GEO-SPACE URBAN DESIGN

GEO-SPACE URBAN DESIGN

GIDEON S. GOLANY

Distinguished Professor of Urban Design
The Pennsylvania State University

TOSHIO OJIMA

Professor of Architecture
Waseda University, Tokyo, Japan

JOHN WILEY & SONS, INC.
New York Chichester Brisbane Toronto Singapore

This text is printed on acid-free paper.

Library of Congress Cataloging-in-Publication Data

Golany, Gideon.

Geo-space urban design / Gideon S. Golany, Toshio Ojima.

p. cm.

Includes bibliographical references and index.

ISBN 0-471-16252-3 (cloth : alk. paper)

1. City planning. 2. Underground architecture. I. Ojima, Toshio, 1937- . II. Title.

HT166.G598 1996

307.1'216—dc20 96-22125

Printed in the United States of America

10 9 8 7 6 5 4 3 2 1

DEDICATED TO THE ENGINEERS, ARCHITECTS, AND SCIENTISTS OF JAPAN FOR THEIR INGENUITY IN DEVELOPING MODERN BELOWGROUND SHOPPING CENTERS.

CONTENTS

PREFACE

In recent years, the rapid growth of major urban centers has become a serious concern for contemporary urban designers. Increases in land prices, congestion, administrative deficiencies, neighborhood deterioration, and countless other issues associated with this growth have made it essential to study the past and present experiences of urban design professionals, because they can provide valuable lessons for developing alternatives to expanding urban frontiers.

For more than two decades, I have studied the many physical, social, and economic problems resulting from present trends in urban design, and combined that knowledge with my extensive research on various indigenous uses of belowground space. By compiling information from both the past and the present, I have developed a comprehensive plan that addresses contemporary design needs for the construction of what I consider to be the optimal future city. Part of this research involved comparing other possible pioneering concepts—such as mountain cities, cities that float above or below sea level, and space settlements—to each other and to geo-space. While each one of this diverse array of city plans has its distinct advantages and disadvantages, it is quite clear to me that while the geo-space city is but one of many design options, it is perhaps the most feasible.

Although my research on geo-space started in the 1950s, it

wasn't until the energy crisis of the 1970s that I immersed myself in specific details related to the realization of such a futuristic design concept by conducting thorough research on case studies in historical belowground space development in China and Tunisia, as well as modern Japanese shopping center designs. I used these examples as a laboratory to shape my overall comprehensive concept for the geo-space city of the future. During my stay in Tokyo in 1990, specific ideas for this concept of a geo-space city continued to develop.

My research since that time has focused on refining comprehensive urban designs that incorporate not only diversified use of land within the city, but also a fusion of the conventional supra-space city with the new geo-space city. There exists also the challenge of encouraging the general public to overcome biases toward living and working underground. In order to accomplish this type of integration and meet the psychological needs of the inhabitants of such a city, innovative designs and practical construction methods are essential. Some of these needs can be met by utilizing sloped sites and designing structures conducive to fluid movement between supra- and sub-space, which allow for maximum direct sunshine and openness while not losing sight of the overall city scale and aesthetic.

I view comprehensive geo-space urban design as a revolutionary, yet realistic, response to many existing urban problems. Geo-space city design is a landmark contribution to urban design. To the best of my knowledge, this book is the first of its kind in the English language.

I would like to give special thanks to Dr. David Shirley, Senior Vice-President for Research and Graduate Studies at PSU; Professor Neil Porterfield, Dean of the College of Arts and Architecture; Dr. Edward Williams, Associate Dean for Research and the Graduate Program; and Professor Michael Fifield, Head of the Department of Architecture at PSU for their continuous encouragement and financial support. Appreciation is extended to my colleague, Professor Peter Magyar, who reviewed some of the drawings in this volume and provided valuable comments. I would also like to thank my editorial assistants, Ms. Ruth Vastola and Ms. Claudia

Link, who helped in the editing of my work, and Linda Gummo and Karen Bair, department secretaries, for their friendly help throughout this project. Last, I would like to thank Ms. Yue Li, Ms. Christin Yun, Ms. Li Hong, Mr. Chen Xie, and Mr. Shijun Huang, who helped to prepare the final illustrations from my drafts. They all deserve thanks and appreciation for their assistance and dedication.

G. S. Golany
The Pennsylvania State University
University Park, PA

INTRODUCTION

Urban agglomeration within major cities around the world has created an urgent need for innovative solutions to many economic, social, environmental, and physical problems facing metropolitan centers today. During this century, many responses to this need have yielded design concepts that focused primarily on vertical and/or horizontal aboveground expansion of a city. Although diverse uses of belowground space have existed in scattered locations throughout the history of civilization, a wide-scale movement toward utilization of geo-space as a solution to urban expansion has yet to be fully considered. The reason for this neglect has evolved in part from highly deficient design practices in the past—designs ill-equipped to meet the norms and standards of contemporary societies in developed countries. Therefore, geo-space has generally been associated with an unpleasant environment and images of darkness, dampness, and disease.

The recent movement calling for the use of belowground space favors innovative technological design, sunshine penetration, efficient ventilation, and multipurpose land uses. *Geo-Space Urban Design* is a product of this movement. It describes practical methods for developing belowground space in order to complement, rather than replace, existing supra-space cities or new towns. It suggests comprehensive and integrative design ideas for all aspects of a city, including transportation, residences, delivery systems, circulation, health, and the many other planning aspects that urban designers need to address.

The most amenable locations to a geo-space city environment are those that have one, most, or all of the following conditions or needs:

- **Land preservation:** countries that need to preserve agricultural land from rapid urban expansion, such as in Holland, Israel, and Japan.
- **High urban land prices:** countries where the requirement of space for urban growth has become acute and land prices have skyrocketed, making geo-space construction economically necessary, such as in urban United States, England, and Japan.
- **Defense:** countries where space is an essential factor for defense and where geo-space in mountains or rolling hills can respond adequately to security needs, such as in Israel and China.
- **Efficient urban scale:** cities where urban physical expansion has resulted in overextended and inefficient urban utility networks; the geo-space city would bring a significant reduction to the sprawl of the network and make it economically and logistically efficient. Examples of networks that have expanded to a level of inefficiency are the many large metropolitan cities throughout the world.
- **Response to stressful climate:** cities located in a stressful hot-dry climate, such as the Middle East, southwestern United States, or northern Africa or Australia, or a cold-dry one, such as central Canada or Siberia, are subject to climatic discomfort and high consumption of energy.

The introduction of the geo-space city concept raises a complex of independent and interrelated questions that need to be addressed by a variety of professionals, such as urban designers, engineers, developers, architects, and legislators. Only in this way does it become possible to realize a comprehensive city plan that effectively considers *all* aspects of construction and habitation. Technological, logistic, legal, territorial, environmental, social, economic, safety and security, and, perhaps above all, psychological considerations are all part of successful implementation of the geo-space design concept.

Like any other pioneering idea, geo-space urban design will require time for adjustment and application. Yet, the growing complications facing our expanded cities will soon bring to light the necessity for visionary solutions to very real problems.

1

GEO-SPACE CITY LAND USE

G. S. Golany

The subterranean habitat is the oldest form of shelter known to humans. Even when man moved to aboveground housing, he maintained partially belowground space for storage and living because of economic circumstances. Due to tremendous urban expansion, acute space shortages, and skyrocketing urban land prices, however, an increasing interest has emerged recently in earth-enveloped architecture. The potential benefits of such a habitat have led to the exploration of a design concept of a thoroughly modern underground city. There are two levels of issues and problems associated with such a city design: construction and usage. The demands of construction can be met with existing modern technology. However, usage encompasses socio-psychological attitudes that are less easily addressed. While humans have the sophisticated technology to overcome the first problem, urban designers, engineers, and architects need to address the socio-psychological issues using innovative, unconventional designs. They need to study and synthesize ancient vernacular design practices with modern theories and introduce distinctive, innovative, and multidisciplinary concepts that suit modern norms and standards.

Historically, the use of indigenous belowground space has been associated with the low-income and homeless populations in rural areas around the world. This association has created a negative image, regardless of the clear benefits to be gained by the use of such space. However, once the geo-space design becomes associated with the availability of advanced technology and accepted by the middle and upper classes, its image will improve and its positive benefits will become obvious. It is the reality of today that technologically advanced countries are the ones that have taken serious interest in this endeavor. Some nations are already using geo-space to different degrees within their existing cities, pressured by the shortage of urban space, soaring land prices, urbanization intensity, and other factors.

Underground human settlements have existed mostly in rural centers. These settlements arose in order to meet a variety of needs, among which were to save precious agricultural land, for defense, to escape religious persecution, or for protection against harsh climates. The most widely recognized of these geo-space cities are the Roman city of Bulla Regia in northern Tunisia; the underground salt mine city network of Wieliczka, near Krakow in southern Poland; Yan'an city in China; and the many Byzantine communities in Cappadocia, central Turkey; and Nabataean settlements in southern Israel and Jordan. Subterranean villages have also evolved for many millennia in southern Tunisia, southern Italy, central Turkey, northern China, Spain, and North America.

TYPOLOGY

Since the early use of the cave many millennia ago, humans have developed a variety of earth-enveloped spaces for shelter and other uses. Accordingly, numerous terms have been adopted, especially recently, to describe these diverse forms of dwellings. Primarily, the terms refer to the different degrees of spatial relationships to the soil and, consequently, the thermal performance patterns, the adjustment to local environmental conditions, and the availability of building materials (Fig. 1.1). *Geo-space* is a term that refers to the space built within the earth. In this book, the term geo-space is used to indicate earth-enveloped subterranean spaces. Recently, however, a number of terms describing a variety of forms of earth-enveloped space usage have been used interchangeably. Among them are earth-sheltered, semi-belowground, subsurface house; subterranean or belowground space; and geo-space. This author would like to take the liberty of clarifying this terminology and these definitions. Also, to properly inform the reader and the researcher, it is necessary to typologize all such varieties of human settlements explored throughout the history of civilization, their origin, and their distinctive features (Fig. 1.2).

Earth-Sheltered Habitat

The term *earth-sheltered habitat,* or *supraterranean earth-sheltered habitat,* is now commonly used in the United States to describe a habitat constructed mostly above ground and enveloped by a layer of earth, usually about one-half meter in thickness (Fig. 1.2*a*). This method is currently used to respond to energy consumption demands, especially in harsh climates where energy consumption is high for cooling or heating. The earth-sheltered habitat functions primarily as a thermal insulator between the indoor and outdoor air temperatures. Its process of heat retention is very limited, due to the relatively small thickness of the earth enveloping it, yet it is effective in minimizing diurnal heat gain and loss. This method of earth-sheltered living space was used by native Americans in the arid regions of the southwestern United States. Its ancient form of construction has been used in regions of stressful climate, such as hot-dry areas of the Middle East, the Mediterranean region (Iran, Turkey, Italy, and Spain) and in northern China. A good example of this construction form is the indigenous house of Jerusalem (Fig. 1.2*b*).

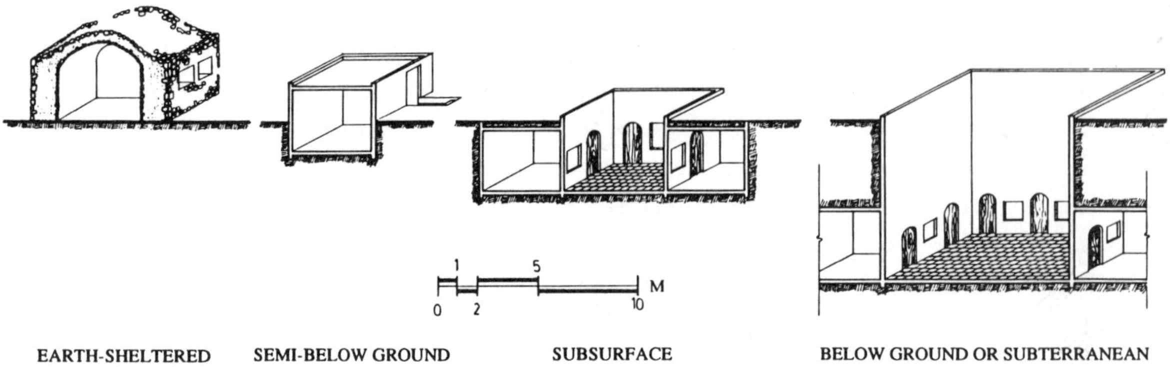

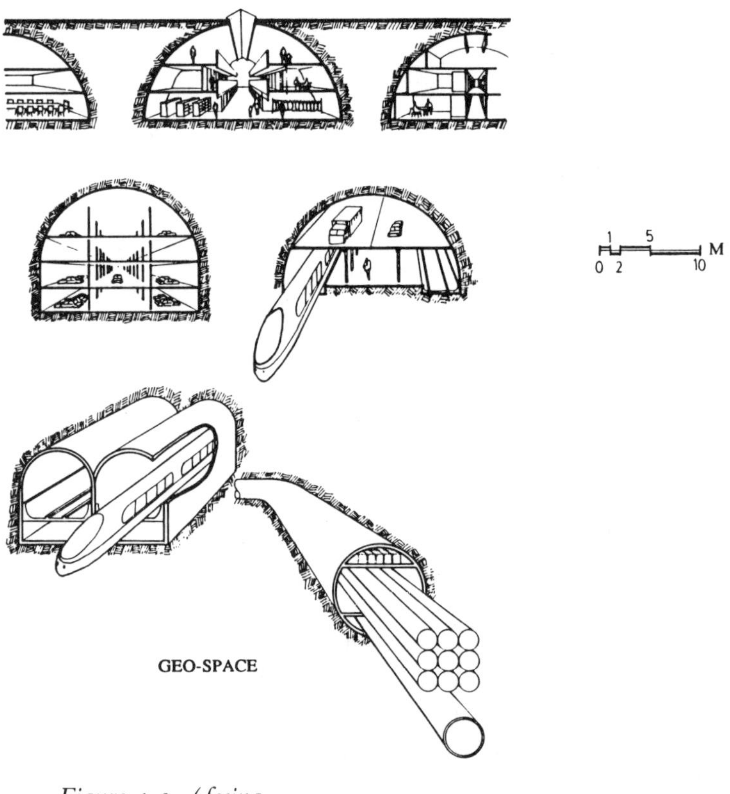

Figure 1.2. (facing page) Typology of human-made earth-enveloped habitats. This principle has been explored throughout the history of human settlements.

Figure 1.1. Terminology of earth-enveloped space units (cross sections). Terms are coined according to different degrees of depth of the space unit's relationship to the surface soil.

1. EARTH-SHELTERED (SUPRA TERRANEAN) HABITAT

A. EARTH-ENVELOPED HABITAT (RECENT AMERICAN)

B. INDIGENOUS JERUSALEM HOUSE

2. SEMI-BELOW GROUND

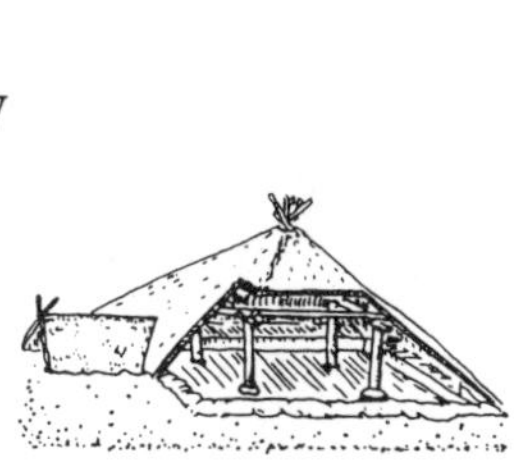

C. NEOLITHIC (CHINA AND JAPAN)

D. TERRACED (MEDITERRANEAN)

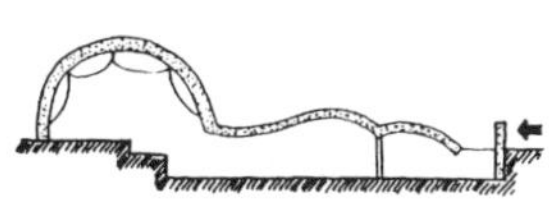

E. IGLOO, ESKIMO

3. SUBSURFACE HOUSE

F. ROMAN SUMMER VILLA (NORTHERN TUNISIA)

4. BELOW GROUND (SUBTERRANEAN)

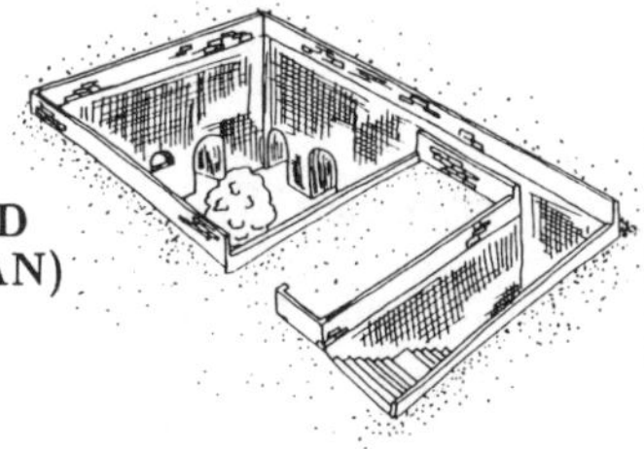

G. PIT TYPE (TUNISIAN AND CHINESE STYLE)

H. CLIFF TYPE (CHINA)

I. NEST TYPE (CAPPADOCIA, TURKEY)

5. GEO-SPACE (JAPANESE CONCEPT)

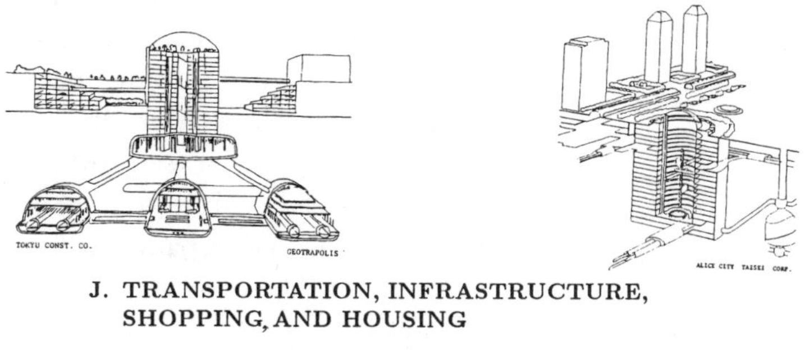

J. TRANSPORTATION, INFRASTRUCTURE, SHOPPING, AND HOUSING

Semi-Belowground Dwellings

A *semi-belowground dwelling* is a unit constructed partly below and partly above ground. This is one of the most common human-made housing forms and was used in Neolithic village communities in China, Japan, and other places of the ancient world (Fig. 1.2*c*). It is still in use in rural communities in Africa. The basement form, which also falls within this category, is commonly used throughout the world. A similar functional form in northern China and in southern Tunisia is the terraced cliff, where part of the house is built below ground and part is built above (Fig. 1.2*d*). Another example is the Eskimo winter igloo and summer semi-belowground house (Fig. 1.2*e*).

Subsurface House

A *subsurface house* is a shallow belowground level dwelling with a short distance between its ceiling and the soil surface, usually about one-half meter or less. This subsurface type of house was used in ancient times by the Romans in the city of Bulla Regia in northern Tunisia (Fig. 1.2*f*). The Romans built a large number of peristyle subsurface summer villas to escape the intense heat of north Africa. This design form was most likely used by the native Berber tribes of North Africa before the Romans arrived. In some modern houses, basement units are built subsurface style.

Belowground or Subterranean Space

Belowground or *subterranean space* has been the most common form of earth-integrated space used throughout history. Subterranean describes spaces developed below ground at a reasonable depth—usually about three meters from the ceiling to the soil surface. Because of the soil thickness, the belowground space is usually created by the "cut-and-use" method, in which no building materials are required. This method is used on flat topography for a pit-type design (Fig. 1.2*g*), as well as for a terraced form on cliffs (Fig. 1.2*h*). It is typically developed in limestone or in tufa because cutting is relatively easy, as in Cappadocia (Fig. 1.2*i*), or in loess-type soil, which is firm and holds its shape when dry, as was the case in the cave dwellings of northern China.

Geo-Space

The term *geo-space* is currently used by the Japanese to describe space that is fully submerged deep within the earth.[1] Japanese designers have introduced some innovative and pioneering concepts of geo-space forms for multipurpose human activities at depths of 50 meters or more (Fig. 1.2*j*). The renewed interest in the use of earth-enveloped space of this type originated primarily among the technologically advanced countries of the world.

Another example of deep geo-space usage is the Rannfast project on Norway's southwestern coast. The project comprises two undersea fjord crossings, and provides mainland connection for all of the major islands in Rennesoy County. The longest tunnels among this network are the Byfjord Tunnel (5,860 m) and the Mastrafjord Tunnel (4,405 m). Since the Byfjord Tunnel opened in November 1992, nearly 2,400 cars have crossed it every day.[2] Tunneling, however, is a new frontier in geo-space and its use is expected to be accelerated with further advancements in technology.[3]

PROS AND CONS OF GEO-SPACE USAGE

One of the prime objectives of the geo-space city is to preserve nature and enhance aesthetic and social space in the supra-ground conventional city. Thus, the design of the geo-space city is intended to significantly improve the conventional city.[4] Different reasons can lead each country to accept geo-space city development.[5] Some of those reasons include saving land for agriculture; providing shelter and security for defensive needs; protecting against harsh climates; obtaining a comfortable ambient temperature; easing the stress of high land prices; achieving compact land uses and efficient use of urban land in order to contend with contemporary urban expansion; decreasing the length of urban utility networks; enhancing the health and attractiveness of supra-urban space;

or creating a more livable city with plenty of greenery and free from motor vehicles.[6] In the United States, for example, the geo-space system has been explored by the middle and upper classes in recent times in response to innovative housing needs.

The application of advanced technology, along with innovative design, construction, and operational processes, has the potential to enhance the use of geo-space. The primary goals of the geo-space city, referred to in this book in connection with flat or slope locations, are to upgrade the quality of existing urban life, to eliminate or diminish some of our urban problems, to enhance the natural environment, to coordinate a new belowground development, and to introduce a visionary, yet realizable, urban pattern to be implemented in future urban centers.[7] The land-use discussion in this chapter refers primarily to urban nonresidential land use.[8] Residential land use will be discussed in the following chapter.

Urban expansion can occur in three different directions, either simultaneously or separately: horizontally, vertically upward, or vertically downward. Throughout the 19th century into the 20th century, cities expanded in a vertically upward pattern in the form of high-rise buildings. Twentieth-century growth has also been characterized by horizontal expansion, with the dispersion of the cities into the suburbs to create single-family homes. Since both approaches have been explored extensively, it is our belief that the city of the 21st century will expand again in the vertical direction, but this time it will expand downward into the geo-space city. This development will take place not only out of necessity, but also from the recognition of geo-space potentialities and, most importantly, due to its positive developmental effect on the conventional supra-space city.

Almost all types of conventional supra-space land use are adaptable to the geo-space city.[9] Using geo-space is more beneficial for some land uses than others. The vertical downward expansion of the geo-space city would reduce the sprawling network of the urban infrastructure. Our research findings indicate that using geo-space can shorten the total combined length of all utilities and infrastructure, including roads, telephone, TV, electricity, water, and sewage systems, by 75 percent.[10]

The geo-space city eases much of the intense land-use pressures associated with the conventional city center.[11] At the same time, it enhances the physical and social environments of the city core by converting roads to pedestrian use and by introducing extensive green space and natural vegetation. It also controls—if not reduces—land price speculation, provides a pollution-free environment, reduces energy consumption for cooling and heating, and increases safety in the downtown city area. Like any complex design endeavor, there are advantages and limitations to the geo-space city enterprise. In any case, it is necessary to be aware of its features in order to enhance its strengths and minimize its weaknesses.

Significant Benefits of the Geo-Space City

We have defined the geo-space city concept as the extensive and diversified use of belowground space and the comprehensive fusion of this space with the existing or planned supra-space city. Geo-space offers a new urban dimension that will significantly affect the conventional supra-space city. This innovative urban form complements the supra-space city but does not replace it, and provides a new vision of land use.[12]

Psychologically, logistically, and economically, it is more efficient to develop a complete city exclusively on and within a slope rather than to use a flat site for geo-space housing construction. A slope accommodates needs of privacy and lighting (chapter 2). Yet it is still feasible to use flat sites for many of the other city land uses. While the geo-space city is a tradeoff, involving both gains and losses, the significant benefits of such an alternative are numerous. The following discussion illuminates some of these important benefits.

Climate. Geo-space living is an effective way to cope with outdoor climatic conditions—especially harsh ones. Its weatherproofing provides a comfortable ambient environment in both winter and summer. It consumes less energy and reduces heat gain and loss.[13]

Land cost. Generally, the price of land is reduced, either because of its dual use or because of initial construction on low-priced sloping land. In fact, land savings can be significant, since belowground construction frees space for aboveground artery lanes and other purposes, such as playgrounds, recreation areas, or other buildings. The insulated quietness of the geo-space habitat enables such spaces to be built even near airports. Design and building costs can be minimized when a large production takes place or when, geologically, lack of stratification enables the cut-and-use method of construction. Also, the limited need for exterior building materials and the low level of required maintenance will reduce its price. There is no need for periodic re-roofing (every 10 to 15 years in aboveground construction), and only minimal repair is needed to fix the damage caused by wind, hail, rain, snow, or other natural forces.

Environment. The geo-space city preserves the land and the environment and saves plenty of space for green and open areas. It provides privacy, quiet, and a relaxing environment. As such, it is ideal for creative work, such as writing, music, composition, painting, and sculpture.

Safety. The geo-space city can be safer than the aboveground environment if it is properly designed. It protects against storms, tornadoes, and radioactive fallout. It is fireproof, and there is minimal potential for a fire to spread to neighboring structures. Accordingly, insurance rates are expected to be lower than for the conventional aboveground structures. Geo-space living can also be made safe against destruction by earthquakes if protective design measures are taken. In addition, geo-space does not have problems associated with temperature fluctuation, such as freezing and cracking of water pipes.

Sleeping and tranquility. Although specific scientific research has not yet been thoroughly conducted in belowground living spaces, logic would dictate that the significant reduction of noise and the increase in privacy and intimate atmosphere are conducive to tranquility and, by extension, better sleeping conditions. Interviews conducted with people who live below ground in Australia revealed that they felt that the geo-space environment in which they lived was more tranquil and healthier than living in an above-ground space, especially for children. For example, one mother indicated that her children had never slept as peacefully as they did in the geo-space habitat.[14]

Industry. Geo-space offers diversified land use with special advantages. The temperature stability of geo-space is most suitable to industries that require seasonal or perennial thermal stability, such as the film and wine industries. The author's research indicates that the minimal thermal fluctuation that exists in geo-space facilities speeds the healing of external wounds after an operation. It seems that the wet- and the dry-bulb temperatures, along with their stable synchronization, reduce surgical patients' recovery times by 20 percent or more.[15]

Refrigerated storage. Due to its cool summer temperatures, geo-space can be used for large-scale refrigeration and central food storage, especially for citrus fruits and vegetables. Such storage centers are currently used in Cappadocia, Turkey; in Ohya, Japan; and in Chongqing, China. In the Middle East, as well as in China and many other places of the ancient world, geo-space was used extensively for grain storage. Grain storage pits used in China during the T'ang Dynasty (seventh to tenth centuries) were recently discovered in Luokou and in Luoyang city (Henan Province), where more than 50 percent of the grain was still in good condition.[16] Modern methods of geo-space storage are now being researched in the United States, with the potential for large-scale development.

Educational facilities. Another effective use of geo-space found in the United States today is that of educational facilities, such as libraries, schools, and bookstores. In universities, where space is precious and the expansion of an existing well-established library is most often necessary, the nearby geo-space is an ideal solution, which does not interfere with the existing environment. (While thermal stability is an asset here too, because it helps preserve the quality of the books, there is also a need for low relative humidity.) Some American universities that have expanded their libraries into geo-space are the University of Illinois, Harvard University, Johns Hopkins University, and Cornell University. About 100 modern, earth-integrated schools have also been developed in the United States.

Offices. As with educational facilities, a large number of offices, public gathering places, business spaces, and religious centers have been constructed within geo-space in the last two decades.[17] This new movement toward earth-sheltered usage was initiated in response to the energy crisis, the need to obtain ambient thermal conditions, or environmental protection. Today, this movement has gained momentum in most of the developed countries of the world.

All in all, the geo-space city responds positively to the elements of the conventional supra-space city. In addition, the geo-space city stimulates designers and business people to employ innovative ideas to bring an effective measure of rejuvenation to the city. Table 1.1 summarizes the pros and cons of the geo-space city.

Ultimate Goals

The ultimate goals for construction of the geo-space city are to solve some of the existing and anticipated problems arising from the supra-space city.[18] Ultimately, we view the geo-space city to call for:

1. The relocation of most, if not all, of the supra-space urban transportation network, especially that in the city center, to the geo-space level. The ultimate purpose of this is to open the urban supra-space for pedestrians, to introduce natural landscape into the city, to separate human activities from vehicles, and to allow relaxation in a natural environment.
2. The penetration of the natural environment deep into the core of the supra-urban environment.
3. The movement of all the city's infrastructure to the geo-space level, with construction fully and comprehensively organized into a sophisticated operation.
4. The design, implementation, and operation of a totally automated underground delivery system for the city.

TABLE 1.1.
Summary of Pros and Cons of the Geo-Space City

PROS	CONS
HISTORICAL	
• Use of the geo-space in large scale has proved to be feasible and can offer clues for future development • Historical designs provide impetus for innovation in contemporary practice to meet modern norms and standards	• Although individual historical cases have achieved their ultimate goals, their norms of design cannot be an example for our contemporary times, which require sunshine, natural light, ventilation, and other environmental needs
LAND USE	
• Introduces natural environment deep into the city when moving urban arteries into the geo-space • Preserves the beauty of nature • Provides dual land use for supra- and geo-space • Enhances integrated land use to fuse the existing supra-space city with the geo-space city • Creates proximity of land uses for daily needs • Offers more comfort to the users through mixed integrated land use • Total separation of transportation significantly reduces the human friction with transportation network • Shortens all urban utility networks • Consumes fewer building materials • Floors have increased load capacity • Compact land use reduces energy consumption • Improves agricultural and poultry production	• Intensifies land use and may introduce congestion • Congestion may lead to loss of some privacy, which may not be acceptable to some cultures
SOCIAL	
• Enhances social interaction among age groups through the introduction of mixed and proximate land uses • Enhances social urbanity • Encourages social integration among heterogeneous groups and reduces isolation	• Cultural bias may cause acclimatization difficulties among heterogeneous groups • May cause possible claustrophobia and psychological constraints, which necessitate special design • Difficulty in eliminating psychological resentment against belowground space • Requires adaptability

(*continued*)

TABLE 1.1. *(Continued)*

PROS	CONS
ECONOMIC	
• Dual land use reduces urban land cost • Slows speculation of city land prices • Compactness brought to the geo-space city reduces the complexity of the infrastructure and expenditure in its design, construction, and maintenance • Increases housing options and employment • Heat gain and loss are minimal; lower energy consumption • Provides low-cost refrigeration • Increases labor productivity by comfortable and stable air temperature • Lessens speculation of city land uses	• Increases costs of pumping water • Initial investment costs are high • Land prices may escalate after initial development is made • New design may be costly and require research • Costs during construction may be high • Illumination may increase maintenance costs • Possible need for blasting may increase costs • Geological and soil mapping costs are involved • Extensive excavation may require costly reclamation of land surface to refit it for agricultural and other uses
TRANSPORTATION	
• Increases the usage of nonpolluting mass transportation, such as subways • Reduces the use of private transportation • Shortens commuting time	• Creates a major change in the urban system • Carries a risk of transportation vibration from heavy vehicles
SAFETY	
• Provides protection from manmade and natural disasters • Water utilities will not freeze, burst, etc. • Increases resistance to fire • Minimizes flood risk due to slope structure for dwellings • Protects against earthquakes with responsive design measures and proper site selection • Protects against tornadoes, hurricanes, and electrical storms	• Has fire evacuation difficulties • Geological faults increase risk of impacts from earthquakes
ENVIRONMENT	
• Minimizes impact on the natural environment • Creates a pleasing and generous relaxing urban environment affiliated with daily city land uses • Provides a livable urbanity, yet active city • Provides a stimulating environment for	• May introduce some environmental constraints and require adjustments • Noise levels, generated from gathering in public spaces, may increase • Has potential exposure to radon • Special indoor landscape is needed for some spaces

(continued)

TABLE 1.1. *(Continued)*

PROS	CONS
creativity of writers and artists	
• Provides overall privacy in housing	
QUALITY OF LIFE	
• All transportation is moved below ground • Provides a wide natural environment in the city with pleasing green spaces • Eases mental pressure on individuals and groups • Brings proximity to diversified land uses • Provides cleaner and fresher air • Introduces wide, safe pedestrian networks throughout the city, totally separated from motorways • Reduces transportation noise and air pollution, is quiet, and provides safety	• In a closed environment, air pollution may become a problem unless passive or active ventilation is designed or electric cars become a common reality
HEALTH	
• Comfortable ambient temperature supports relaxation and mental vitality • Quietness stimulates creativity • Minimizes visual and audible distractions • Reduces the post-surgery healing period by 20 percent • Tranquility reduces stress	• Increases dampness, especially in humid regions • Potentially induces claustrophobia • May require passive or active ventilation design
CLIMATIC COMFORT	
• Increases weatherproofing against extreme and stressful climate • Introduces stable seasonal and diurnal temperature and is beneficial for health and for some industry • Resists temperature fluctuation and contributes to a comfortable ambient environment • In extreme cold climates, survival is still assured when electricity or heat is interrupted • Reduces impact of wind significantly	• Risk of cover by dust storms in some regions
MAINTENANCE COSTS	
• Lowers maintenance costs • Extends durability of structures • Housekeeping is reduced • Fire insurance rates should be lower • Proximity of land uses reduces utility costs significantly	• Increases costs of pumping water to the sloped dwellings and pumping waste from the pit units

5. The implementation of a cohesive plan of land use into integrated and segregated patterns of geo-space, acting as an integral part of the supra-space city. Consequently, the restructuring of the supra-space city land-use pattern should conform to a human scale, with the pedestrian network dominant throughout the city.
6. The introduction of sloped geo-space habitat design principles and the integration of those principles with the flat lowland city without placing subterranean housing on the flat land.
7. The relocation of all windowless structures into the geo-space.

It is not realistic to expect the implementation of the geo-space city to take place within a short period of time. As a revolutionary concept in urban design, its merit and strength lie primarily in its long-range response to existing urban issues. From legislative, economic, design, implementation, and other urban developmental points of view, it is easier to construct a geo-space city first as a new city built from scratch, rather than to integrate it with an existing city.

In all advanced regions of the world, the trend toward urbanization is usually accompanied by changes in the landscape and in the overall physical, social, and economic environment. The irreversible pattern of urbanization will continue to dominate our environment and, with it, increase the consumption of space, goods, high technology, and the standard of living, all of which lead to drastic changes in socioeconomic values.[19] Any planning system should consider these changes and the forces behind them. Under these circumstances, the role of the urban design system becomes critical and essential. Urban design is conceived as a synthesis of the natural and the artificial environments, where all contributing forces shaping the city structure are carefully considered. The focus of urban design is the people and their physical three-dimensional environment. The goal of urban design is to create a continuous, pleasing environment for all the residents in the entire city.

This work introduces the belowground city, and its fusion

with the aboveground city, as a design concept. The marriage of the aboveground city with the geo-space city should bring a rejuvenation of the supra-space city. The fusion of the two cities introduces to the aboveground city a healthier environment, a renovated cityscape, a closer proximity of residences to places of work, a safer pedestrian network throughout the city, an integration of the natural environment with the heart of the city, a relaxing and tranquil social environment within the dynamic urbanity, and an environmental setting free of air and noise pollution.

This geo-space city project is to be treated comprehensively by considering most, if not all, aspects of the city rather than a general conceptualization. In order to develop the geo-space city model, historical lessons, as well as existing practices and experiences, have been studied.

The theme of the geo-space city endeavor is a long-range comprehensive vision, which introduces an unconventional yet applicable concept.[20] Until this century, urban growth expanded horizontally and vertically upward. The geo-space city introduces a new downward vertical dimension to urban growth. People now have the technological ability and the courage to implement these solutions to benefit humanity and society in general, not simply to create another beautiful but flawed historical architectural monument.[21]

Although the geo-space city concept seems visionary, radical, and revolutionary, it is this author's belief that, over time, belowground space usage has increased steadily and will continue to do so. Research and study of this subject, and, most importantly, the need to ease our urban problems, will bring this project to further realization. To meet contemporary standards, the design of geo-space, especially for day-to-day living, necessitates treatment different from that of supra-space design.

National Urban Policy

Our observations, and findings of urban design and regional planning in many regions of the world, indicate that there is a strong correlation between the size of a given country, population density, and level of technology on the one hand, and the collective consciousness and the need for a comprehensive national urban growth policy on the other. Illustrating this correlation are the cases of England (new town development), Holland (polder reclamation and regional development), Japan (technology and design consciousness), and Israel (land-use regulations and population redistribution). These countries are small in size, have high population densities, and are limited in natural resources, but they have been able to establish a cohesive and comprehensive national urban growth policy and effective spatial urban design.

Serious use of geo-space in any given country necessitates a comprehensive study of its feasibility, its prospects, and its social, spatial, economic, and environmental ramifications.[22] Clarification of large-scale goals and objectives, formulation of a national development policy, establishment of codes by which geo-space should be regulated, incentives to be offered by the central government to the private and the public businesses, and issues surrounding transportation are also important factors in the implementation of any national movement toward widespread geo-space construction efforts.

BASIC LAND-USE POLICY

The use of geo-space is intended to intensify and to optimize land use in the city, while enhancing the social, economic, and health conditions of the residents.[23] There is an increase in the economic and social benefits when geo-space development is carefully monitored.

The integration of geo-space land use with the supra-space city is a complex issue. The dual use of the land in its supra- and sub-levels has an economic implication in areas where land prices are subject to substantial increases. To achieve optimal use of urban geo-space, its integration with other design concepts should be considered. Such urban development should be viewed as tripartite in nature. This notion is called the "trinity concept" by this author. The three elements involved are slope site, compact urban form, and geo-space. A trinity concept of urban design and construction calls for the integration of belowground space with compact form and with slope site selection.[24] The fusion of these three design elements is accomplished in order to upgrade and enhance the geo-space concept to meet our modern and future design norms and standards. Indeed, these three elements have been used in human settlements extensively throughout history. The difference here is that in the past, they were treated separately rather than integratively. Even the combination of two identical land-use elements, or two related land-use elements of the trinity concept, above and below ground can ease or eliminate problems traditionally associated with belowground space use. The association of the three principles should be viewed, analyzed, and implemented as a cohesive integral unit that can also offer a variety of design principle applications to the existing city itself. The urban designer can use each of the three principles either side-by-side in isolated form (one principle), in semi-integrated form (two principles), or in its fully integrated form (all three principles together). In this way, the effective use of subterranean space can respond comprehensively to some of the more pressing urban problems.

Another approach to the geo-space land-use pattern is the introduction of a grid artery system within the geo-space and its integration with that of the existing supra-space land use. The geo-space constructed cavities would mirror existing built-up patterns in supra-space, such as the grid transportation network. Also, active or passive aboveground open spaces, parks, arteries, and the like would correspond with geo-space land uses. These should include routes for pedestrians, vehicles, and subways, along with areas for parking lots. Integrative above- and belowground land use is necessary to achieve structural soundness, accessibility, flow of pedestrian circulation, or safety. For reasons such as orientation, this type of mirrored design pattern will prove to be beneficial, primarily for pedestrians (Fig. 1.3).

Compact Land Use

The compact form of urban land use has been commonly used throughout the history of human settlements, often

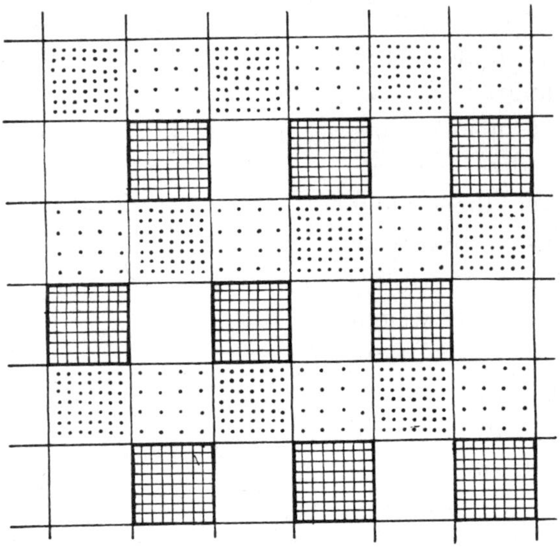

A. ONE PATTERN WITH ALTERNATED LAND USES

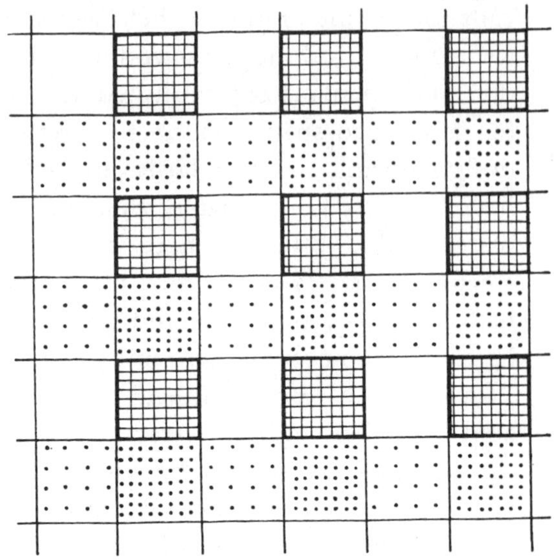

B. ANOTHER PATTERN WITH DIFFERENT ALTERNATIONS

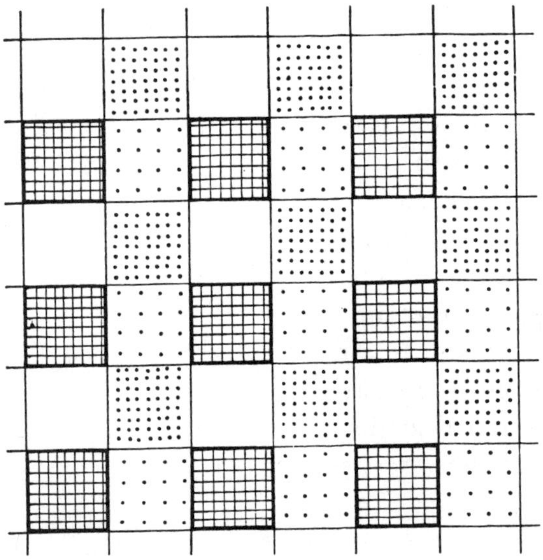

C. THIRD PATTERN WITH DIFFERENT ALTERNATION

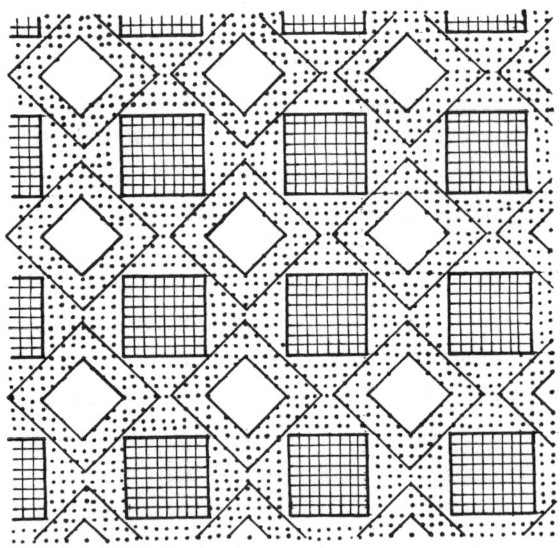

D. A COMBINATION OF ALTERNATIONS

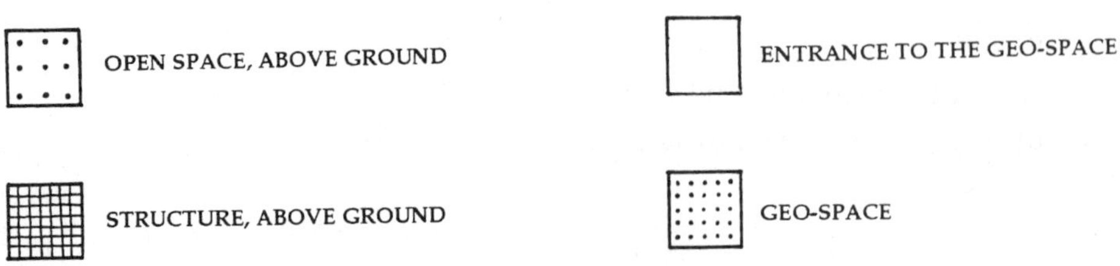

without particular regard to topography—using either flat, hilly, or sloping sites. Its development evolved for a variety of reasons, including defense, conservation of arable land, protection from harsh climates, economizing on labor for city wall construction, or territorial cohesiveness and social identity. In some arid and semi-arid climates, the compact city evolved as the dominant form over many millennia. It was also used for many centuries in temperate climatic regions, such as those of Europe and the Mediterranean. During the present century, the compact city was reintroduced in the search for the optimum urban design by city planners, architects, and social scientists, including Ebenezer Howard, Frank Lloyd Wright, Le Corbusier, and Paolo Soleri.

The compact land-use city differs from the dispersed style typical of American cities in that it also introduces a sharp demarcation between its physical existence and the immediate, natural, and open environs. This inherent contrast contributes both practically and subconsciously to a strong awareness of the natural environment and a sense of community cohesiveness.

Discussions on the compact land-use city are usually associated with the idea of a high density of people per unit. However, it is necessary to distinguish between the density of people per habitat unit and density of habitat units per city. Compact city land use should not simultaneously combine an increase of people living in each habitat unit with an increase in the number of habitat units per city. A higher density of habitat units per city, however, will not diminish the quality of indoor privacy and should be encouraged. A minimization of privacy, an increase in noise, limited movement, and disruptive behavior within the population will more likely result from increasing the population density per habitat unit in a given city than increasing the number of habitat units in the city. Instead, a compact city should introduce proximate dwelling units and minimized passive open space. Careful design of the dwelling units and the introduction of other land uses can improve the physical and psychological environment.

A compact city is more likely to be accepted by a homoge-

Figure 1.3. (facing page) Grid and linear patterns of land uses for the geo-space city, combining below- and aboveground cities.

neous group than by a heterogeneous one, since the former enjoys better communication and tends to accept the intimacy that is found in a closely knit area. Also, the compact city does not detract from personal or family space, which is the most essential space of the urban dweller.

Adaptability to space conditions depends on the existence of shared human elements, such as origin, culture, and lifestyle. Because homogeneous groups share common values, they communicate better than do heterogeneous groups, and are, therefore, usually more tolerant in a high population density arrangement. The priorities of space use also differ among cultures and between one family and another, and usually relate to the size of the family and the age and number of children. Basically, indoor personal space is more important for individuals and families than is the outdoors in the street, marketplace, or other public spaces. Similar priorities exist outdoors in relation to the usage of public space, shopping centers, cultural centers, or educational institutions. Within the compact land-use city, it is assumed that personal contacts in shopping centers and other public areas of activity occur more frequently than in the conventional city because of the proximity of these centers to the neighborhoods.

However, it is likely that different types of societies would react differently to living in a compact city, and the mutual contact established by proximity may not be completely acceptable to some. Social, economic, and habitation norms are not static, but rather are subject to fluctuating social expectations. To meet these future changes, the urban planner/designer should create a flexible plan.

The compact city has been presented here within a tripartite concept of urban design that is integrated with urban geo-space. It is through their shared interaction that the concepts gain meaning, are more likely to be realized, and are mutually reinforced. The urban planner/designer can offset the potential negative effects of high unit density by incorporating the needs of its residents for abundant open spaces and green areas, safety of movement for children and adults, proximity to services, and distance from sources of noise and pollution.

The major contribution of the geo-space city is increased space. By virtue of its new dimension, the geo-space city enforces compact land use by establishing vertical and horizontal proximity of usage. This reduces commuting time, speeds delivery of services, and establishes a city free of supra-space transportation and its associated environmental damage. Table 1.2 offers a comprehensive listing of the pros and cons of the compact city form. The study of the history of indigenous compact city land use requires further research

TABLE 1.2.
Pros and Cons of the Compact City Land Use

PROS	CONS
PHYSICAL	
• Provides proximity of land use for daily needs • Significantly reduces transportation networks • Shortens utility length • Increases use of pedestrian networks • Provides proximity to nature	• Can be congested, resulting in a loss of some privacy
ECONOMIC	
• Conserves land for agriculture and other uses • Reduces energy consumption • Provides efficient land use • Shortens commuting time • Saves time in design, construction, and maintenance	• Escalates land prices due to increased building density
SOCIAL	
• Increases social propinquity and interaction • Increases interaction among age groups • Improves urban self-image • Increases use of public institutions • Encourages pedestrian movement	• Requires adaptability and takes time
HEALTH AND WELL-BEING	
• Reduces pollution • Protects against stressful climates	• Potentially increases noise generated from public spaces
ENVIRONMENTAL	
• Decreases land consumption • Creates pleasing urban environment	• Innovative design is necessary and can be costly

and understanding from social and economic points of view. However, we view the introduction of the compact land use as a tradeoff between social and economic ends.

As will be discussed in the following chapter, the trinity concept and its compactness will be adjusted to the habitat part of the geo-space city in slope topographical forms. This basic form introduces a diversity of subforms that best suit our needs (Fig. 1.4).

In conclusion, the need to employ the geo-space land use is motivated by many factors that exist in contemporary cities. These include escalating land prices, shortage of space for expansion, increase in urban land consumption, land-use dispersion, transportation congestion, length of utilities development, urban design inefficiency, and high maintenance expenditures. The introduction of the compact city responds positively to most, if not all, of these problems, especially when it is associated with geo-space slope usage.

The compact city has the potential to reduce rush-hour traffic and commuting time, to encourage people to walk short distances, and to establish a safe and comfortable environment. A compact residential area describes concentrated and unified buildings in close proximity to each other and in conjunction with consolidated land uses.

Urban subterranean city construction is an innovative venture that requires that the many conventional approaches to urban design be reexamined. These include the urban design concepts of site selection criteria, physiography, orientation to available natural sunlight, macro- and microclimatological systems, geological limitations, drainage patterns, accessibility, and environmental constraints.

Subterranean cities have many benefits and some constraints, but these issues must be studied with long-range considerations in mind. There is an acute shortage of theory and modern practice in such endeavors, and the lack of published work demands, both economically and practically, decisive research in this area. Underground land uses are a revolutionary concept compared with conventional cities. Underground construction responds to a complex group of issues: protection from increasingly harsh climates, rising en-

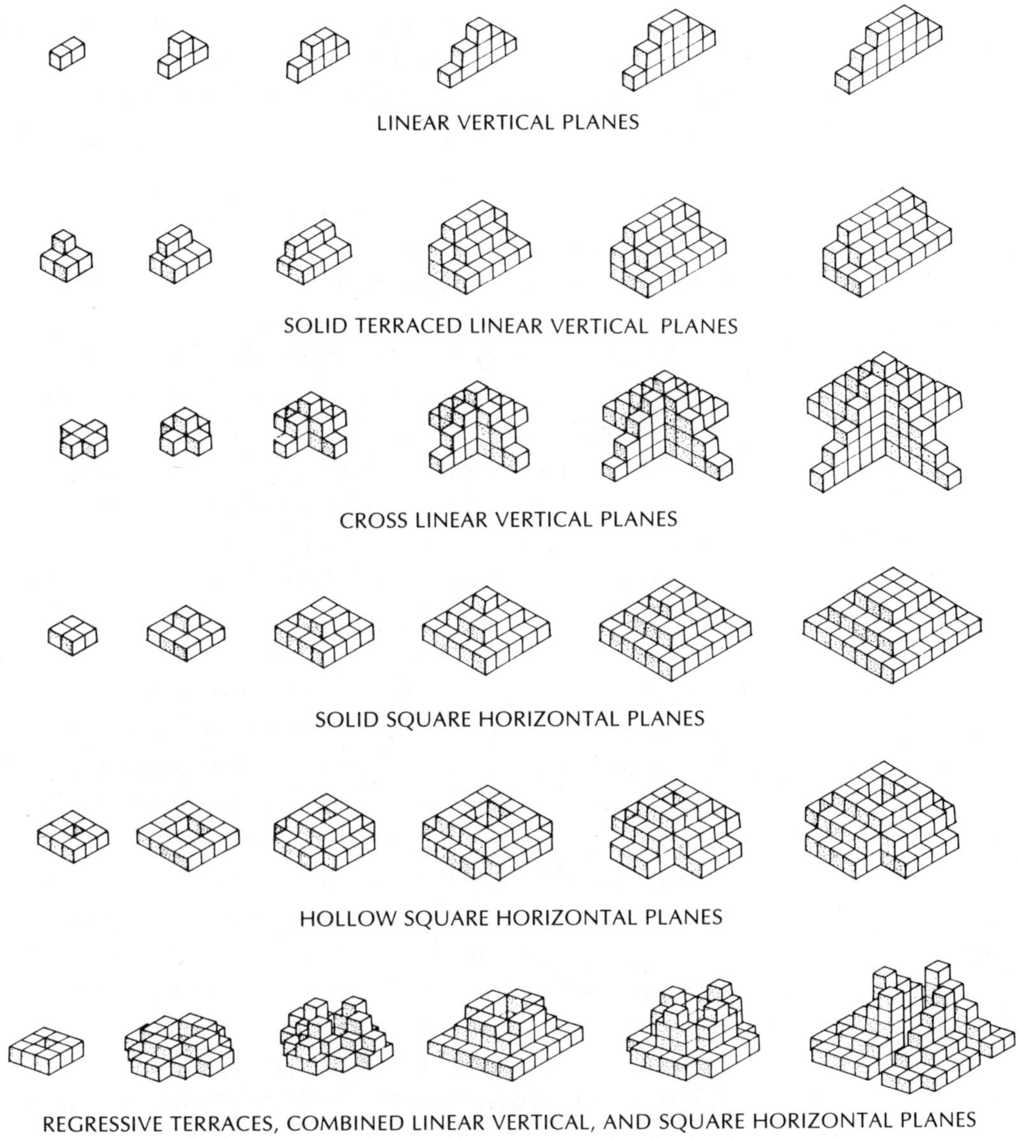

Figure 1.4. Progressive geometric models of horizontal-vertical terraced slopes. The models are introduced in a variety of possible slopes for compact physical urban form.

ergy costs, construction costs, maintenance costs, land-use types, safety concerns, health conditions, and environmental damage.

Segregated or Integrated Land Use

The fusion of supra-space and geo-space cities intensifies the land use, and therefore necessitates the introduction of principles that call for integrated and segregated land uses. Segregated land use is the designation of a single land use in a given urban zone in order to isolate its reciprocal impact on the other urban land uses. Integrated land use is the mixture of different types of uses that have common interests and interrelated needs, at least from the point of view of the user in a zone. Integrated land use combines uses that have no negative reciprocal impact on segregated land use. As such, integrated land use is best applied to a compact city.

The land uses of the geo-space city should be carefully selected so that related uses, such as habitats, educational facilities and cultural centers, or nonpolluting industry, storage, and transportation centers can be easily grouped and consolidated. Integrated land uses offer an opportunity to minimize the distance between a place of residence and work, to introduce variations in the human landscape, to offer multiple cityscapes, to introduce a choice of nearby shopping centers, to establish greater proximity to schools and therefore greater safety for children, and to introduce urbanity to a neighborhood that otherwise would have been characterized by monotonous, segregated housing. Integrated land use can serve a diverse blend of age and gender groups, and can significantly enrich the geo-space community.

Segregated land use is the allocation of a single, dominant type of land use in a designated area, such as an exclusively industrial or residential area. Segregated nonresidential land uses are those often considered to have a negative impact on their adjacent land, through air and noise pollution, large quantities of waste, or some other nuisance. This type of land use typically generates heavy traffic for loading and unloading materials for heavy industry, processing plants, large storage warehouses, food refrigeration, or agricultural production or packing. Yet, segregated land use is mostly dominated by daytime population and is unsecured at night.

However, in a geo-space city, it is necessary to integrate residential land use into the geo-space city, as well as to maintain conventional supra-space housing. Basic land-use principles to be adopted are:

1. Geo-space housing should be exclusively located on the slope and not on a flat site or low land. This principle does not conflict with the development of conventional aboveground housing on a nearby flat or low land.
2. Most nonresidential geo-space land uses should be selectively located on both the slope and the flat land as well as above and below ground.
3. Land uses that do not introduce negative impact on the residential area, such as educational, cultural, or commercial facilities, should be integrated with residential land use. Such integration will balance the volume of the daytime and nighttime population.

Nonresidential land uses include public gathering spaces, parking areas, and circulation spaces, and should provide safety exits and entrances and traffic thoroughfares. All these activities require special spaces and illustrate the types of land uses that can be integrated into the combined slope geo-space and supra-space. The fusion of the geo-space and supra-space is possible, but not without technological difficulty and innovative design.

DEPTH AND SOIL THERMAL PERFORMANCE

The thermal performance of soil is one of the strongest elements that influence geo-space use. Generally, in the past, interest in the study of the soil thermal performance process was limited to two specific zones of the soil. The agronomist's interest has justifiably and typically concentrated on the very shallow depth that affects vegetation. Mining engineers, however, have been concerned with a depth of hundreds of meters. But with the dawn of modern geo-space usage, geo-space designers have become interested in the zone ranging from zero to 10 meters in depth.

The soil thermal process differs significantly from that of aboveground space. In general, we can state that down to a depth of about 10 meters, the soil depends on the sun for heat. As the depth increases, the effect of the heat on the earth increases. However, the thermal behavior of the soil depends not only on the intensity of the source of thermal supply, but also on the soil composition and density (rock, sandy, alluvial, etc.) and the degree of water containment.

During the day, solar heat influences the soil at a shallow depth generally ranging between 5 and 7 centimeters, but there is a continued seasonal movement of this heat to a depth of around 10 meters. The basic rules for thermal performance are:[25]

- Between the soil surface and a depth of 10 meters, there is a seasonal thermal fluctuation, which decreases at deeper measures and increases toward the surface. No diurnal temperature fluctuation occurs.
- Around a depth of 10 meters, the seasonal temperature becomes stable.

In short, the thermal performance of the soil has two distinct functions. First, it acts as an efficient insulator against heat gain and loss, and second, it acts as a thermal retainer. The author's findings show that it takes one season (approximately three months) for the air temperature to travel to a depth of around 10 meters.[26] The mass of the earth retains thermal consistency. Thus, the summer air temperature will reach geo-space at a depth of 10 meters by the winter season and the winter air temperature will reach that depth by the summer, when it is needed most.

The discovery of those terra-thermal zones, between the land surface and a depth of 10 meters or more, could offer promising prospects for industries or even human comfort needs during future geo-space development. Certain industries require constant seasonal and diurnal temperatures for the most efficient production, such as for storage or human post-surgical recovery. Thus, these zones will have broad economic, social, and health implications.

LAND USE AND DEPTH

The geo-space city calls for the use of belowground space to an almost unlimited depth. Primarily, depth in slope construction is defined as a horizontal dimension; depth in flat construction is primarily defined as a vertical dimension. Using a combination of both dimensions can significantly enrich city land use. Our geo-space concept is based on the integrative usage of flat and sloped sites.

This broadened definition of depth presents us with a perceptual dichotomy. In the visual sense, depth perception within geo-space is limited to immediate surroundings, while depth perception of the traditional supra-space city encompasses a wider horizon. For example, the segregated land uses in the geo-space city, such as industry, are perceived only by those who use it as a workplace. Others are even less visible, such as the rapid delivery network and some rapid transportation systems, which are automated. The reality, however, is that geo-space depth has a great expansive potential because there are virtually no limits to horizontal and vertical depth.

The location of the geo-space land uses is related primarily to the vertical and horizontal depth of the soil. Proper designation of land uses fuses the supra-space city with the geo-space city as one cohesive urban organism. Accordingly, the suggested basic principles regarding depth and the geo-space city land-use distribution pattern are:

1. The shallower the depth of the geo-space, the more intensive and extensive the diurnal human activities of this space will be. Also, because of the shallow depth of the geo-space, the integration of the geo-space land use with the supra-space city will be strengthened.
2. The greater the depth of the geo-space, the higher the increase of automated land-use activities and the lower the degree of human diurnal activities should be.
3. In slope geo-space, where depth is primarily a horizontal dimension, most, if not all, land utilization is most

likely to be confined to a single level. Diverse land uses related to intensified human activities are to be integrated with the slope habitat within or adjacent to the shallow depth. The slope geo-space habitat at the shallow depth has direct contact with nature and the natural source of sunshine and light. Land used for nonresidential daily activities can be located deeper in the earth. Examples of such nonresidential land uses are theaters, art centers, educational institutions, shopping centers, sports arenas, and the like, which are occupied by people diurnally.

Accordingly, the soil depth, at least in generic terms, should be divided into three basic land-use groupings: shallow, medium, and deep (Table 1.3). Land use and depth, however, are to take into account the intensity of human activities and the possible introduction of automated systems.

Aboveground space should gradually move its three major conventional land uses—transportation, utilities, and the delivery network system—to the geo-space level. In so doing, more space for greenery and the natural environment, for pedestrians, and for leisure will be created. The aboveground space will continue to be used for most of the other conventional land uses.

TABLE 1.3.
Basic Grouping of Land Uses and Human Activities in the Geo-Space City as Correlated to Depth

LEVEL/DEPTH (1)	HUMAN ACTIVITIES (2)	SEGREGATED/ INTEGRATED LAND USE (3)	LAND USE (4)
		SUPRA-SPACE LEVELS	
ABOVE GROUND: Height: unlimited; depends on technology	Highly intensive and extensive diurnally	Selective integrated mixed and segregated land use	• Removal of the following to below ground: • Transportation network (especially from the cores of the city), except for emergency cases • Delivery system • The infrastructure network

(*continued*)

TABLE 1.3.
(*Continued*)

LEVEL/DEPTH (1)	HUMAN ACTIVITIES (2)	SEGREGATED/ INTEGRATED LAND USE (3)	LAND USE (4)	
			• Increased urban features: green open space, natural environment, bodies of water, children's playgrounds, rest areas, open markets, art displays • Expanded pedestrian network into all parts of the city as a comprehensively integrated network • Most of the traditional land uses will continue to be represented	
		GEO-SPACE LEVELS		
SHALLOW DEPTH: 0 to 10 meters	Intensive and extensive diurnally	Selective integrated mixed and segregated land uses	• Habitats (exclusively in the slope) • Pedestrian network • Hotels (for transition, short stay) • Surgery and recovery rooms • Conferences • Light infrastructure • Restaurants • Cultural centers, theaters, museums • Offices	• Entertainment centers • Special recreation • Sporting activities • Educational activities, libraries • Creative environment • Spiritual activities • Light transportation • Parking • Shopping centers • Public gatherings • Limited transportation network
MEDIUM DEPTH: 10 to 50 meters	Selective daily human activities	Segregated land use	• Subways, highways, streets • Parking • Infrastructure network • Automated delivery system	• Refrigeration • Energy storage • Nonpolluting industry • Storage (short-range)
DEEP: 50 meters and more	Very low human activities/very high automated technology	Segregated land use	• High speed transportation (intra-urban) • High speed automated network delivery systems	• Energy storage • Special storage (long-range) • Heavy infrastructure

Whether related to slope gradient (horizontal depth) or to the flat gradient (vertical depth) sites, there is a clear distinction between depth, its relationship to human comfort, and allocation of land use (Fig. 1.5).

Shallow Depth Zone

The depth between zero and 10 meters, for our purposes, is called the shallow depth zone and is characterized by insulation from outdoor temperatures, by increasing seasonal temperature stability, and by cool temperatures in the summer and warm temperatures in the winter. The shallow zone also experiences seasonal temperature fluctuation. As depth increases toward 10 meters, temperature stability increases, while at 10 meters, temperature is stable seasonally. Because this range is immediately below the soil surface, feelings of claustrophobia (a common concern when discussing below-ground space) are minimalized. Therefore, it is suggested that this depth level should be the zone that accommodates the most human activities, because it is most comfortable physically and psychologically. The shallow depth will accommodate intense diurnal human activities similar to those on the ground surface. The aboveground land uses adjacent to this shallow zone should be fully integrated with it. Examples of such integrated land uses are entertainment, cultural, and sport centers; creative activities, such as painting, sculpture, or music; and some light infrastructure limited to transportation networks, shopping centers, and public gathering halls. It is this similarity of activities that will fuse the supra-space and the geo-space cities (Fig. 1.6). The integration of similar land uses is essential to the success of the geo-space city concept.

The pattern of soil thermal performance within the shallow depth zone is suitable for mixed land uses. Such a pattern ensures the increase of diurnal human activities, since a wider variety of land uses will attract more people throughout the day and evening. The shallow zone can accommodate two to three vertical floors, which should increase the activity level. The careful combination of elements such as restaurants, en-

Figure 1.5. Schematic divisions of vertical soil depth in a flat site and horizontal depth in a slope site. Note that the density of the geo-space land use is reduced as the depth increases.

tertainment centers, and cultural and educational activities will achieve this goal.

Although it is preferred to transfer urban supra-transportation to the medium level (discussed below), it may also be possible to move some of it to this shallow level. This would be feasible only if transportation at this level would not interfere with other land uses (Fig. 1.7). From the human point of view, the land-use design of the shallow zone is the most sensitive and necessitates careful work by a team of specialists, including architects, artists, social scientists, economists,

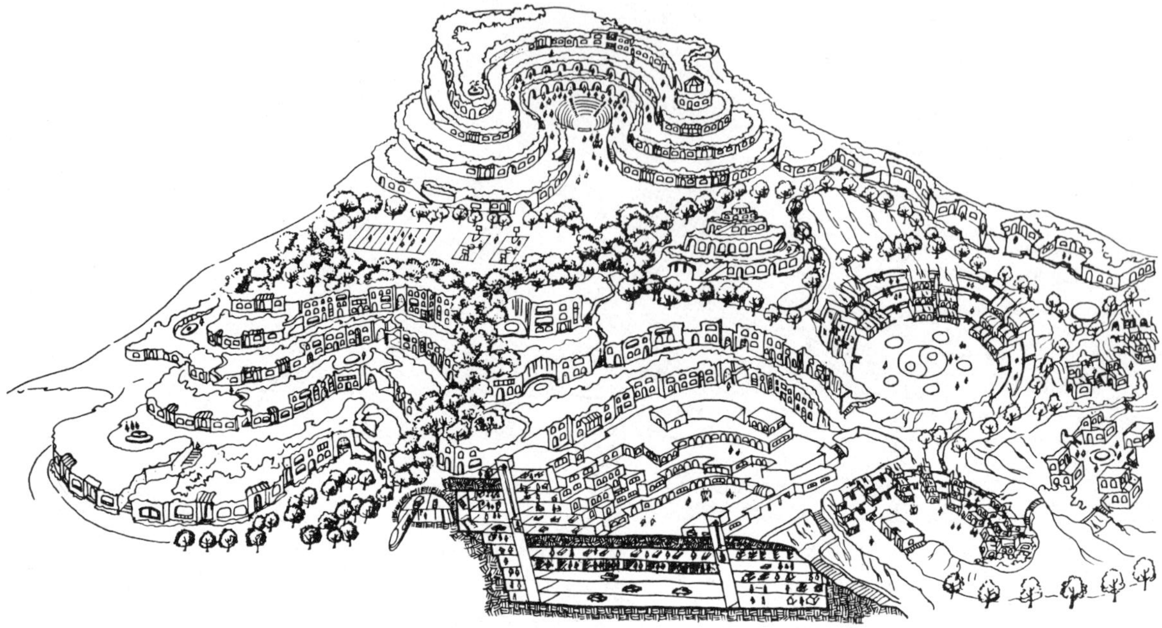

Figure 1.6. Schematic layers of vertical and horizontal geo-space city land use. The levels may be vertical (or almost flat form) and horizontal (slope form) for supraterranean city land use.

health experts, physiologists, and environmentalists, and headed by urban designers.

Medium Depth Zone

The medium depth land-use zone falls generally between a depth of 10 to 50 meters. Composition and geological structure within this zone will vary from site to site. Typically, the temperature is relatively stable, thus providing suitable space for specific types of industry. This is the zone where segregated land-use types can be introduced, although some human diurnal activities may still be adopted here (Fig. 1.8). This zone should also accommodate the subway network, all types of vehicles, highway and city streets, railway stations, and parking areas, as well as industry and storage. The link between this and the shallow depth zone is by stairway and an extensive network of elevators. This medium depth zone should also accommodate nonpolluting industry. Within the 40 meters of depth, 8 to 12 vertical floors can be built, assuming that each floor will occupy three to five meters in height from floor to ceiling.

Figure 1.7. An urban transportation network and some of the utilities introduced at a shallow depth. The integration of the transportation and utility networks into the geo-space city has a significant positive impact on the social, health, safety, and environmental conditions of the supra-space city (Japanese model).

In a valley example of a medium depth zone, gravel and other deposits cover the base rock. Such is the case of Tokyo. This type of geological stratification has an impact on construction, since the base rock provides a strong foundation. It also has a legal implication based on the fact that different countries extend land ownership to different depths and a change in depth may signify a change in ownership.

Deep Zone

The deep zone, around 50 meters and more in depth, is a unique zone that should feature few, if any, human activities and more automated, programmed systems. This zone holds segregated land-use types.

This author expects that by the early 21st century, the use of nonpolluting (electric) vehicles will become common, and, therefore, this zone will provide high-speed inter-urban transportation, heavy-duty infrastructure network, high-speed cargo delivery systems, large long-range storage spaces, and other limited-use activities.

In short, depth usage should be correlated with thermal

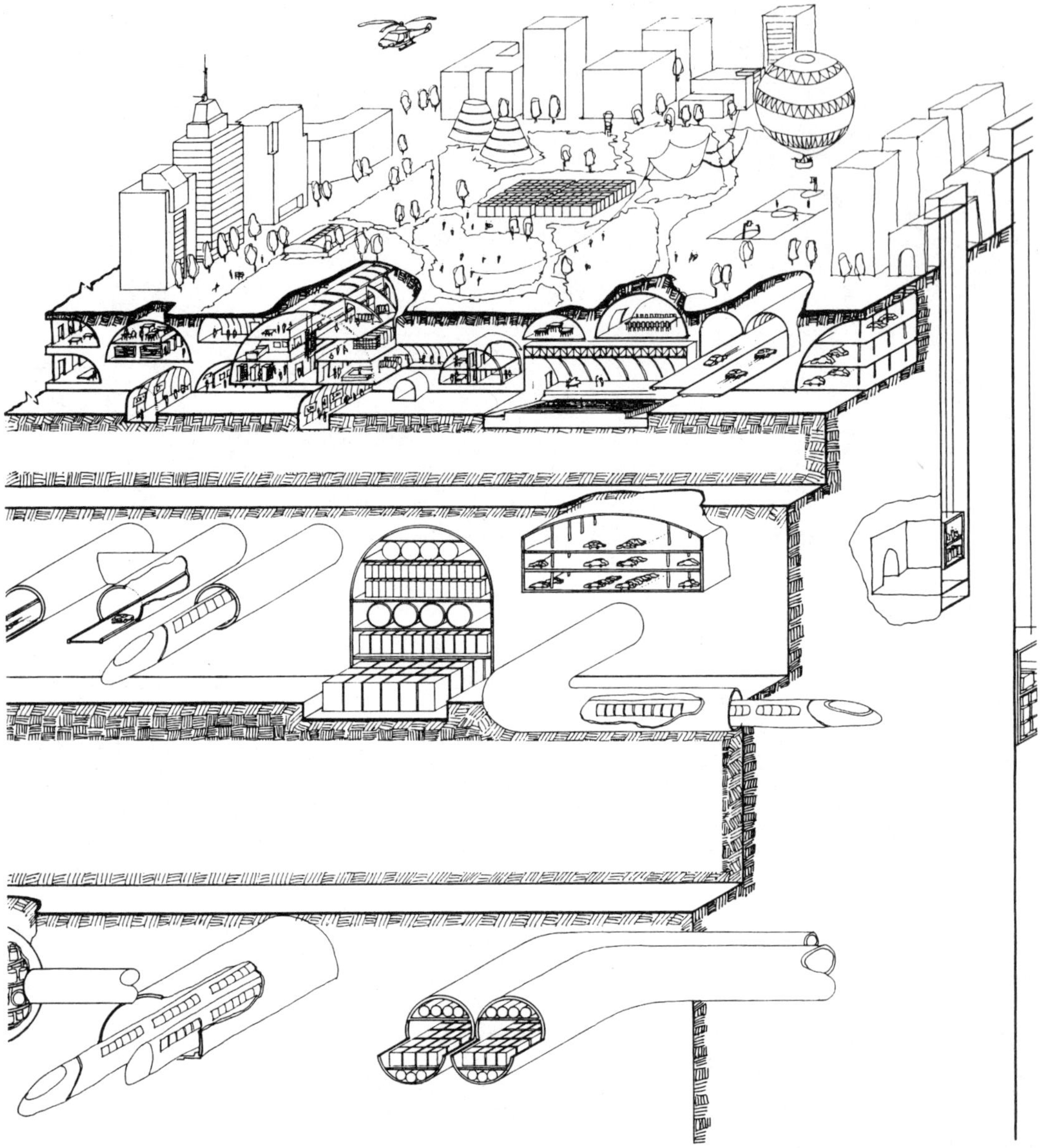

Figure 1.8. Cross-sectional worm's eye schematic perspective of land uses. Examples of the three geo-space zones along with an example of the aboveground land use.

soil behavior, geological stratification, soil composition, and, most importantly, the degree of adaptation of human activities. The basic rule is that the greater the depth, the lower the degree of human activity. These conditions are related to physical and psychological comfort.

URBAN TRANSPORTATION AND DELIVERY SYSTEM

One basic theme of the geo-space city, as presented here, is the gradual transferral of the supra-city urban transportation and delivery system to different belowground levels within the city. The idea is that the released aboveground space will be used for extensive planting of natural vegetation, for safe pedestrian movement, and to free the aboveground space from environmental nuisances associated with conventional transportation. The newly renovated aboveground environment will enhance the integrated social and natural environment and further increase urbanity in the city. The aboveground environment will also be adjacent to shopping centers and other commercial operations both above and below ground and be linked to the new belowground transportation network. Revolutionary changes to the environment and the social climate of the urban society, such as these, require thorough investigation of technological aspects of urban transportation in a practical form.

Urban Transportation

Transportation transferral to the medium zone should take place first in the downtown area of the city, followed by a gradual movement to the city's periphery.[27] Although the city center will be the most difficult to negotiate technically, it is also the most urgent because of high land prices, heavy congestion, pollution (air and noise), and, finally, the need for renovating and modernizing deteriorating portions of the city center.[28] Once transportation has been moved from the center of the city, a great deal of aboveground space is opened up for freer movement and introduction of new land uses and social interaction. It will be one of the safest aboveground places within the city and will enable the natural landscape to flourish within the heart of the city. The importance of these benefits should not be overlooked, as many big-city dwellers currently do not have the opportunity or resources to interact with the natural environment. Such changes to the profile of the city center in a major metropolitan area will revolutionize the lifestyle of its inhabitants in that the overall quality of life will be greatly enhanced.

The first renovative steps, geological survey and legislative regulation, are to be done simultaneously. The latter is a difficult one, as it is time-consuming and requires tough political negotiation. In some countries, such as Japan, landowners control the deepest point into the ground and the highest point into the air. In others, this control is limited to a specific number of meters. In either case, private enterprise is most likely to step forward and get involved to control belowground space development. Here, the free market economy will help to induce the private developers and bring about extensive geo-space urban transportation utilization if the public sector will support it by mapping, planning, and developing public utilities.[29] Legislative changes would be necessary when digging starts at a deeper point. The lower depths would be primarily for transportation, which would require a lengthy line along a given area and would involve debating legislative rights with more than just a few landowners. However, it seems to this author that under those circumstances, a joint public and private enterprise would be desirable.[30] In any case, the public would not be freed from up-front investment needed for underground mapping, designing, and construction of the infrastructure. Basic investments will give incentives to the private investor to enter into this new frontier enterprise.

Other major concerns when developing geo-transportation involve the familiarity with innovative geo-space design practices, advanced construction technology for different types of rocks and soil, and all the logistic processes associated with such construction. However, some countries, such as Japan, Sweden, Great Britain, France, the United States, and Canada, have now gained a remarkable advantage in geo-space urban transportation construction.[31] Our experience with existing tunneling technology has entailed extensive construction and intensive use.[32] Countries such as Japan have gained significant knowledge of subway transportation networks with the Honshu-Hokkaido undersea connecting tunnel and tunnels for trains. Great Britain and France obtained experience during the construction of the tunnel underneath the Strait of Dover, and Sweden through its subway, tunneling, and mining construction. Tunneled geo-space technology promises public urban delivery and passenger movement for the geo-space city into the 21st century.[33]

Another important issue related to belowground transportation is the pollution and health of people using these systems and their spaces. The ultimate goal is that all transportation will be nonpolluting.

Zero-Pollution Vehicles

One promising nonpolluting motor vehicle is the electrically fueled car furnished with electric loaded batteries, which has been in the experimental stages for the last decade. Another type is the car run on electricity supplied by solar energy. This car would be especially efficient in areas where solar energy is dominant year-round, such as in a hot-dry climate. The third promising nonpolluting car is the newly tested Honda from Japan. This car, called the HRK model, is expected to replace conventional gasoline fuel with hydrogen, which is produced from water. It can reach a speed of 150 kilometers (or 92 miles) per hour.[34] The United States has required that all U.S. auto makers produce nonpolluting cars by the year 1998.

Although General Motors (GM) did not produce the first electric car in the world, it can claim to have introduced the best experimental electric car to date. At the very least, it was the first to create an electric-powered vehicle that outperforms its gasoline-powered equivalent. California has passed a law mandating that at least 2 percent of the cars sold in the state produce zero emissions in 1998 and 10 percent produce zero emissions by the year 2010. Based on forthcoming laws like California's, GM is motivated to meet this standard. The California law forces companies to conduct special research on the development of electric cars. Both Ford and Chrysler have also responded to this law and are developing electric-powered versions of minivans, while the European and the Japanese car manufacturers have focused mainly on converting existing passenger cars (Table 1.4). GM's goal is to design the electric car completely from the ground up for higher performance, rather than converting an existing model.

Many automobile companies are conducting intensive research focusing on improving the electric car's acceleration and overall performance through special styling design and technology. The new car must be more efficient and attractive than the conventional gasoline car to convince the public of the potential of this new concept. GM's Impact experimental model has been tested since mid-1993 in New York, Long Island, Los Angeles, Washington, DC, Atlanta,

TABLE 1.4.
Progress of Experimental Electric Vehicles to Date

VEHICLE	BATTERY	RANGE	DATE DUE	COMMENTS
GM Impact	Lead-acid 1,100 lbs.	70 miles EPA city; 90 miles EPA highway, 100 miles at a steady 55 mph	1998	0 to 60 mph in eight seconds. Regenerative braking.
BMW E-1	Sodium-nickel-chloride 440 lbs.	165 miles	No date set	Four-passenger platform; 78 mph top speed; aluminum and plastic body panels; 1980 lb. curb weight.
Chrysler T-Van	Nickel-iron About 1,800 lbs	80 miles in mixed driving conditions	1997	Available as of April 1993; 0 to 60 mph in 25 seconds, 65 mph top speed.
Ford Ecostar	Sodium-sulfur 780 lbs.	100 miles at 45 mph; 200 miles at 25 mph	1998	75 mph top speed; 900 lb. payload.
Nissan FEV	Nickel-cadmium 444 lbs.	100 miles at a steady 45 mph	No date set	Roof-mounted solar cells help augment charging; 0.19 drag coefficient; regenerative braking system; FEV is Nissan's test bed for new EV technologies.
Peugeot 106	Nickel-cadmium 572 lbs.	75 miles at 50 mph	1995 in Europe	Peugeot's electric vehicle based on the existing 106 platform; rear wheel drive; 55 mph top speed; on-board charger in the trunk.

*Dan McCosh, "We Drive the World's Best Electric Car." *Popular Science,* January 1994, p. 58.

Houston, Phoenix, San Diego, Sacramento, San Francisco, Ft. Lauderdale, and Harrisburg. At least 1,000 ordinary people will have use of this car for two weeks to one month to evaluate its performance.[35]

The belowground transportation networks are expected to include nonpolluting vehicles, including subways, low- and high-speed trains, light and heavy cars, parking, car services, elevators, horizontal passenger movers, and automated cargo delivery systems. Geo-space parking lots have been in use for a long time and are expected to multiply.[36]

The application of the geo-space arteries within nonresidential land use is more likely to develop a linear pattern along the aboveground original traffic movement, especially at the parts that need to be linked and integrated with the existing supra-space city network. Similarly, the pattern of the geo-space parking area may follow the aboveground one. The aboveground main lines of network will have much impact on geo-space linearity. Also, the geo-space arteries below ground will require points of reference with the aboveground existing streets and roads of the city. At the supra-space level, the former street land will be converted to green space covered with a natural environment and become exclusively a pedestrian zone. All the aboveground traffic will be directed underground. Also, it is common that most streets are legally owned by the public, which will make their subspace also publicly owned.

Automated Delivery System

The geo-space automated cargo delivery network will replace the aboveground urban, and probably some interurban, cargo systems. The heavy-duty long-range delivery network will occupy the deep zone. Short-range light cargo delivery will use the medium or shallow depth zone.

The basic concept of the automated delivery system is the distribution of goods in special containers from aboveground stations by vertical elevators. The stations will be connected to underground nodes of horizontal networks, which will be again linked to vertical elevator networks leading back to the aboveground stations. Both the vertical and the horizontal integrated networks will be computerized and programmed to work automatically, free from human involvement at the belowground levels. This concept is being researched now by Japanese institutions.[37]

LAND-USE FUSION OF SUPRA-SPACE AND GEO-SPACE

For practical reasons, the passage that connects the surface of the supra-space city with the shallow geo-space environment should allow swift movement and easy access from inside and out, should be naturally lighted, and should be wide enough for the movement of masses of people without obstacle. The stepped passage itself could contain shops and other land uses. Since both the ground level of the supra-space city and the geo-space shallow zone are designed to be active diurnally, their fusion should involve careful design of this transitional passage.

The supra-space and the shallow geo-space should have a mixture of similar and complementary land uses. The supra-space level should be mixed with greenery, vegetation, and some shops, offering diversions for the pedestrian throughout the day and into the late hours of the night. Similarly, in the upper level of the geo-space city, people will be bustling in restaurants, cultural and entertainment theaters, sport arenas, and educational centers.

Like the flat city, the sloped city should occupy supra-space as well as geo-space. This dual usage would intensify the overall land use and support compactness, offer greater opportunities for investors, and stabilize land prices.

The land-use fusion pattern is the key to merging the supra-city space and the geo-space city. The most significant area lies between the zones of the two cities that are adjacent to each other—that of the supra-city surface level and that of the shallow geo-space ranging between zero and 10 meters below the surface. By virtue of being the most active two zones of the city, they should have a common and related land-use character and be used almost diurnally. They have an extensive mix of complementary and similar land uses in close proximity to each other. It is this proximity that must be carefully examined when designing the land uses of these neighboring zones.

The land uses selected for each zone should be continuous with, compatible with, or complementary to the land uses of each adjacent zone so that they are nearly integrated. To achieve a fusion of the two zones, an active human environment should be established on both levels to promote a dynamic and livable urbanity, the psychological importance of which should not be overlooked. Educational, cultural, entertainment, and shopping facilities should have lengthy utilization periods during the day and night. By combining these types of land uses, the two cities will become fused. Another key issue to realizing this fusion is the design and construction of a comprehensive downtown center in the geo-space city.

Comprehensive Downtown Center

A downtown area offers residents and visitors the first impression of a city. In most cities of the world, it is the downtown area that needs revitalization most and attracts the greatest attention of the public and private sectors. Due to increasing land prices and the tendency of urban centers to attract large populations, it is essential to integrate the downtown supra-space and geo-space land uses together cohesively. But there are few tools available to help achieve this goal. One possibility is to design a layout and then to erect a comprehensive, yet unique, downtown urban center. A comprehensive downtown center refers to the inclusion of a diversified all-day center, which will add to a lively and pleasant urbanity. Moreover, the construction of a comprehensive center will entice private investors to contribute to this goal.

As this author explained earlier, all transportation and delivery systems will be removed from the supra-space city and placed in the geo-space city, providing safety, security, freedom from traffic, and a green environment in the downtown part of the city. It is in this part of the city that implementation of the geo-space city will be most feasible and become a pioneering endeavor of subterranean land use.

The comprehensive downtown center should include land use for commercial, cultural, entertainment, leisure, recreation, sport, social, and educational activities, as well as for restaurants, offices, some selected manufacturing, overnight accommodations, and many other such businesses. Such activities would attract all age and gender groups. It would combine medium- and high-rise buildings for diversified uses, along with a sunken patio (Fig. 1.9). The sunken patio centered in the downtown complex would be large in scale and would operate as the main entrance to this part of the geo-space city (Fig. 1.10).

The pedestrian circulation movement in this proposed comprehensive downtown center would contain three levels, including the above and below ground (Fig. 1.11). The parking lots and all other vehicles would be exclusively within the geo-space level, and most likely at a medium depth (10 meters or more). The overall pedestrian circulation would move horizontally on at least three levels, two of which would be above ground and at least one below ground. The first level would be at the fourth floor of all the buildings surrounding the center. At this level, all the roofs would be flat and would be linked by bridges to provide open pedestrian space on the roofs. The fourth and fifth floors would be designated for restaurants and leisure activities, which could use the roofs, floors, and bridges between these floors for their expanded activities. However, this roof-bridge level would also create a horizontal layer for human circulation throughout all the supra-space buildings of the center. The second horizontal level of pedestrian circulation would be on the ground surface, which would now be free of any vehicles and provide a fully natural environment. This plan would allow the ground level, too, to provide a tranquil, pleasant, and safe urban atmosphere. It would create a rich, full, and lively urban setting and continuously attract people with varied interests (Fig. 1.12). The third horizontal movement of pedestrians would be at the geo-space level, where much of the daily activity would take place (Fig. 1.13). At this level, the pedestrians would have access to every building located above ground, since all these buildings will also meet underground. In short,

Figure 1.9. (facing page) Schematic land use of the downtown comprehensive center and its relation to supra-space and geo-space cities.

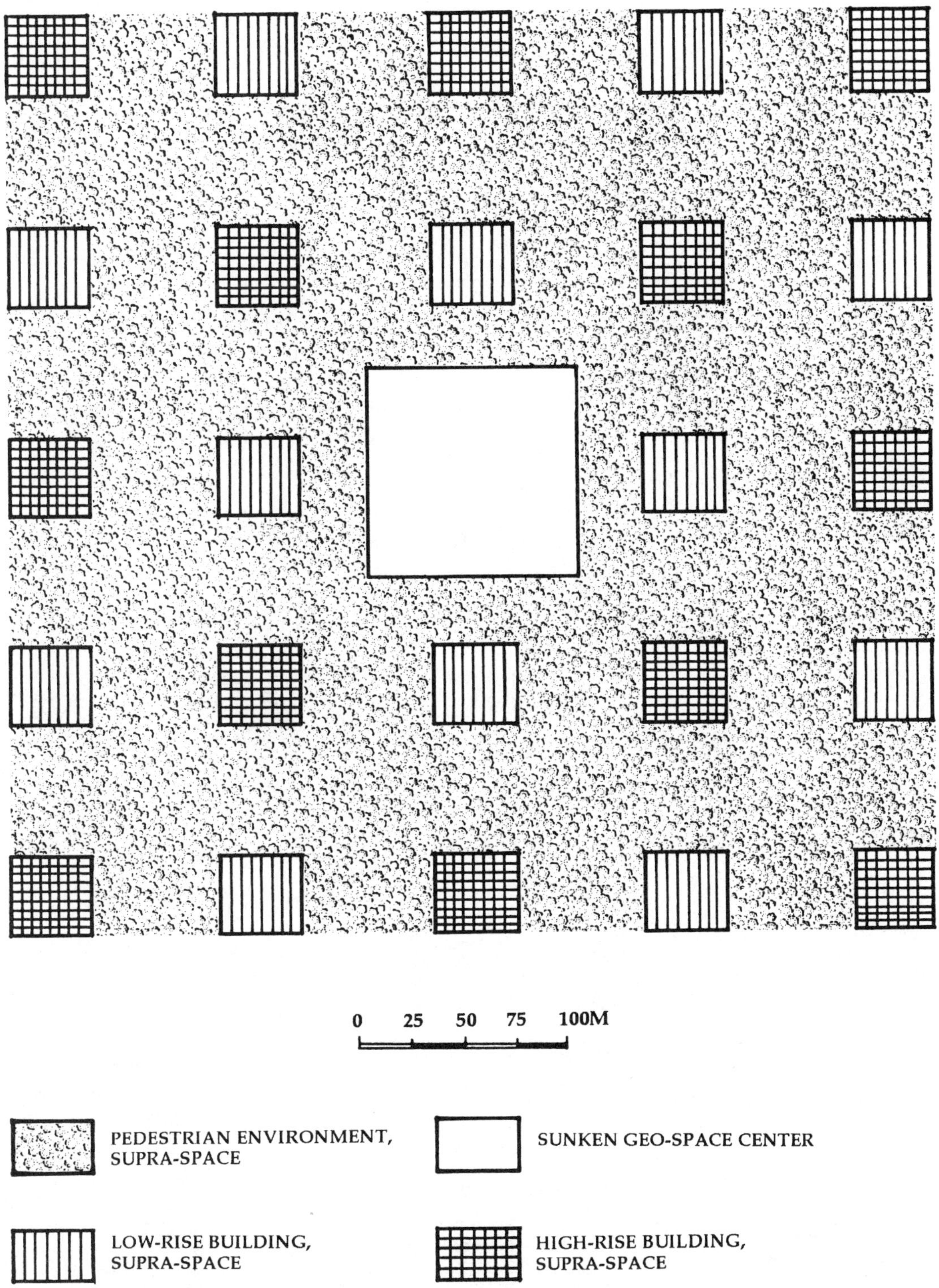
0 25 50 75 100M
PEDESTRIAN ENVIRONMENT, SUPRA-SPACE
SUNKEN GEO-SPACE CENTER
LOW-RISE BUILDING, SUPRA-SPACE
HIGH-RISE BUILDING, SUPRA-SPACE

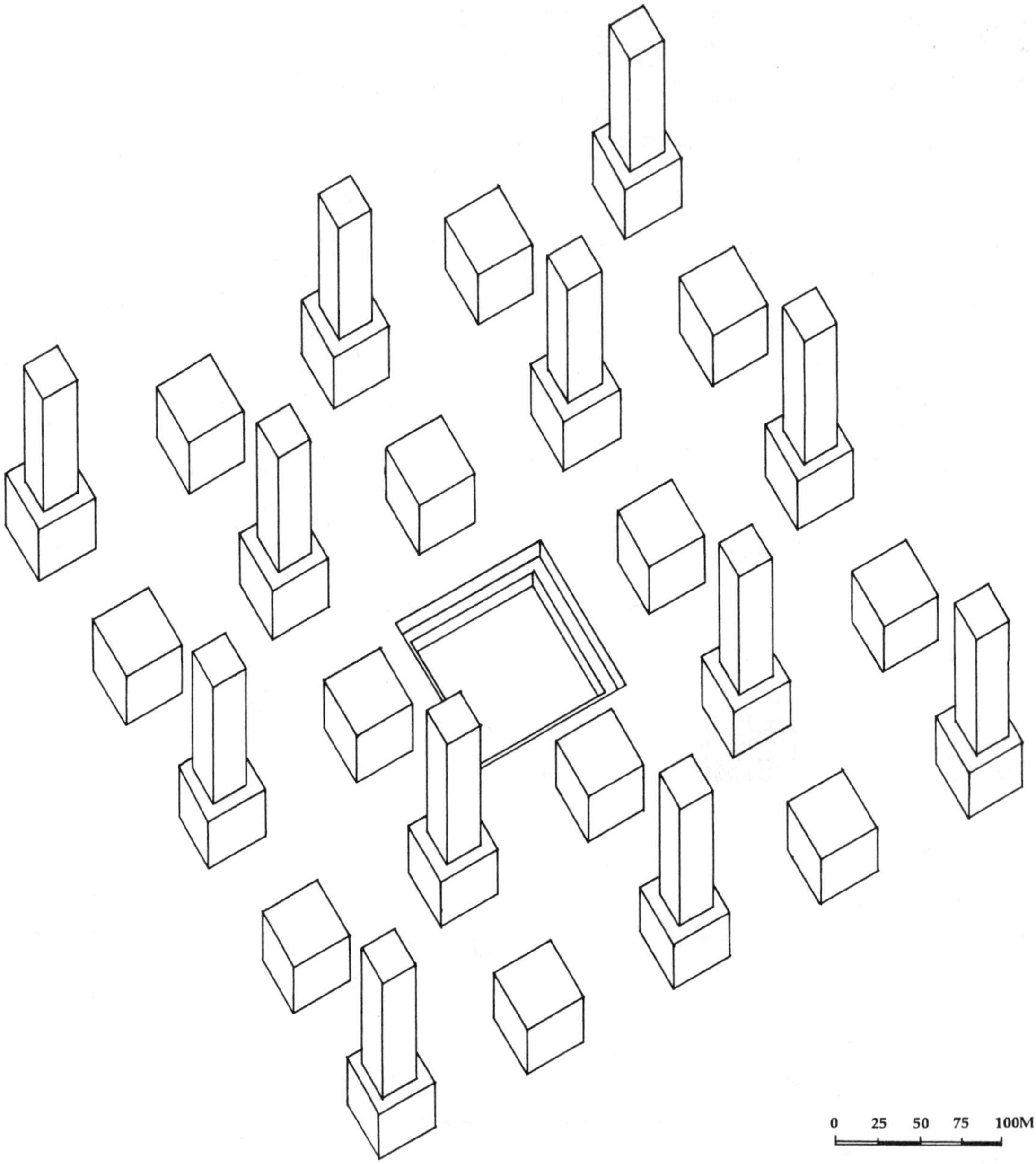
0 25 50 75 100M

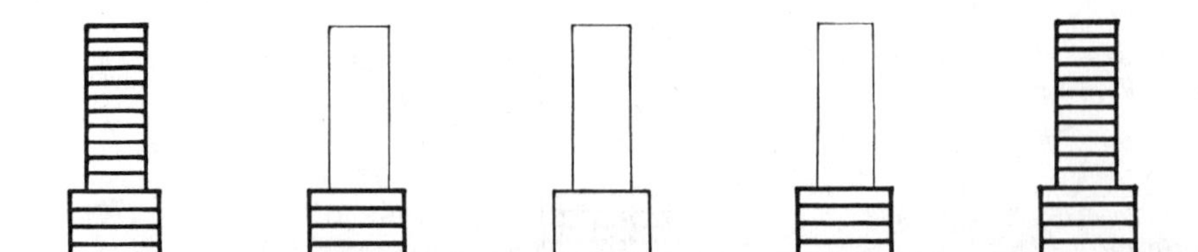

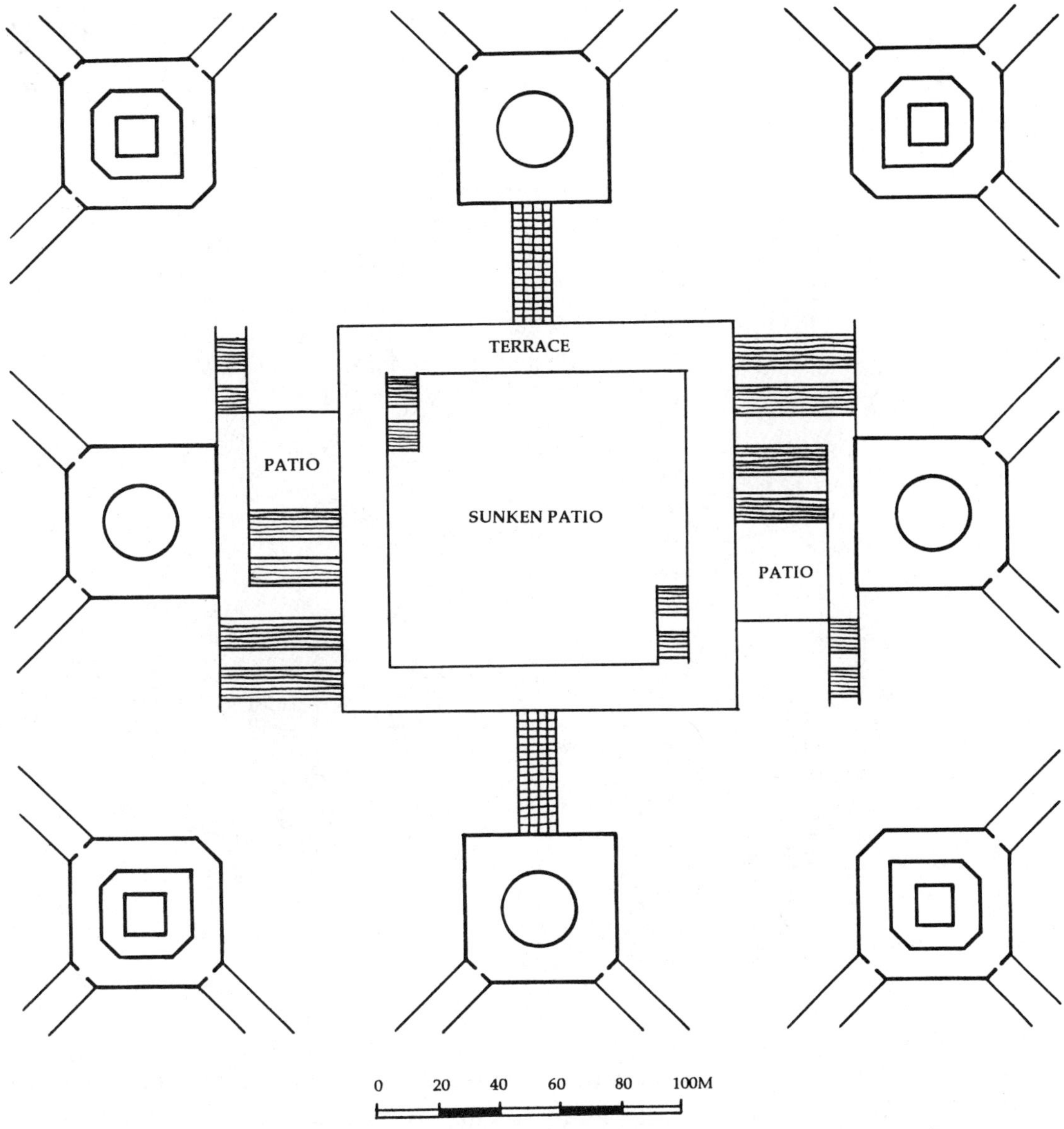

Figure 1.10. (facing page) Schematic perspective and cross section of the downtown supra-space and geo-space.

Figure 1.11. Overall plan of the sunken pit geo-space and the high- and low-rise supra-space buildings surrounding it. All the supra-space buildings are also connected to the geo-space levels.

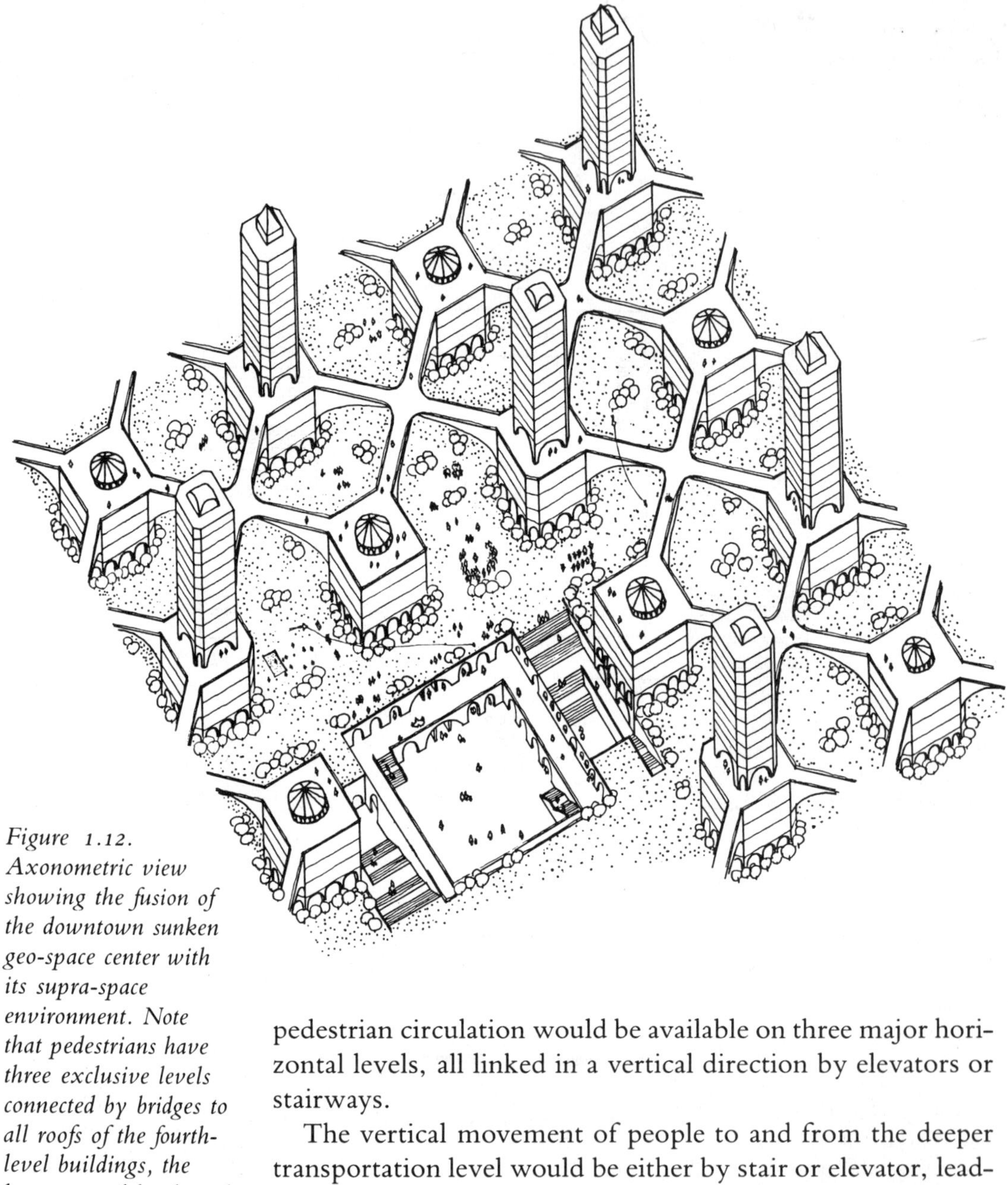

Figure 1.12. Axonometric view showing the fusion of the downtown sunken geo-space center with its supra-space environment. Note that pedestrians have three exclusive levels connected by bridges to all roofs of the fourth-level buildings, the large ground level, and the geo-space level. All levels are free of motor vehicles, and parking is beneath the sunken pit-patio.

pedestrian circulation would be available on three major horizontal levels, all linked in a vertical direction by elevators or stairways.

The vertical movement of people to and from the deeper transportation level would be either by stair or elevator, leading from the parking lots and vehicle routes throughout all levels to the top of each building located above ground. Traffic circulation would take place on levels allocated for subways, motor vehicles, bicycles, and parking. Transportation would be below the level of human activities within the geo-

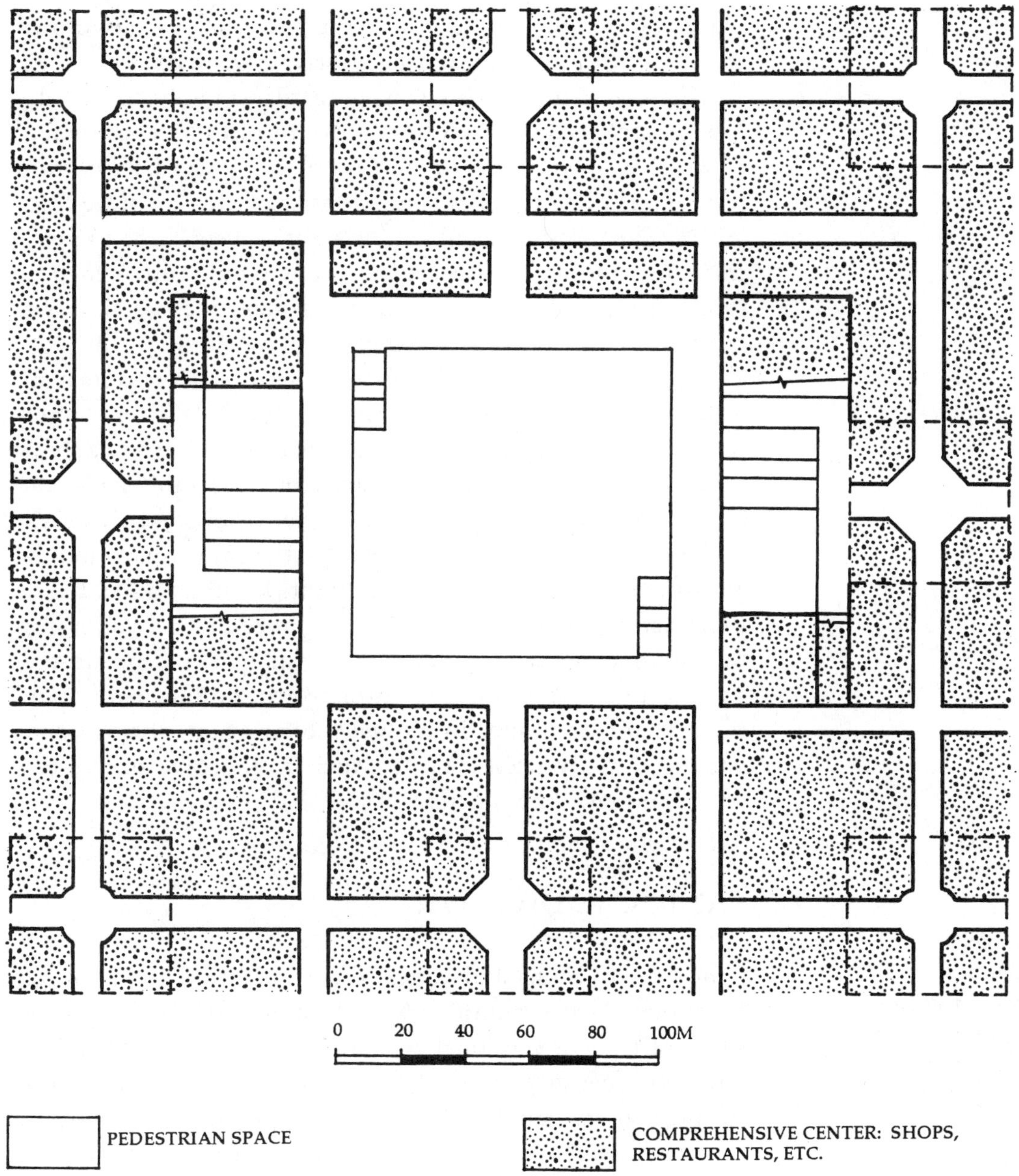

Figure 1.13. Geo-space level illustrating the pedestrian circulation pattern around the comprehensive sunken pit center.

space. However, people who would live in this downtown area would most likely not need private vehicles.

The core of the comprehensive center would be marked by a large sunken pit built of regressive terraces. The pit would be largely open at its top and narrower toward its bottom (Fig. 1.14). The terraces could consist of two or more floor levels. Their design would allow light and sunshine to penetrate into the geo-space center. Light would also penetrate through alternate openings made of glass tiles or skylights built at the ground surface level.

The aboveground buildings surrounding this sunken pit would most likely house apartments, hotels, offices, restaurants, and schools. The geo-space levels would provide shopping centers, restaurants, sport activities, assembly halls, and cultural and educational centers, as well as other daily needs.

The ground level, located between the supra-space and the geo-space levels, would be lit at night. It would likely be the most active space, holding free, organized social gatherings throughout the day and into the night. It would be large in size, of prime location, and attractive in design, and would provide freedom and diversified means of circulation and a variety of daily needs, all of which would give this level a social and city-like quality.

LAND-USE DIVISION

Figure 1.14. (facing page) Axonometric view and cross section of the sunken terraced downtown center. It provides a focal landmark of the geo-space network of commercial, social, and educational activities. Each terrace is 10 meters high in order to allow air movement and possibly provide space loft.

Land-use divisions and patterns in conventional cities differ from one place to another. A discussion of the land-use divisions of the conventional city and that of new towns, constructed especially in Great Britain after World War II, contributes significant lessons to the proposal for a geo-space city. However, the land-use division of the conventional city seems to represent the common and more realistic pattern than that of the new towns.

Both technically and practically, it is feasible for the geo-space city to adopt most of the conventional city's land uses. The following land-use types are commonly used in the supra-space and could be used in the geo-space city.

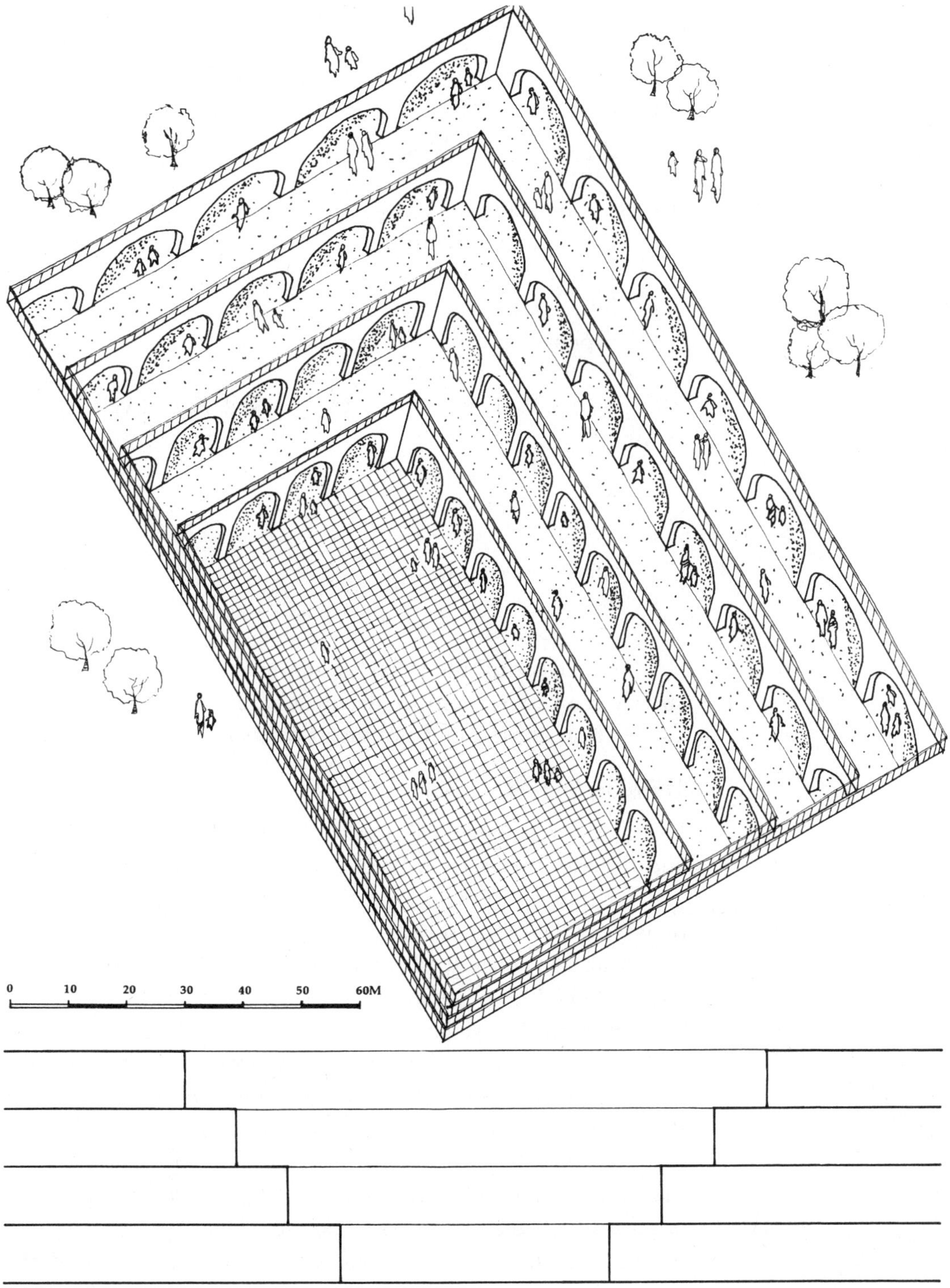
0
10
20
30
40
50
60M

Used above and below ground:

- Habitats (discussed in chapter 2 of this book)
- Commercial buildings (including shopping centers)
- Industrial buildings
- Offices
- Some health services
- Educational facilities
- Entertainment
- Some agriculture
- Open space and playgrounds

Suggested land-use types to be used exclusively below ground:

- Transportation
- Utilities
- Delivery systems
- Storage

When analyzing the types of geo-space usage, it is important to consider them without windows to witness their proper functioning and pollution control.

Windowless Structures

Except for residences, all types of urban structures can exist and function properly without windows. A large portion of urban land uses in the conventional city fall into this category of windowless structures.

Windows typically allow for sunshine, ventilation, and air circulation. They establish visual contact with the outdoor environment. However, some contemporary buildings in our large urban centers are intentionally designed without windows.[38] Such types include entertainment buildings, educational facilities, cultural centers, restaurants, industrial buildings, shopping centers, parking facilities, and many others (Table 1.5). Windowless environments can facilitate productivity and reduce outside interference.

To accommodate geo-space structures, almost all of the

TABLE 1.5.
Survey of Urban Windowless Environments Implementable in Geo-Space

AGRICULTURE AND FOOD

- Bakeries
- Catering services
- Meat-packing plants
- Meat retail and wholesale shops
- Mushroom growing
- Raising chickens
- Shops, department stores of all types
- Slaughterhouses

BURIAL

- Cemeteries, mausoleums
- Funeral homes
- Pet cemeteries

BUSINESS SERVICES

- Auction houses
- Automobile rental agencies
- Building maintenance and repair
- Data processing offices
- Distribution services
- Employment services
- Equipment, other rental services
- Firefighting services
- Moving companies
- Packing and shipping
- Pet grooming services
- Police headquarters
- Post office
- Rental agencies, real estate
- Résumé writing services
- School supplies
- Some office buildings
- Travel agencies

CREATIVE WORK

- Art galleries
- Artist display shops
- Artists' painting supplies
- Crafts and handiwork
- Meditation centers
- Music stores
- Sculpture
- Sewing and fabric shops
- Writing supply stores

EDUCATIONAL FACILITIES

- Art schools
- Bookshops
- Classrooms: schools, universities
- Cultural centers
- Exhibition halls
- Lecture halls
- Libraries
- Museums of all types
- Music schools
- Newspaper facilities
- Photography shops
- Publishing houses
- Research centers

ENTERTAINMENT-SPORT CENTERS

- Bars and taverns
- Bowling alleys
- Boxing arenas
- Broadcasting centers
- Cinema (movie theaters)
- Football, basketball, tennis centers
- Gambling casinos
- Gymnastics centers
- Music shops
- Opera and concert halls
- Party halls
- Pool halls
- Public gathering halls
- Skating rinks: roller, ice
- Social clubs of all types
- Sports arenas
- Sports clubs of all types
- Swimming pools
- Table tennis, indoor courts
- Tennis and badminton courts
- Theaters
- Video rental shops
- Wrestling arenas

FINANCIAL CENTERS

- Banking services
- Banks
- Brokers, financial services
- Clearinghouses
- Insurance companies
- Real estate offices
- Trust companies

INDUSTRIES, FACTORIES

- Automobile manufacturers
- Building equipment and supplies
- Building manufacturers
- Carpet industry
- Cleaning industry
- Clothing industry
- Commercial industry
- Dumping areas
- Film industry
- Food industry
- Furniture manufacturers
- Processing plants
- Recording industry
- Tailors
- Textile industry
- Wine industry

MEDICAL CENTERS

- Animal hospitals
- Clinics
- Health care equipment
- Hospitals, surgery rooms
- Human service organizations
- Medical laboratories
- Pharmacy and drug centers
- Physicians' offices
- Rehabilitation centers
- Shelters
- Social service organizations

(*continued*)

TABLE 1.5. *(Continued)*

PERSONAL SERVICES

- Aerobics centers
- Barber shops
- Beauty shops
- Chiropractors
- Gymnasiums
- Health spas
- Laundry service
- Masseurs
- Physical fitness

PRINTING SERVICES

- Advertisement centers
- Bookbinders
- Copy and photo services
- Drafting equipment
- Printing and page layout

SHOPPING CENTERS

- Alcoholic beverage stores
- Clothing stores
- Furniture stores
- Garden and lawn supplies
- Grocery stores
- Hardware stores
- Household items
- Pet shops
- Post offices
- Restaurants
- School supplies
- Supply shops
- Surplus stores
- Thrift shops
- Trade centers
- Wholesale shops

STORAGE

- Agricultural products
- Equipment
- Furniture storage
- Garages
- General storage
- Grain storage
- Refrigeration
- Snow removal equipment storage

TRANSPORTATION

- Automobile display and sales
- Auto repair and services
- Bicycle shops
- Car washes
- Parking garages and lots
- Parts supply
- Railway, bus, taxi stations

WAREHOUSES

- Building materials
- Personal storage

WORSHIP CENTERS

- Churches and synagogues

traditional functions of the window can be replaced using technical methods. Ventilation and air circulation can be channeled into every section of the building regardless of its location above or below ground. Residential land use is an exceptional case, as natural ventilation and sunshine should be channeled directly into the building. However, it is possible to construct residential geo-space. As will be discussed in chapter 2, specific technology for accommodating residential needs exists and homes can still receive full sunshine, good ventilation, and other environmental necessities if built on a slope and not on a flat site.[39]

Natural light and sunshine can be introduced deep into the geo-space structure through a prism or by refracting light through a set of mirrors, which will reflect the light deep into the ground.[40] The U.S. Underground Space Center in Minneapolis, Minnesota, is constructed 33.5 meters (110 feet) deep below ground and still receives natural light and sunshine. In addition, the dynamic rhythm of the outdoor landscape can also be introduced into the subterranean indoor space through audiovisual images projected on a screen-window.

Research conducted on windowless schoolrooms has indicated that windowless classes are more conducive to focused attention.[41] Normally, outside activities can be very disruptive. While this research did not provide definitive or conclusive answers, it did indicate that windowless environments may affect the learning achievements of some youngsters. Interestingly, the teachers unanimously preferred windowless classrooms and were pleased to have more wall space for displays: "The children were no longer distracted by outside events when the classrooms became windowless."[42] In another study, conducted at Abo Elementary School in Artesia, New Mexico, the results showed improved achievement among the students in belowground classroom settings.[43]

Commercial Centers

The contemporary practice of constructing geo-space shopping centers has been successful in some technologically advanced countries, such as Japan and Canada. Although these shopping centers are typically located within a flat site, they can be suited to slopes as well. Except in the United States, shopping centers are usually located at the center of a city or at the center of a neighborhood. Most of Japan's 76 belowground shopping centers are a by-product of the need to connect subway stations, and are usually developed on flat sites. Their design and construction are remarkable, and they have efficient operating systems. Similar successful experiences have been achieved in Toronto and Montreal, Canada.

Geo-space shopping centers within flat sites face different challenges than those on sloped sites and require special design treatment. In most cases, they are integrated within an existing urban structure in need of renewal, or are located on a new site integrated with urban supra-space high-rise buildings. In either case, an innovative urban and architectural design is essential to meet the new environmental conditions. These shopping centers can be designed in both flat and sloped topography. In addition to careful site selection, an imaginative design is needed for each case in order to take advantage of the full potential of natural light and sunshine. In both locations, issues of accessibility, safety, ventilation, parking, circulation, and environmental conditions need to be resolved using unconventional designs. The choice of site has a significant impact on the design quality of each of these issues.

Sites with sloped topography have greater potential to resolve these design issues than do flat sites. A commercial center on a slope oriented southward can receive much more natural light and sunshine. If the shopping center consists of multiple floors, it can have the unique advantage of having two entrances, one at the low level and the other through a shaft-type form (Fig. 1.15). This design concept provides

Figure 1.15. Geo-space commercial center cross section. The center is located on a sloped site with two entrances—one at its lower level on the slope and the other at its upper level—to provide ventilation, easy circulation, alternate accesses, safety, and reasonable amounts of natural light and sunshine.

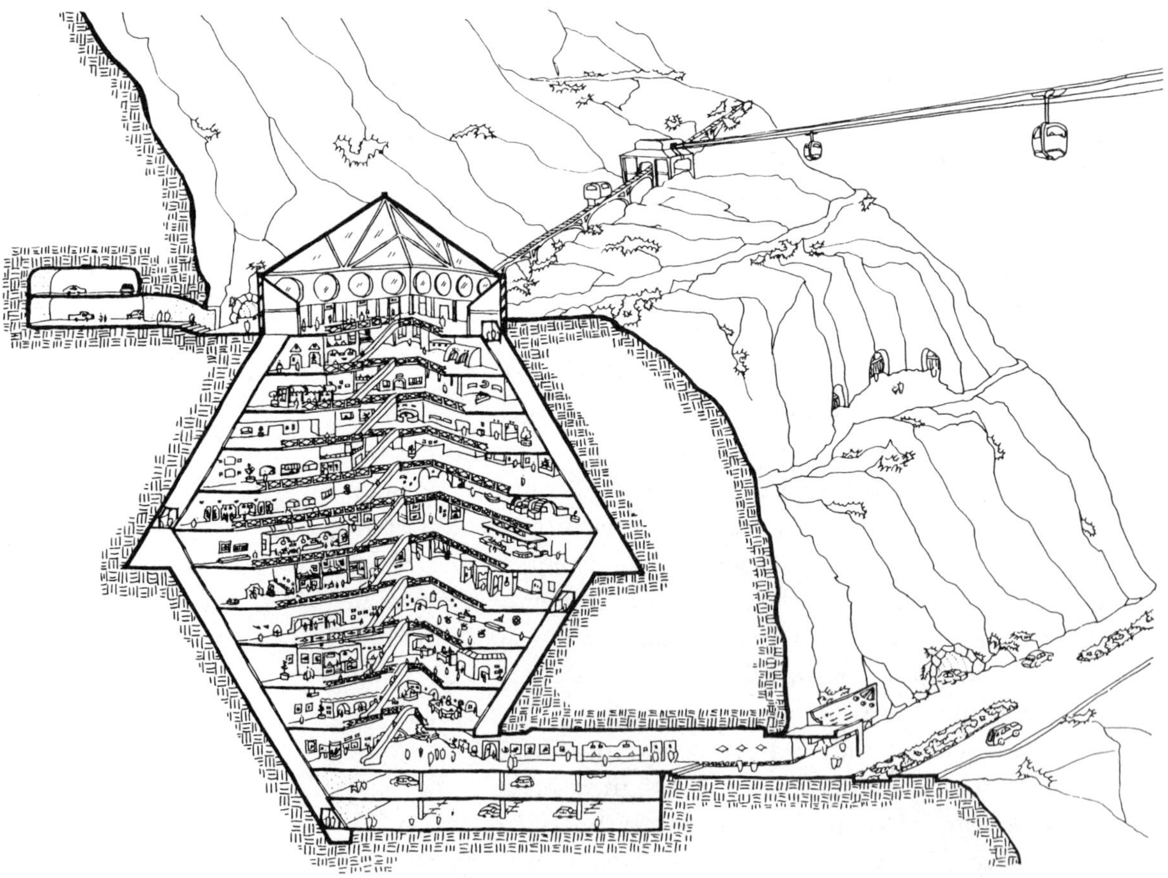

adequate but mild ventilation, two openings for light, easy accessibility, and safety. Such a design can allow additional entrances at each floor level to increase accessibility from different levels of the neighborhood and to permit emergency evacuation. Commercial, entertainment, and other socio-economically consumed land uses constitute the bulk of the activities in the urban core area. However, the slope geo-space section of the city should receive its share of these land uses and even establish its own town centers.

It is feasible to integrate a slope neighborhood shopping center near belowground subway and motorway transportation networks (Fig. 1.16). The proximity of these three elements will enhance the residents' quality of life.

A linear geo-space commercial center is suited to the slope geo-space habitat complex. Such linear shopping centers can also be designed to receive plenty of natural light and sunshine and to accommodate parking (Fig. 1.17).

Another type of slope geo-space shopping center is the terraced form, which blends with the natural configuration of the topography (Fig. 1.18). With this design, each geo-space slope has a comfortable orientation and direct exposure to sunshine, ventilation, and a view of the lowland. Other facilities will enjoy this special positioning because the whole complex is open to the sky.

A commercial shopping center or commercial park can also be designed at the border site between a geo-space slope city and a flat city. In this case, it can serve different types of neighborhoods or different related land uses. Many historical cities flourished economically and grew rapidly due to their location at the border of two contrasting geographical regions.

Geo-space shopping centers located on flat land, however, can be integrated with adjacent supra-space buildings. The network of shops will expand to form a complex within the geo-space city (Fig. 1.19). The geo-space pedestrian circulation will be intertwined with the shops to form an even larger network and create an urban feeling within the geo-space city. Similar circulation patterns can be introduced to form one large cohesive unit.

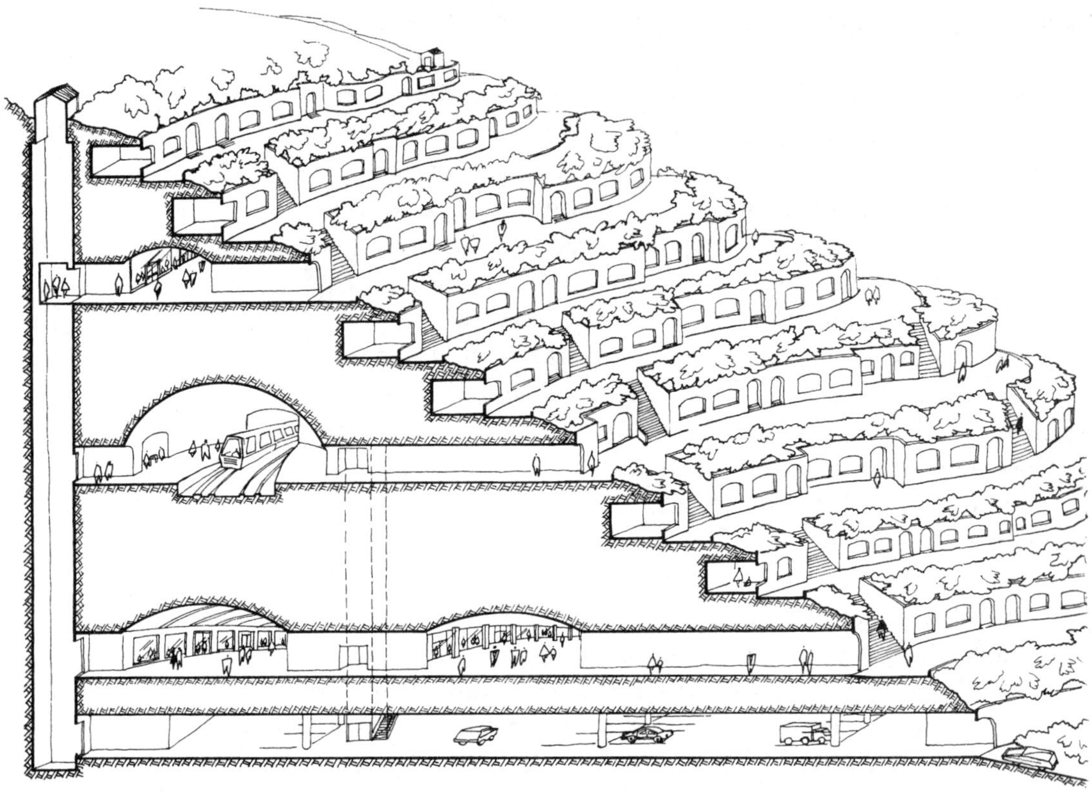

Figure 1.16. Geo-space linear commercial centers. The centers are designed to be adjacent to slope and geo-space dwellings with nearby geo-space transportation.

One basic design principle calls for the creation of a large pit on a flat site around which the commercial center expands horizontally, either in a circumferential, linear, crossing, radial, or polycentric form (Fig. 1.20). Natural light and sunshine are essential elements of the commercial center and can still be provided from the pits and other openings for those geo-space commercial centers located on flat land. The pit form allows the penetration of light through its wide opening, but the form and the materials (glass) of the building can further facilitate the penetration of natural light and sunshine (Fig. 1.21). Also, it is possible to construct pit-type shopping centers on flat sites with aboveground high-rise office buildings and form an interconnected network among them. This complex offers a network for pedestrian circulation on two levels without vehicle interruption.[44] The pit design can take a variety of forms, such as the quadrant, circle, hexagon, or

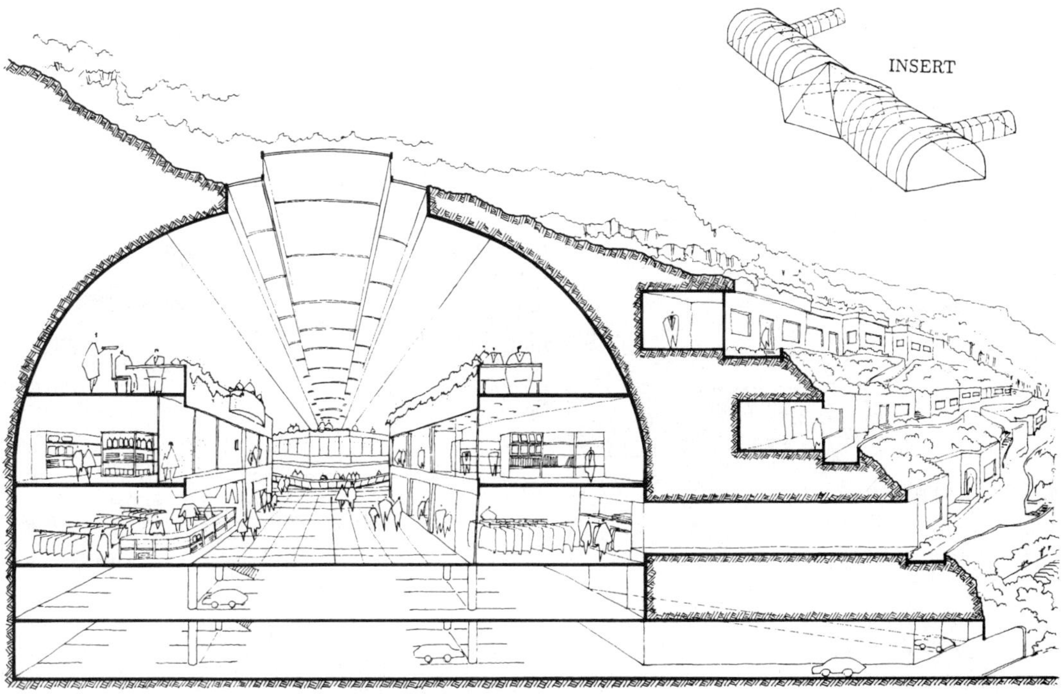

Figure 1.17. Sketch of a linear geo-space shopping center built within a sloped residential neighborhood. The shopping center has air circulation, natural light, and a parking area.

other multisided styles, and can act as a focal point for the geo-space commercial center or geo-space urban center.

The electronics revolution, the increase of urban land prices, and the increasing need for parking space support the fragmentation of some single industries into numerous activities located at different sites. Production, management, storage, and distribution no longer need to be tied together at one site, and the slope geo-space can accommodate one or all parts of an industry. With this fragmentation, the integration of many focal points of different forms has the potential to enhance an urban geo-space atmosphere. To do so, the pit unit and its surrounding shops need to be integrated with the geo-space land use of the adjacent buildings (Fig. 1.22).

Industrial Development

The geo-space urban design industry needs to be divided into polluting and nonpolluting categories. The type of pollution should be further categorized according to air, water, soil, or

Figure 1.18. Cross section of a terraced geo-space shopping center in a slope geo-space city.

noise. Polluting industries usually need a large, flat site, and should not be transferred into the slope geo-space city. Yet, it is possible to apply conventional industrial park development to the slope geo-space when such development combines geo-space and supra-space usage. In this case, certain features must be carefully considered, including accessibility, loading and unloading, parking, storage, the possible inclusion of a management office, and the placement of machinery and heat-producing elements in the supra-space (Fig. 1.23). Non-polluting industry or light industry can be integrated into the flat or slope geo-space. Generally, light, nonpolluting industries are of small size, require limited land area, and compete with other types of land use at the city center. Contemporary light industry is not restricted to any one area and, therefore, is common to many cities. Some types of light industry, such as the precision industries, require sites free of vibration; geo-space is an ideal location for such industries.

Based on our research findings, the comfortable thermal conditions and stable humidity of geo-space can improve labor productivity. In any case, a comparative study of labor productivity in the geo-space and supra-space environments

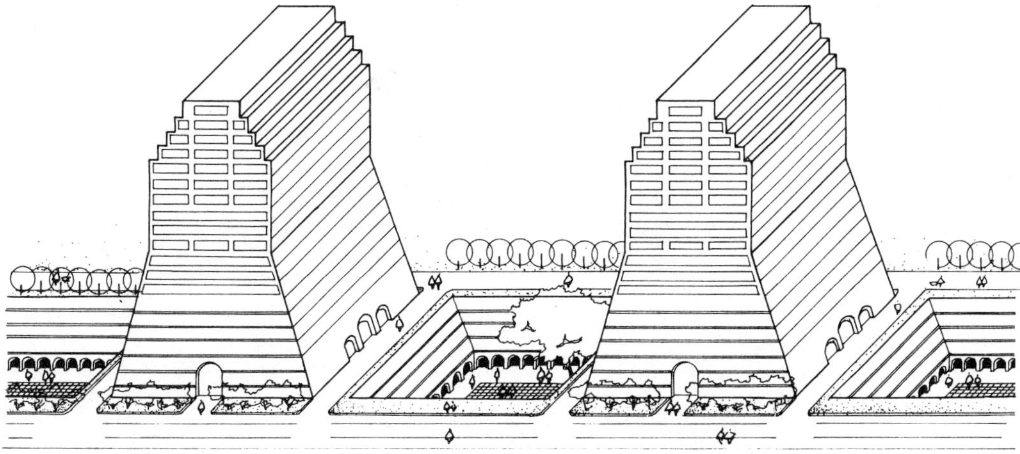

PERSPECTIVE

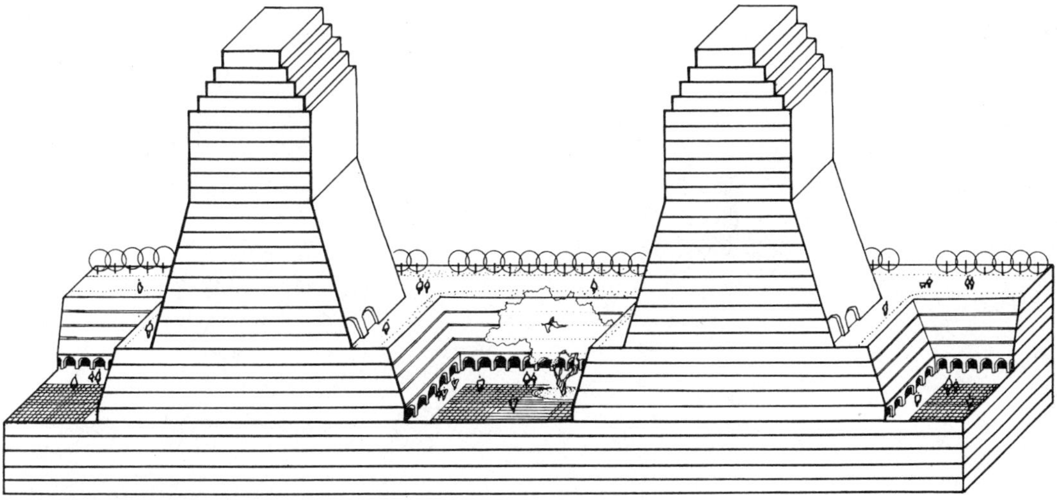

CROSS SECTION

Figure 1.19. Alternating subterranean and supraterranean shopping centers on flat topography. The shopping centers are integrated into buildings that provide for the safe flow of pedestrians and traffic circulation.

is needed. The stable seasonal humidity and thermal performance of the geo-space meet the needs of certain other industries, such as film, wine, and agricultural production.

Many large, old cities that developed on plains are often adjacent to sloped sites. Tokyo, Rio de Janeiro, San Francisco, and Paris are located in such sites and can potentially use those slopes. Slopes alone cannot attract the development of light industry, however, because of the limited space requirements and the cost of developing and operating terraced

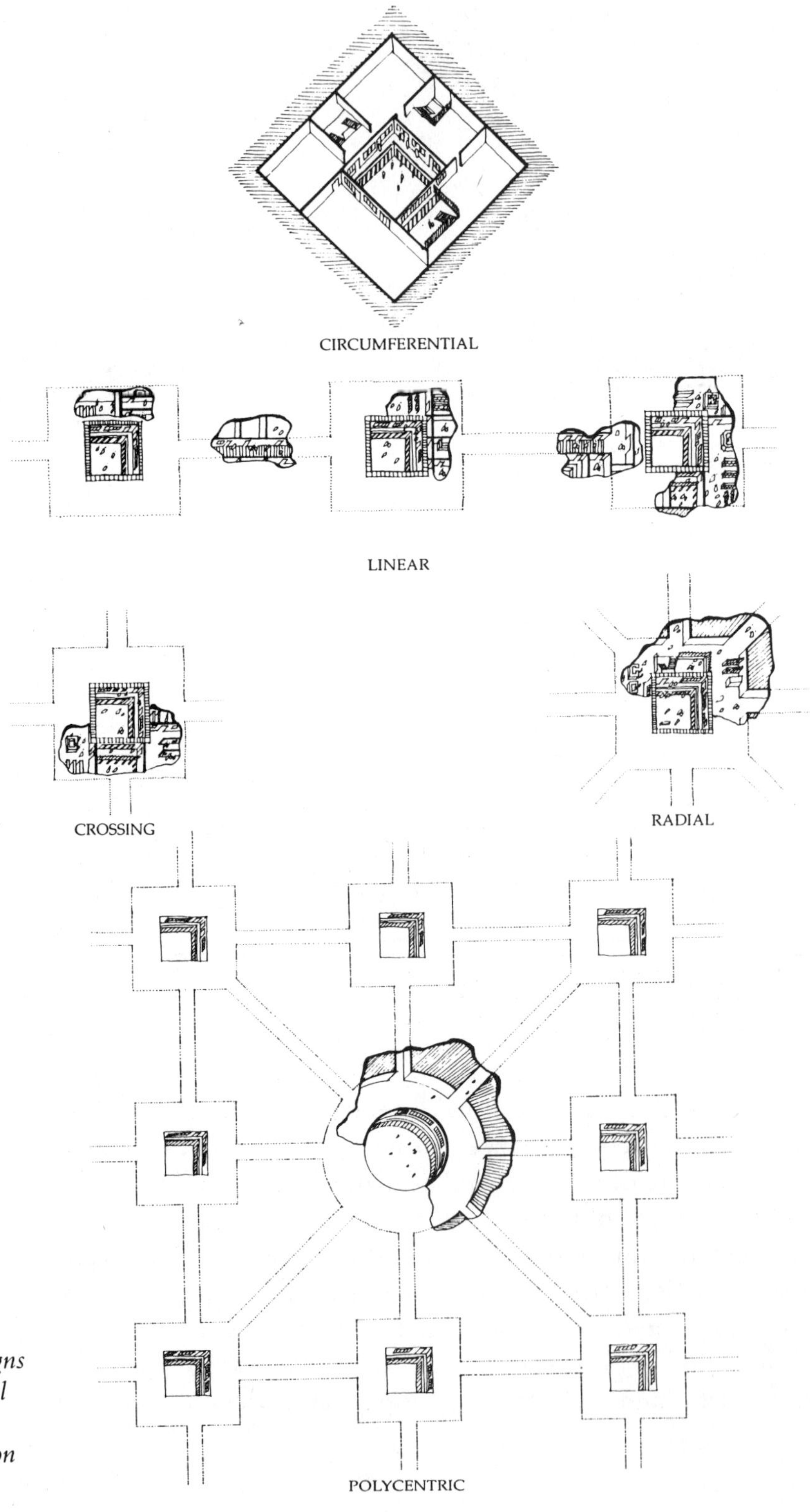

Figure 1.20. Various complex pattern designs of pit-type commercial centers with their potential expansion on flat topography.

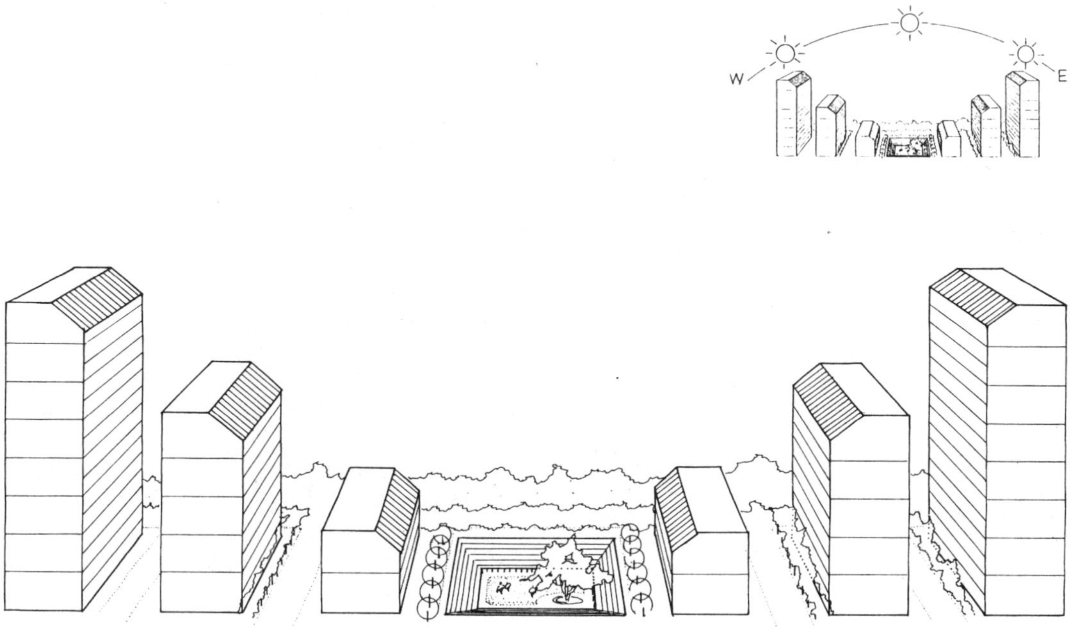

Figure 1.21. Geo-space shopping center on flat land combined with its neighboring buildings. All centers are designed to receive maximum sunshine, light, ventilation, and safe traffic movement.

forms. Yet, with the combination of slopes and geo-space, the relatively low price of land can meet the industry's needs. It seems that slopes can be suitable for large-scale housing, while industry can be completely adapted to geo-space. In some cases, the slope is the only means for future urban expansion, and commercial land use of the geo-space would support the integration of the slope with the flat land uses of the city. This land-use pattern will also reduce the distance between places of residence and places of work.

The major concerns of an industry constructed fully below ground are the transportation and the supply of raw materials, the delivery of products, the accessibility of labor and emergency evacuation systems, comfort, ventilation, and light. But the adjacent slope geo-space neighborhood will use a transportation network of subway, highway, and parking areas that will be provided below ground and can serve both industrial and residential areas. Geo-space city industry should also introduce easy transportation circulation for loading, unloading, and parking. The provision of alternative pedestrian circulation with the outdoor network is essential for emergency needs.

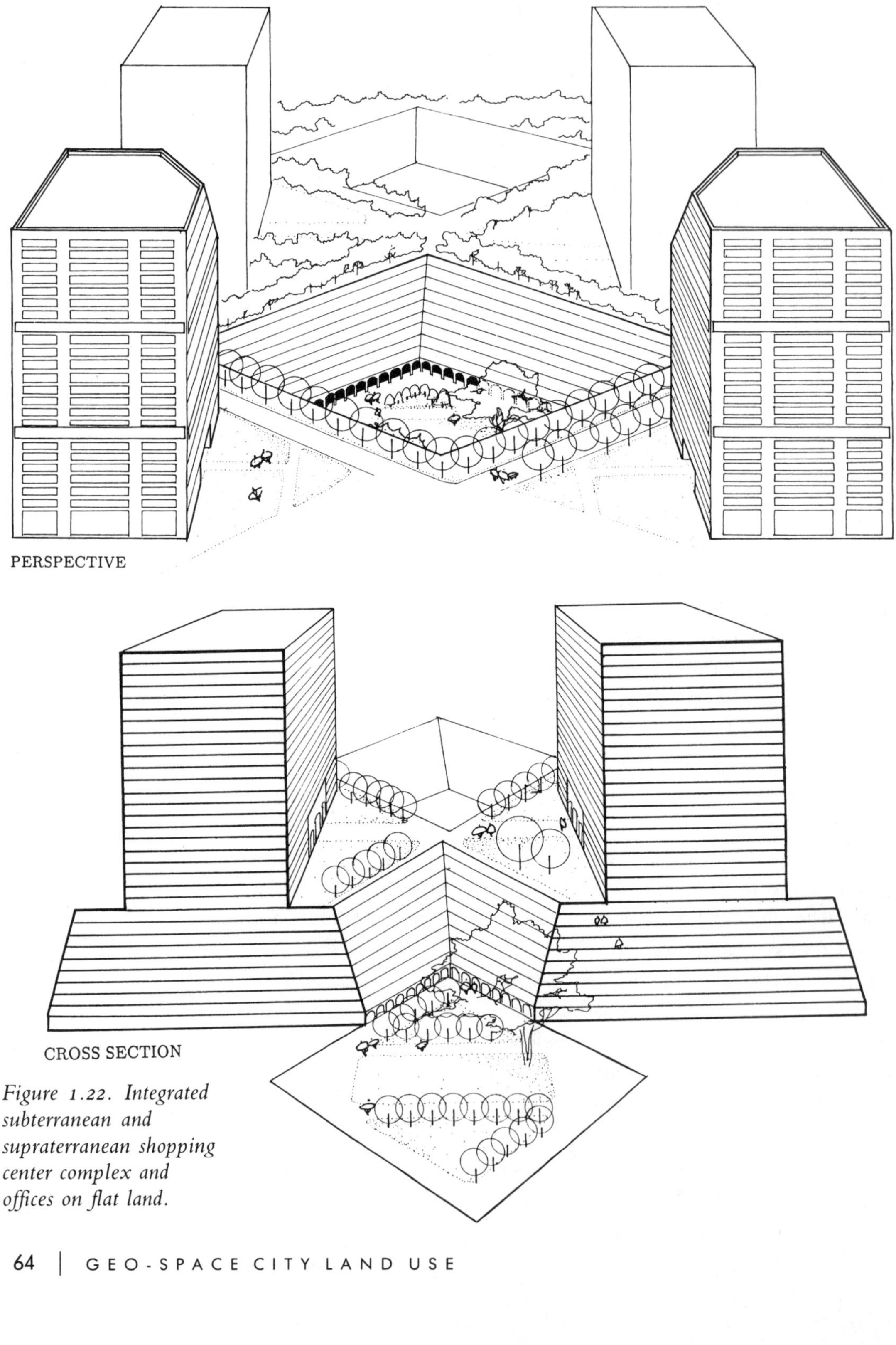

Figure 1.22. Integrated subterranean and supraterranean shopping center complex and offices on flat land.

Figure 1.23. Cross section of an industrial complex in a slope geo-space city. The center has access to a major road and parking lots.

Educational, Cultural, and Entertainment Land Use

Urban slope development introduces a new dimension to the landscape and cityscape within the geo-space city complex. It provides an open view of the city, the lowlands, and the natural surrounding environment. For some urban dwellers, the open slope landscape is perceived as a counterbalance and a complement to the confined space within the geo-space environment.

Educational facilities for all ages need to be located on and within the slopes and therefore have the potential to become the social center of the slope geo-space community (Fig. 1.24).[45] The design and development of such facilities should combine geo-space and supra-space land usage, and, in so doing, these spaces will become a visible part of the slope cityscape.[46] Some types of educational facilities, such as kindergartens, require large and direct exposure to the outdoors as a stimulus to children.[47] The quality of most educational facilities, however, will not suffer from a windowless environment.[48] In fact, students often produce better results in windowless classrooms because the children's concentration is less likely to be diverted.

The view of the slope geo-space cityscape, especially after dark, would be attractive and likely to draw many people to

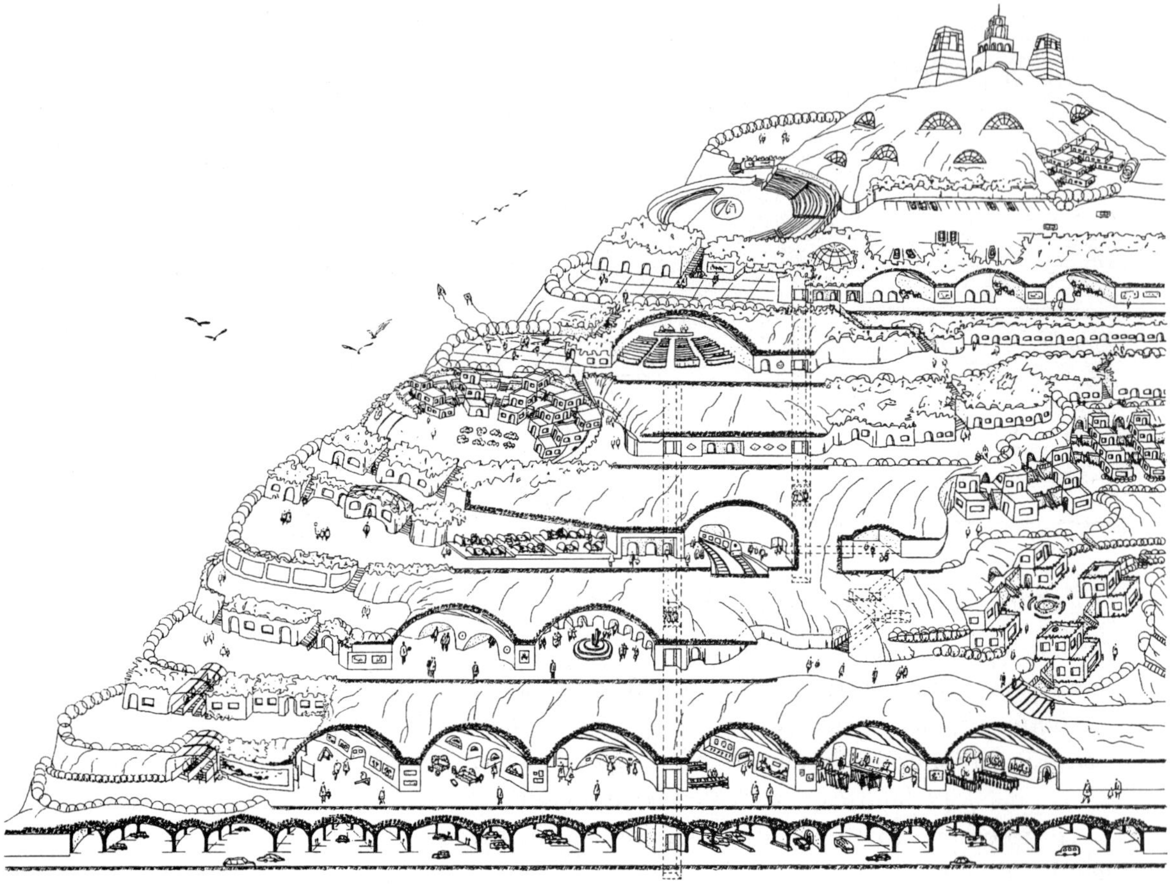

Figure 1.24. Integrated land use on the slope. Combined perspective and cross section (centered) of housing, shopping center, entertainment center, and cultural center at the slope terraced part of the geo-space city, and a university campus at the top of the hill. The centers, as well as the housing, have a view of the lowland.

the area, justifying the development of a downtown center of its own. This draw enables such a location to become a hub of day and evening activity, both cultural and entertainment. Such a center would have a mixture of land uses combining shopping, restaurants, indoor sporting events, concert facilities, and so on.[49] To remain lively and dynamic, the land use should be diverse and respond to the daily and nightly needs of diverse groups of people.

HEALTH, HAZARDS, AND SAFETY

Beyond the technological elements of geo-space design and construction lie two issues associated specifically with the human condition: health and safety. If properly designed, belowground space can address some aspects of these concerns for human beings.

Health

Human health is greatly affected by the environment. Ambient temperature and relative humidity patterns are an element of the environment that governs human health and body development. Change in this environmental element directly influences the physiological functions of the body and its metabolic processes. Belowground space forms an ideal thermal ambient condition for physiological comfort and development.[50] The relatively constant environment helps regulate the nervous system and maintains a balance in heat production and heat dissipation of the body, which in turn helps establish normal physiological functions, promotes effective metabolic rates, and helps the body resist disease.

The findings of one study revealed that in a hospital built in Shanghai, China, where the surgery and the recovery rooms were built below ground, the time required to heal external surgery wounds was reduced by around 20 percent.[51] It seems that the diurnal stability of both the temperature and the humidity sped up the healing process. An added benefit was the quiet of the subterranean rooms, which provided tranquility for the patients.

The constant temperature and the comparatively stable relative humidity had positive effects and possibly increased the longevity of the human body, according to the findings of the research conducted in China.[52] Basically, the belowground space is cool in the summer and warm in the winter.[53] The health benefit of living in geo-space can be summarized in seven items discussed expertly by Huo Yan:[54]

1. **Upper respiratory tract infections.** The stability of the temperature and humidity in geo-space initiates lubrication in the respiratory tract and muscles of the trachea and lungs. When the respiratory tract is in this healthy state, the coughing reflex is able to expel dust

and germs and eliminate infection by bacteria. Consequently, the risk of infection is reduced.

2. **Rheumatism.** Contrary to many popular beliefs, it is not proven that people living below ground are prone to developing arthritis or rheumatism. These conditions develop when air is cold and damp. The climate in geo-space, however, is moderate and humid. It has been found that people in warm, rainy seasons do not have a higher incidence of arthritis or rheumatism than do people living in colder climates. The stability and suitability of the temperature and the relative humidity of underground space can also reduce the incidence of acute rheumatic fever, a disease related to upper respiratory infections. It affects all parts of the body, including many organs, skin, and joints, and is accompanied by pain and fever. The findings of Huo Yan shows that the climate of geo-space, in fact, lessens the occurrence of acute rheumatic fever.
3. **Infections of nasal mucosa.** In a cold-dry environment, the nasal passages are subject to irritation or bleeding and are plagued by spasms. This can reduce one's immunity to bacteria, viruses, and inflammation of tissues, and lead to general discomfort. When living in geo-space, where the relative humidity is higher, the mucosal lining is lubricated. Consequently, the mucous membranes and blood vessels retain their elasticity, blood circulation, and metabolic functions, and increase resistance to colds, drafts, or contaminated gases.
4. **Skin protection.** Skin nutrition is connected to the metabolic functions of the hypodermic glands, which are directly influenced by the weather. In cold-dry weather, the capillaries of the skin are in continual contraction and obstruct blood circulation. This disturbs metabolism inside the tissues, leading to frostbite (injury resulting from cold). Also, cracked skin and bleeding can occur, especially when relative humidity is very low and the temperature is extremely cold. The higher relative humidity and ambient temperature of

the geo-space provide an environment that makes the skin smooth, soft, and resistant to infections.

5. **Noise irritation.** Geo-space is effectively insulated from outdoor noise. In an urban setting, air and noise pollution are intense. Geo-space provides a noise-proof, tranquil environment, which can reduce stress. Moreover, it is an ideal space for peaceful sleep, and can stimulate creative impulses. Prolonged noise irritation can impair the regulative function of the nervous system. In short, geo-space reduces the negative effects of noise on human health.
6. **Radioactive substances.** It is necessary to measure radioactive substances in geo-space. In Chinese cave dwellings built within the loess soil, measurements indicated no traces of radiation or illness by cancer. On the other hand, the Swedish Radioactivity Protection Research Institute has warned that in the upcoming 30 years, living in houses above ground may bring about 1,100 additional cases of cancer. Houses constructed from stone, bricks, and mortar emit some degree of radioactivity.
7. **Good health and longevity.** A survey was conducted in Shanxi Province of China, where 20 percent of the population (5 million people) live in belowground dwellings. In Dayang village of Linfen County, of four families living in old cave dwellings, three people were over 80 years old and one person was over 90 years. Members of the family were in good health.[55] Similarly, in two villages of Yangjiayu Township near Taiyuan, 800 members of 200 families were physically examined. Life spans, death rates, and causes of death among the elderly over the past three generations from these villages were reported. Huo Yan collected data from 364 residents with no family history of genetic diseases who also live in caves dug in the soil. The findings indicated that no arthritis, skin disease, or infectious disease existed, even though 66 people smoked and 13 people drank alcohol. Among the 104 other residents, 72 of them lived in cave dwell-

ings that were built more than 100 years ago. Ninety-nine of them felt that the dwellings were comfortable and all felt that the dwellings were free from dampness; 66.2 percent lived to age 65 or more.

In addition, the geo-space environment provides quiet and tranquility. It is therefore therapeutic for those who need such an environment for fighting insomnia, who suffer from neurasthenia, or who have nervous tendencies. In deep geo-space, a person could sleep longer and more deeply. In addition to the stable temperature and humidity, geo-space provides fresh and clean air free of dust and germs.[56]

The stability of the temperature and the humidity provided by the geo-space, especially in a hot-dry climate, creates a comfortable environment where breathing becomes easier and the skin becomes soft and supple.[57] Body temperature is regulated by the central nervous system through physical and chemical regulation. Chemical regulation controls body heat, which, in turn, regulates metabolism. However, this chemical reaction may be dictated by external physical forces. When ambient temperature is comparatively high and heat dissipation is hindered, body heat production is reduced. On the other hand, when ambient temperature is low, the dissipation decreases, causing an increased body heat production. Thus, physical regulation involves heat dissipation from the body by conduction, convection, radiation, and evaporation. Body growth depends significantly on the metabolism of the substances in the body for energy production and consumption. Extreme ambient temperature and relative humidity cause body discomfort, infection, and even inflammation of the respiratory tract by disturbing the process of evaporation and heat dissipation. In short, most physiological regulatory functions for human comfort and efficient body metabolism are best regulated between temperatures of 10 degrees and 20 degrees C and 30 percent and 75 percent relative humidity.[58]

The findings of this author's field survey on belowground housing in Tunisia and China point out that, in the summer season, people entering dwellings from the outdoors are susceptible to colds. This type of cold is related to external factors and is not caused by bacteria or influenza viruses, which lead to inflammation and infection. The external cold factor can be solved by creating a transitional space between the outdoor and indoor environments.

There may possibly be some contributing factors that a geo-space environment could offer to improve the physiological condition and longevity of the inhabitants. Further research is needed to substantiate the benefits of such an environment.

Hazards and Safety

The geo-space city, especially its sloped parts, is subject to natural hazards and therefore requires special design precautions to ensure safety.[59] Geo-space land use on a slope would reduce the effect of these hazards compared with aboveground usage. Still, an awareness of those hazards and the development of preventive measures are essential.[60] The important slope hazards are debris flow, landslides, slope failure, spread of fire, and extensive erosion.[61]

Debris flow occurs when rock fragments roll down the steep slopes. When debris is mixed with soil, gravel, and water, the downward movement is accelerated, acting like a powerful avalanche.[62] A landslide is the downward movement of a large quantity of soil mixed with rocks, supported by the water runoff that occurs along all types of sloped gradient.[63] Landslides usually occur in soft or clay soil, even with low velocity and small gradient degree. Slope failure occurs suddenly on steep slopes with a small scale, and is often hazardous. Preventive measures include installing retaining walls and proper drainage.

One method of preventing or easing the abovementioned hazards is the planting of tree zones, especially at the upper part of the slopes. In the case of fire, geo-space structures are more likely to inhibit the spread of fires than aboveground structures because, in a sloped site, fire tends to move toward the upper part of the slope. As the smoke spreads, inhalation and heat become critical issues in the geo-space neighborhood. Another preventive measure is to allow for more entrances and exits in every geo-space unit, especially in public buildings and those that accommodate large gatherings.

From the safety point of view, the geo-space city, especially one located on flat land, should have similar pedestrian circulation patterns to the aboveground city.[64] Thus:

1. No belowground unit should be developed as a closed unit, or with limited circulation of movement.
2. The geo-space must have a wide range of arteries free of obstacles for quick mass movement with frequent exit alternatives.
3. The subterranean pattern of arteries for regular daily pedestrian movement should be established similarly to that of the aboveground pedestrian network. The similarity of the two patterns will familiarize the people with the design pattern of the circulation for swift use in case of emergency.[65]

LEGISLATION

In many technologically advanced countries, where planning legislation and city zoning rules and regulations are well established, there is a tendency to be rigid. These rules would need to be reexamined and reevaluated in light of new designs, which integrate slope geo-space and the supra-space city. In most countries, legal land ownership makes reference to the land surface's horizontal dimensions of width and length, with no reference to ownership limits of the vertical dimensions of depth and height.[66] In Japan, for example, the full right of land ownership extends all the way to the center of the earth and to the heavens above. In other countries, there may be a legal reference to the ownership of the natural resources found within the land, such as water, gas, petroleum, coal, or other minerals, which are owned by the public.[67] In still other countries, such as the United States, landowners retain rights to all the mineral resources of the soil. For some time, in Israel, olive trees were owned by a person, while the land on which they grew was owned by others.

The rapid expansion of geo-space construction raises the legal aspect of this new endeavor, especially in technologically advanced countries. During the 1980s, Helsinki, Finland, developed approximately 300,000 m^3 of subsurface space annually. There was a lengthy process involved in acquiring approval for the development of such space and the geotechnical and map information system used in the planning of subsurface developments. Also, recommendations were issued by a committee that has investigated the current practices of subsurface planning and building permit acquisition in Helsinki.[68]

Geo-space usage in a given country should clarify the ownership right of depth, whether it is public or private or joint, and who should own the space rights created within the soil itself.[69] The legality of this issue becomes vital when debating construction of tunnels for subways and other transportation, as well as for the infrastructure needed for the efficient development and safety of a city. It is clear that the introduction of geo-space transportation arteries in the city of Jerusalem, interspersed within the mountainous morphology, can dramatically enhance the city's transportation efficiency and the supra-space environment, as well as the quality of life. Yet, in Japan, most of the 76 geo-space shopping centers were constructed underneath the existing supra-space roads or their cross sections in order to bypass the nonexistent legal rights of geo-space private land ownership. Also in Japan, the majority of the subways are limited to construction underneath existing roads, resulting in lengthy lines and seriously hampering their efficiency.

The public should have input on the legal decision of the development of geo-space transportation, with or without private enterprise. The following items should be legally reexamined within a given country:

1. The legal ownership of the border limit (depth of the geo-space) and the supra-space should be determined. The role of the private landowner and the extent of the depth of his ownership are factors to be considered. Currently, the Japanese are contemplating the possi-

bility of limiting the land ownership of the private holders to a depth of 50 meters. This limit is mentioned since the bedrock in the Tokyo Bay is at a depth of 50 meters.

2. It is this author's opinion that, in legally complicated cases, the owner of the underground space should seek a partnership between the private and the public sectors (local or the central government) as a joint venture.
3. When building in geo-space, the rights and property of the landowner above the ground should be protected under all circumstances. The construction should not result in any future damage or nuisance to those structures, such as pollution of the soil, vibration, or other damage.
4. Geo-space construction should have its own codes and regulations to suit environmental, safety, and health needs in regard to earthquakes, soil pressure, landslides, water leaks, dampness, circulation flow, air movement, light, and ambient thermal comfort (as related to thermal performance and humidity). Any other needs should also be legally regulated.
5. Geo-space development would necessitate the enactment of rules, regulations, and norms concerning construction quality and land-use regulations. The latter should make reference to the segregated or integrated land uses so that environmental quality is enhanced.

The use of the geo-space in a comprehensive and holistic land-use approach is a pioneering concept. It would therefore require an innovative and revolutionary approach to enact a new legal system to protect all parties concerned. It is this author's opinion that private entrepreneurs, if given the incentives, will be a driving force behind geo-space city development. It would also be necessary for the general public to be active in the rules and regulations of construction and to invest in the construction of the basic infrastructure, without which the private entrepreneurs will not participate.[70]

CONCLUSION

The concept of belowground land use is new and requires careful consideration. Diversification of land use is extremely important for success of the geo-space city; however, it may have limited support among general lay persons. It is the responsibility of the urban designer, along with experts such as architects, social philosophers, humanists, politicians, and other potential members of an urban development team, to develop comprehensive and diversified plans that meet the needs of the general public. They can accomplish this, at least in part, by defining problems in our existing urban settings above ground, then introducing realistic yet visionary solutions below ground. The geo-space city is a promising and realistic solution for some of the problems of our congested cities.

We already have some basic experience in the construction of geo-space, which has been developed in this century for limited types of uses. The current trend in cities seems to indicate continued, and even growing, usage of geo-space. We certainly have the technology and the ability to design and construct a geo-space city. A basic limitation is the lack of public awareness and shortage of disseminated research on the social, economic, and environmental implications of geo-space city development.

Our comprehensive approach to the geo-space city necessitates, first, that it must be integrated with the existing supra-space city. Second, it is essential to envision the land uses in their totality, rather than to focus exclusively on one type, such as parking or storage, as has been done conventionally. Third, the development of the geo-space city should be functional and practical for daily mobility. Thus, the supra-space land use, located on the ground surface, and the geo-space located close to the surface should create a flow of pedestrian movement with similar or complementary functions offered to the consumer. Fourth, development of the geo-space city, like any other urban design project, is a tradeoff by which benefits and advantages are gained, but other things are lost. Urban designers and architects seek out utopian or visionary solutions intellectually. In this way, they are often able to introduce pioneering and innovative solutions that may have not been commonly accepted in the early stages of planning, but were adopted over time by the general public.

The geo-space concept is a product of urban necessity. Yet, its implementation will consume a great deal of time. The stronger the pressure caused by accelerated urban problems, the quicker geo-space will be realized. This chapter suggests that development be viewed broadly rather than having the design of belowground spaces in cities continue piecemeal. It is easier to correct mistakes in construction above ground than it is to do so below ground. An overall urban design of the geo-space city is essential for short- and long-range future development, and for the social and economic benefits of this comprehensive approach.

NOTES

1. Tetsuya Hanamura, Mitsugu Shioiri, and Yoshihiko Kouno, "Underground City Development and a Case Study," *Urban Underground Utilization '91, Proceedings* (Tokyo: Urban Underground Space Center of Japan, 1991): pp. 143–48.
2. Tor Geir Espedal and Gunnar Nærum, "Norway's Rennfast Link: the World's Longest and Deepest Subsea Road Tunnel," *Tunneling and Underground Space Technology,* Vol. 9, No. 2 (April 1994): pp. 159–64.
3. Sir Alan Muir Wood, "The Economics of Tunneling: Problems and Opportunities," *Tunneling and Underground Space Technology,* Vol. 2, No. 2 (1987): pp. 131–36.
4. F. L. Moreland, "Earth Covered Habitat—An Alternative Future," *Underground Space,* Vol. 1, No. 4 (Aug. 1977): pp. 295–307.
5. Eugene Raskin, "The Underground City," *Underground Space,* Vol. 3, No. 5 (March/April 1979): pp. 227–28.
6. Kenneth Labs, "The Architectural Underground—Part I," *Underground Space,* Vol. 1, No. 1 (May/June 1976): pp. 1–8. "The Architectural Underground—Part II," *Underground Space,* Vol. 1, No. 2 (July/Aug. 1976): pp. 135–56.
7. Yoshiyuki Uchino, "Utilization of Underground Space of Ginza Area," *Urban Underground Utilization '91, Proceedings* (Tokyo: Urban Underground Space Center of Japan, 1991): pp. 140–50.
8. David Martindale, "Why American Business Is Going Underground," *Underground Space,* Vol. 6, No. 1 (July/Aug. 1981): pp. 21–23.
9. Ichiro Ozawa, Hirai Takashi, and Satoru Otomaru, "A Study on the Method for Areawide Development of Urban Underground Space,"

Urban Underground Utilization '91, Proceedings (Tokyo: Urban Underground Space Center of Japan, 1991): pp. 101–05.

10. Gideon S. Golany, *Earth-Sheltered Habitat, History, Architecture and Urban Design* (New York: Van Nostrand Reinhold, 1983): pp. 162–66.
11. Nobuhiko Nakano, Keiichi Shintani, and Junko Fujita, "Planning of Underground Space Utilization in Tokyo," *Urban Underground Utilization '91, Proceedings* (Tokyo: Urban Underground Space Center of Japan, 1991): pp. 108–12.
12. Nishida Yukio and Nobuyuki Uchiyama, "Subject of Utilization of Underground Space on Urban Development," *Urban Underground Utilization '91, Proceedings* (Tokyo: Urban Underground Space Center of Japan, 1991): pp. 151–54.
13. Gideon S. Golany, *Design and Thermal Performance, Below-Ground Dwellings in China* (Newark, DE: University of Delaware Press, 1990): pp. 23–139.
14. Sydney A. Baggs, "The Dugout Dwelling of an Outback Opal Mining Town in Australia," in *Underground Utilization: Human Response and Social Acceptance of Underground Space,* ed. Truman Stauffer (Kansas City, MO: A Geographic Publication, Department of Geosciences, University of Missouri), Vol. 4 (1978): pp. 573–99.
15. Gideon S. Golany, *Urban Underground Space Design in China, Vernacular and Modern Practice* (Newark, DE: University of Delaware Press, 1989): p. 94.
16. Id., pp. 28–54.
17. Masato Ujigawa, Hiroshi Muto, Masahito Yasuoka, Kotaro Hirata, and Yoshio Tsuchida, "An Experimental Study on the Improvement of the Underground Office Environment," *Urban Underground Utilization '91, Proceedings* (Tokyo: Urban Underground Space Center of Japan, 1991): pp. 505–09.
18. Birger Jansson, "City Planning and the Urban Underground," *Underground Space,* Vol. 3, No. 3 (Nov./Dec. 1978): pp. 99–116.
19. E. L. Foster, "Social Assessment—A Means of Evaluating the Social and Economic Interactions Between Society and Underground Technology," *Underground Space,* Vol. 1, No. 1 (May/June 1976): pp. 61–64.
20. Kunihiko Oshima, Satoshi Ihara, and Hisao Hakuba, "Vision and Subjects of Urban Underground System," *Urban Underground Utilization '91, Proceedings* (Tokyo: Urban Underground Space Center of Japan, 1991): pp. 183–88.
21. Tadashi Tanaka, Takako Kyogoku, and Ikuo Hanatani, "A Planning of Underground Space by Civil Engineers," *Urban Underground Utilization '91, Proceedings* (Tokyo: Urban Underground Space Center of Japan, 1991): pp. 157–62.
22. Yasushi Nakajima, "Perspective Underground's Space View for Resources and Assignment—Case Study of Oya Area Utsunomiya City," *Urban Underground Utilization '91, Proceedings* (Tokyo: Urban Underground Space Center of Japan, 1991): pp. 167–71.

23. Royce LaNier and Frank L. Moreland, "Earth-Sheltered Architecture and Land-Use Policy," *Underground Space,* Vol. 1, No. 4 (Aug. 1977): pp. iii–iv.
24. Gideon S. Golany, *Chinese Earth-Sheltered Dwellings, Indigenous Lessons for Modern Urban Design* (Honolulu, HI: University of Hawaii Press, 1992): pp. 144–54.
25. Gideon S. Golany, "Soil Thermal Performance and Geo-Space Design," *Geotechnical Engineering Proceedings of JSCE—Japan Society of Civil Engineers,* No. 445/III-18 (1992–93) (Invited Paper), pp. 1–8.
26. Golany (1983): pp. 61–73. Also in Gideon S. Golany, *Earth-Sheltered Dwellings in Tunisia, Ancient Lessons for Modern Design* (Newark, DE: University of Delaware Press, 1988): pp. 84–108.
27. Junji Nishi, Hideo Igarashi, Keiichi Sato, and Mototo Morita, "Traffic Problems and Underground Utilization in Urban Area," *Urban Underground Utilization '91, Proceedings* (Tokyo: Urban Underground Space Center of Japan, 1991): pp. 223–30.
28. Keshan Zhu and Sishu Xu, "A Promising Solution to Surface Congestion: Using the Underground," *Underground Space,* Vol. 6, No. 2 (Sept./Oct. 1981): pp. 96–99.
29. Heinz Duddeck, "Future Trends in the Structural Design of Tunnels," *Tunneling and Underground Space Technology,* Vol. 2, No. 2 (1987): pp. 137–42.
30. Andre Guillerme, "Urban Underground Network and Terminal Facilities," *Urban Underground Utilization '91, Proceedings* (Tokyo: Urban Underground Space Center of Japan, 1991): pp. 39–45.
31. ITA Report, "Cost-Benefit Methods for Underground Urban Public Transportation Systems," *Tunneling and Underground Space Technology,* Vol. 5, No. 1/2 (1990): pp. 39–68.
32. Masahiro Endo, "Design of the Underground Expressway in Tokyo," *Urban Underground Utilization '91, Proceedings* (Tokyo: Urban Underground Space Center of Japan, 1991): pp. 206–10.
33. Carlo Grandori, "Tunneling and Underground Construction in the Twenty-First Century," *Tunneling and Underground Space Technology,* Vol. 2, No. 2 (1987): pp. 143–46.
34. CNN Television Broadcast News, February 21–22, 1994.
35. Dan McCosh, "We Drive the World's Best Electric Car," *Popular Science* (January 1994): pp. 52–58.
36. Katsuya Matsushita, Toshio Ojima, and Shuichi Miura, "Environmental Study on Underground Parking Lot Development in City," *Urban Underground Utilization '91, Proceedings* (Tokyo: Urban Underground Space Center of Japan, 1991): pp. 234–41. Ikuo Sugiyama, Yoshinori Nakatani, Michinobu Ohashi, and Takao Wakamatsu, "Programming and Design of Underground Car Parking Network," id., pp. 245–50. Koichiro Hamajima and Shinshiro Matsuoka, "Computer Aided Planning Systems for Underground Parking," id., pp. 253–58. Katsumasa Igarashi and Shinji Okumura, "Conception of Underground Parking System," id., pp. 262–65.

37. Takimochi, et al., *The Japan Corridor Plan,* Committee Report (Tokyo, 1990) (in Japanese).
38. Paul B. Paulus, "On the Psychology of Earth-Covered Buildings," *Underground Space,* Vol. 1, No. 2 (July/Aug. 1976): p. 128.
39. Masato Sato, Masao Inui, and Junichiro Sawaki, "Psychological and Behavioral Effects of Visually Closed Office Space," *Urban Underground Utilization '91, Proceedings* (Tokyo: Urban Underground Space Center of Japan, 1991): pp. 493–504.
40. Wybe J. Van Der Meer, "Possibilities for Subterranean Housing in Arid Zones," in *Housing in Arid Lands, Design and Planning,* ed. Gideon S. Golany (London: The Architectural Press, 1980): pp. 141–50.
41. C. Theodore Larson, *The Effect of Windowless Classrooms on Elementary School Children,* Project Consideration (Ann Arbor, MI: Department of Architecture, University of Michigan, 1965): p. 46.
42. Id., p. 55.
43. Frank W. Lutz, "Studies of Children in Underground School," *Alternative Energy Conservation: the Use of Earth Covered Buildings,* Vol. II, proceedings of conference held in Fort Worth, Texas (July 9–12, 1975), Conference Coordinator: Frank Moreland, pp. 71–77. Also, see Phelan G. Allen, "Fremont Elementary School Air Conditioning," in *Underground Utilization: A Reference Manual of Selected Works, Vol. IV: Human Response and Social Acceptance of Underground Space,* ed. Truman Stauffer (Kansas City, MO: A Geographic Publication, Department of Geoscience, University of Missouri, 1978): pp. 562–63.
44. Shuichi Miura and Toshio Ojima, "Environmental Study on Underground Shopping Center," *Urban Underground Utilization '91, Proceedings* (Tokyo: Urban Underground Space Center of Japan, 1991): pp. 471–77.
45. Charles R. Nelson, "Geotechnical Design of the University of Minnesota CME Building," *Underground Space,* Vol. 8, No. 2 (1984): pp. 77–82.
46. Kenji Okuyama, "A Study of Urban Underground Public Space in Japan and the Planning Complexity of Pedestrian's Selected Space Factions in Underground Space," *Urban Underground Utilization '91, Proceedings* (Tokyo: Urban Underground Space Center of Japan, 1991): pp. 460–67.
47. Frank W. Lutz, "Studies of Children in an Underground School," *Underground Space,* Vol. 1, No. 2 (July/Aug. 1976): pp. 131–34.
48. D. N. Carter, "Terraset School," *Underground Space,* Vol. 1, No. 4 (Aug. 1976): pp. 317–24.
49. "Georgetown University's New Underground Sports Center," *Underground Space,* Vol. 6, No. 3 (Nov./Dec. 1981): pp. 142–46.
50. Golany (1990): pp. 23–125. Also, Golany (1988): pp. 91–108.
51. Golany (1989): p. 94.
52. Yan Huo, "The Effects of Cave Dwellings on Human Health," *Tunneling and Underground Space Technology,* Vol. 1, No. 2 (1986): pp. 171–82.

53. Golany (1990): pp. 33–37.
54. Huo (1986): pp. 172–73.
55. Id., p. 173.
56. Id., p. 175.
57. Xin Zhonghua Zashi journal of 1934, as noted by Huo (1986): p. 171.
58. Huo (1986): pp. 171–72.
59. Junichi Oda and Yuki Shuto, "An Analysis of Disaster in Underground Space," *Urban Underground Utilization '91, Proceedings* (Tokyo: Urban Underground Space Center of Japan, 1991): pp. 578–81.
60. Ikuo Watanabe, Shigeyuki Ueno, Masahiro Koga, Kiyoaki Muramoto, Toru Abe, and Tamiko Goto, "Examination on Safety and Disaster Prevention Measure for Underground Space Based on Analysis of Disaster Cases," *Urban Underground Utilization '91, Proceedings* (Tokyo: Urban Underground Space Center of Japan, 1991): pp. 570–75.
61. Golany (1992): pp. 109–15.
62. Tomomitsu Yasue and Isao Tsukagoshi, "Hazard and Disaster Prevention in Japanese Cities: Dealing with Landslides, Slope Failures and Fires," *Regional Development Catalogue,* Vol. 12, No. 2 (Summer 1991): p. 78.
63. Id.
64. Thomas Frost, "Safety Criteria for Underground Developments," *Underground Space,* Vol. 9, No. 4 (1984): pp. 210–24.
65. I. Watanabe et al., "Safety and Disaster Prevention Measures for Underground Space: An Analysis of Disaster Cases," *Tunneling and Underground Space Technology,* Vol. 7, No. 4 (Oct. 1992): pp. 317–24.
66. E. E. Sterling and C. J. Circo, "Legal Principles and Practical Problems in the Two-Tier Development of Underground Space," *Underground Space,* Vol. 8, No. 5–6 (1984): pp. 304–19.
67. William A. Thomas, "Ownership of Subterranean Space," *Underground Space,* Vol. 3, No. 4 (Jan./Feb. 1979): pp. 155–64.
68. S. Narvi, U. Vihavainen, J. Korpi, and J. Havukainen, "Legal, Administrative and Planning Issues for Subsurface Development in Helsinki," *Tunneling and Underground Space Technology,* Vol. 9, No. 3 (July 1994): pp. 379–84.
69. Allan M. Ziebarth, "Potential Problems of Legal Liability in the Design, Construction and Ownership of an Earth-Shelter," *Earth Shelter* 2 (1979–1980), USC Series, Collected Papers presented at Earth Sheltered Housing Conference & Exhibition (Minneapolis, MN: The Underground Space Center of the University of Minnesota, 1980): pp. 133–38.
70. William A. Thomas, "Technology, Law, and Public Perception of Subterranean Space," *Underground Space,* Vol. 3, No. 4 (Jan./Feb. 1979): pp. iii–iv.

2
SLOPE GEO-SPACE HABITAT DESIGN

G. S. Golany

Historically, communities have used geo-space for defense, as a response to a shortage of space, for protection against harsh climates, for environmental advantages, or because of other external or internal forces, such as poverty. With the exception of the Roman subterranean town at Bulla Regia in northern Tunisia, early geo-space usage was lacking in three basic elements. First, it was not a product of a preconceived design and the basic design of the geo-space land use was deficient. Second, it ignored the features of light, sunshine, and effective ventilation. Third, advanced technology was not available to overcome the obstacles and introduce a healthy underground environment. Examination of the historical experience leads us to the conclusion that even though the implementation was improper, the basic concept of geo-space usage was advantageous to society. If we resolve those problems and introduce an innovative design to meet contemporary standards of comfort and acceptability, we will gain from the benefits of geo-space.

As the use of geo-space is quite novel in this century, its design still has many issues that need to be resolved. There is a need for treatment of soil composition, to improve the degree of resistance to humidity and temperature, to determine the structural constraints and proper building materials, and to allow sufficient natural light and sunshine penetration, air circulation, accessibility, and contact with the natural environment.[1] This author's study indicates that with modern technology, innovative design, and management, these structural design problems will be able to be resolved and meet modern standards. The major obstacle that will remain is a psychological one.[2]

PERCEPTION AND PSYCHOLOGY

Many of the problems associated with living in subterranean habitats are not technological ones, but rather are related to the degree of social acceptance of the concept and to the individual perception of the space.[3] There are three dimensions of psychological problems associated with belowground space.[4] These are bias and prejudice, claustrophobia, and self-image.[5]

Bias and Prejudice

The psychological barrier that the public has formed against geo-space usage is understandably one of the most difficult problems. Bias or prejudice has been common throughout the world among all socioeconomic classes.[6] This bias evolved from the real conditions that settlers have encountered throughout the history of civilization. Today, most people have developed images of negative conditions in geo-space, such as darkness, dampness, disease, isolation, infestation with vermin and snakes, and, finally, poverty and backwardness. However, bias and prejudice are the perceptions mostly of outsiders who have not been exposed to belowground usage.[7] The perceptions of belowground space usage have as much to do with the education of the public as with the quality of its historical design.

Claustrophobia

Claustrophobia is defined as an abnormal fear of being in confined, narrow, or closed spaces. People who are claustrophobic and have an inherent tendency to perceive confined spaces as stressful can have their perceptions, but not their fears, eased by special design treatment of belowground space. Basically, closeness is a relative scale to human beings' perceptions; its reference to the design dimensions of ceiling height, width, and depth; and elements of light and sunshine penetration, positioning of space in relation to its outdoor environment, space size, light versus dark colors, or the introduction of natural environment into the confined space. Wise use of these and other design techniques can significantly contribute to diminishing the feeling of claustrophobia, though not cure it.[8]

Claustrophobia is most likely to be found among elderly persons, children, and women in underground spaces.[9] These users must be correctly informed about geo-space in order to handle their claustrophobia. Certainly, design techniques that incorporate appropriate ceiling height, light colors, wide spaces, natural elements, and a direct view to the outdoors would also minimize the feeling of claustrophobia.

Self-Image

Historical development of belowground space has left indigenous users with a low opinion of this environment. When we interviewed users during our field survey research in both China and Tunisia, they expressed negative sentiments about their underground habitats. In Tunisia, the younger generation is in the process of leaving the belowground dwellings to live in aboveground habitats.[10]

There are many reasons why people live underground. In China, for example, 40 million people live in belowground space because they have no alternative living accommodations. In another case, a family willingly accepted to live in a modern belowground dwelling, recently designed for them by this author, which was filled with light and sunshine, and built on a cliff.[11] Similarly, the users of modern Japanese belowground shopping centers, which were designed and developed with high quality, have indicated to us in our interviews that they are very pleased with their environment, accommodations, and the overall design. These centers are extensively used, are growing in number, and employ all the latest technology and facilities. This author considers a positive self-image by indigenous and modern users an essential element that should be taken into account.[12] Designers need to introduce plans that will improve the user's perception through site selection, colors, space dimensions, and forms. In this way, the designer will create a positive attitude toward the use of underground space.[13]

The other psychological dimension is the need for tranquility and quiet, especially for some mentally ill persons. Geo-space provides such conditions for mentally ill people. Several mental hospitals have been placed in the cavities of underground salt quarries near Krakow, Poland.[14]

In conclusion, uprooting bias and prejudice is a lengthy process and requires intensive explanation and education. The approach should be directed to all age groups through different forms of media—visual means and lectures. More than anything else, public opinion can be changed when innovative design responds to the problems associated with belowground usage. The positive self-image of dwellers in underground habitats will also grow when middle-class society actually begins to use belowground dwellings. Improvement of public image of the belowground will happen when people recognize the significant benefits of such dwellings in ambient comfort, health conditions, energy efficiency, environmental quality, and, no less importantly, economic costs.

The most difficult problem, however, is eliminating the negative psychological attitudes we have developed against belowground space usage based on the deficient designs of the past. To eliminate psychosocial problems, the following policy is advised:

1. **Design:** Introduce an elaborate innovative design of urban, large-scale dwellings, which integrates the usage of belowground space with aboveground space. Such a design should meet our modern standards of living and provide a pleasing and healthy environment with none of the negative aspects associated with past

design practices. The design should consider a wide habitat space, with high ceilings, large windows to allow sunshine penetration, light and pleasing colors, direct indoor-outdoor eye contact, and the introduction of the natural environment into the geo-space habitat.

2. **Consumption and use:** Increase belowground space for residential and nonresidential uses and emphasize its advantages so that it can compete in the free-market economy. The estimate is that the potential space within the city for non-dwellings may average between 40 percent and 50 percent of the total city area. Most of those spaces can occupy a windowless environment (Table 1.5). Thus, it is possible to transfer a reasonable portion of our modern city land uses below ground.
3. **Public image:** Develop public education so that people recognize that belowground space can respond adequately and positively to our space needs and meet our modern standards.

INNOVATIVE IMPLEMENTATION METHODS

If we look at the geo-space habitat project comprehensively, scientifically, and rationally and free ourselves from our traditional biases, we may conclude that the advantages outweigh the disadvantages (see Table 1.2, page 23). An innovative design approach to the geo-space city concept fuses the supra-slope and the geo-space designs. One method of design implementation is to continue the practice of hillside slope construction that has evolved throughout history. The other method involves making use of the space below the slope, to be called geo-space habitat design. However, it is necessary to first outline the wealth of practical experience that has been gained throughout the history of human habitat development on the supra-slopes. Subsequently, an investigation should be conducted of the advantages and disadvantages and lessons to be gleaned from the experiences of the supra-space development for the sake of enhancing the geo-space habitat design. Our suggestion is that the habitats of the geo-space be developed exclusively on the slope.[15]

Hillside Supra-Space Settlements

Most, if not all, hillside towns developed through trial and error rather than from a preconceived design. Slope or hillside towns, common throughout history, evolved in the ancient world of the Middle East, the Mediterranean, Europe, and many parts of Asia and Japan. Hillside towns have used slopes very efficiently in gradients ranging from gentle to acute, mostly in compact form with aesthetically and functionally creative and innovative forms. The slope site was selected for its defensive position, climate, efficient use of land for agriculture and other purposes, or for capturing the view of the surrounding landscape.[16] The hillside city has proved to be a successful development, both architecturally and functionally (Table 2.1).

TABLE 2.1.
Slope Hillside Supra-Space Practice, Pros and Cons

PROS	CONS
HISTORICAL VALUES	
• The history of human settlements is rich in experience on slope development and has even a few cases of belowground habitats on slopes.[17] These two models are worthy of study. • Modern slope site revitalization introduces significant, unique quality and is now in practice in Israel and Hong Kong.	• Although the location provided many benefits, the implementation of slope geo-space in many historical examples was associated with negative features including stress, darkness, backwardness, and poverty.
ECONOMIC	
• Saves the lowland for agriculture. • In a site with shallow soil, foundations of the slope geo-space city are solid because the base rock is close to the surface. • At the early stage of slope site development, land prices are low and can offset or ease the higher costs of infrastructure construction.	• The initial investment cost for the construction of roads and infrastructure is higher than that of the lowland. • Requires greater costs for pumping water to higher elevations. • Construction costs are generally higher than on a flat site. • In a site with deep baserock, there is a need for a deep foundation. Such sites are limited

(*continued*)

TABLE 2.1. *(Continued)*

PROS	CONS
	to the construction of single houses and are not adequate for high-rise buildings or heavy structures. • Construction time is longer.
SOCIAL	
• Gives privacy to individual houses and the neighborhood. • Provides inhabitants an unobstructed view with a pleasing feeling. • Increases the value of the houses and the city. • Provides propinquity and a sense of community.	• May discourage pedestrian movement of the elderly, especially uphill.
AIR QUALITY, VENTILATION	
• Is free from air inversion. • Reduces pollution and provides fresh, clean air free of dust. • Provides good ventilation, needed especially in warm-humid climates. • Has fewer odors and noises of all kinds.	• Is subject to atmospheric changes, changes in wind direction, and occasionally strong wind, which may limit construction to sheltered slopes.
HEALTH AND SAFETY	
• Provides unobstructed sunshine. • Creates a healthy environment, especially in highly rainy regions. • Has less automobile traffic than flatland cities and therefore a lower risk of car accidents. • Receives maximum desired light and sunshine deep into the housing unit. • Has a low water table, which keeps the foundation of the house dry.	• May require some walking between parking and residences. • May have limited accessibility.
AESTHETIC VALUES	
• Provides direct indoor-outdoor view. • Provides the opportunity for private land ownership and detached habitats. • Introduces an attractive view to the surrounding lowland. • Introduces spectacular, unusual, attractive, and varied views to the cityscape and enhances the aesthetic quality and local charm,	

(continued)

TABLE 2.1. *(Continued)*

PROS	CONS
contrasted with the uniform landscape of the lowland city. • Introduces a panoramic view upon approaching the city during the daytime and at night. • Introduces a continuously attractive view for pedestrians, especially during horizontal movement. • Blends habitats with the natural surroundings, creating harmony between humans and their natural environment.	
ENERGY CONSUMPTION	
• Slopes bring the option of building at a higher elevation and with it, a decrease in temperature and an increase in breezes, which is advantageous in the humid-warm season.[18] • Consumes less energy for cooling and heating than the lowland. • Slopes allow for increased air movement.	• Consumes higher energy for driving and increases vehicle depreciation. • Possible higher energy consumption for pumping water.
QUALITY OF LIFE	
• Provides a tranquil environment. • Has less traffic noise. • The aboveground part requires careful roof design and regulations in order to maintain an attractive view. Flat roofs can be used for family recreation in warm regions. • The horizon is not obstructed by buildings, as horizons are on the lowland, and keeps inspiring the imagination. • Requires flat roofs that offer space for recreation.	• Development is limited to housing of 1 to 2 stories and is not usually feasible for high-rise buildings. • Requires stairways for vertical movement, which may be difficult for the elderly, pregnant women, and disabled people. • Roofs need special design to enhance the quality of the view.
ENVIRONMENT	
• Drainage is easy and there are no threats of flooding. • Increases proximity and attachment to the natural environment within an urban setting. • Easy sewage disposal. • Enhances the environment for good living. • Offers an integrated, welcomed contrast to the landscape between the artificial cityscape and the natural surroundings with an immediate view.	• Slopes are subject to landslides, especially in highly humid and rainy regions with loose soil. • Slope development is limited to sites where there are supporting soil conditions. Risk of landslides exists if soil is mixed with gravel and the slope has a high gradient. • Threats of landslides and earthquakes require special design and construction measures.

(continued)

TABLE 2.1. *(Continued)*

PROS	CONS
MISCELLANEOUS	
• Optimal integration of supra-space with geo-space land use. • Eases urbanization pressures in the lowland. • Eases pressure on the lowland where there is a need for agricultural land preservation. • Offers tradeoff conditions in which there are significant gains for using slope development.	• Transportation movement is difficult and necessitates special land surveys and careful, innovative design.

The slope townscape integrates the artificial environment with its contrasting natural surroundings in close proximity. In the daytime, the hillside town reflects the light of the sun. At night, the illumination radiates from the street and the interior of the buildings and is the only visible part of the city, making the town quite attractive. In regions where farmers search for more land for food production, the development of the slope town eases the pressure on the lowland, leaving it free for agricultural use.

In most ancient cities, the city center was quite distinct—it capsulized the urban atmosphere of the city. In our modern times, peripheral sections surrounding the city center resemble each other all over the world in an "international" city pattern. On the other hand, the slope town is unique in each of its parts and in its overall environment. Towns were also developed along the tops of the hillside ridges and were grouped with the slope towns. These types of towns and cities evolved along the spine of the ridge and branches on both sides of the hills. Slope city expansion can take three forms: dispersion throughout the hills, high-rise vertical growth, or geo-space development.

Jerusalem is an example of a city developed along mountain ridges. Originally, the master plans of Jerusalem prepared in this century designated the tops of the hills and open spaces for public buildings, the slopes for residential development, and the ravines for green areas. The housing that expanded throughout the hills has grown into confined neighborhoods, each a separate physical entity with its own ethnic identity.

There are also a large number of successful slope cities in the Middle East, Mediterranean area, Europe, and Japan. Others throughout the world were preconceived and planned, such as Haifa in Israel, Athens, and Hong Kong.

Hong Kong is a good example of a modern slope city. Approximately 80 percent of the island is steeply sloped, with gradients greater than 1 : 5. Hong Kong has highly intermixed land use. In an urbanized city like Hong Kong, where psychological stress is typical, residents of the slope town feel significant relief and relaxation when looking out from their slope residences to see a distant view.[19]

Japanese hillside or slope cities were also developed in Nagasaki, Atami (southwest of Tokyo), Yokohama, Kobe, Hakodate, and elsewhere. Due to the need for a deep port for anchorage, important Japanese ports—such as Nagasaki, Yokohama, Kobe, and Hakodate—were usually built on a hillside near the port. These ports were linked with the west, especially Nagasaki to Holland, and the culture of these cities often reflected their western influences.[20] Many Japanese people have grown fond of these cities.

Except in some special cases, most American cities were developed on plains or in valleys. Most contemporary urban expansion throughout the world has developed in a similar fashion to that found in the United States. It seems illogical that people in the past developed cities on difficult hilly sites when technology was less sophisticated than today, while contemporary urban expansion favors flat areas when technology is at its peak capacity.

The decline of the slope city started with the rise of the industrial revolution in Europe at the beginning of the 19th century and continues to the present day. The industries consumed large, flat spaces for the production of raw materials and their storage in warehouses. Also, the new transportation system of railroads was built in the lowlands closely situated to industry, which furthered the decline of the slope city.

In addition to favorable climatic conditions and benefits to human health, the slope town has a distinct spatial charm, which the lowland city cannot offer. In Japan, India, and Europe, some slope sites were even developed for the residences of well-to-do people because of their alluring appeal. This charm is expressed in the building materials (usually local stone), in the street configuration, in the restricted spots of the landscape views toward the valley and the lowland, and the harmonious arrangement of housing units in relation to each other and to the slope elevations. In the slope town, public spaces—especially small, rapidly emerging ones—have distinct symbolic expressions that go beyond their physical form.

In the past, housing in Tokyo was extended to the slopes where the plateaus and valleys intertwine in the Yamanote district. The main roads of Tokyo stretched radially from the center of the city to the plateau ridge where the *Bukechi* (*samurai* family district) was built. The developed slopes connected the low farming land of the valleys with the *samurai* on the plateau.[21] The sea-oriented cities of Nagasaki, Kobe, and Yokohama were also built on slopes at the wishes of the wealthy. The residential district for foreigners built at Yamanote, from the Edo period until recently, is an architectural model for the Japanese. These districts became an exclusive residential area with a view of Tokyo Bay. There are still slopes in Tokyo named *Shiomizuka,* which means "tide-viewing slope."[22]

Recently, the city of Nagasaki has focused on the renovation of slope housing in order to create a more attractive city. The work is driven by a strong vision and by the design of scenic views. Much of Nagasaki has slopes of 15 percent gradient or more.[23]

In short, past hillside supra-space settlements provide excellent evidence from which we can draw design lessons for modern development. This practice introduces health improvements, aesthetics, and other benefits to the users. The fusion of this slope supra-space method with modern geo-space habitat technology, to be discussed in the following section, will further enhance geo-space habitat design.

Significance of Slope Geo-Space Habitat

Like any other design endeavor, the design of a geo-space habitat is a tradeoff of gains and losses from both the economic and social points of view.[24] While a comprehensive list of general benefits of slope habitats can be drawn, this author's focus is on specific uses of slope geo-space housing advantages.

After examining non–geo-space hillside slope settlements, this author discovered that several slope design principles could be applied to the slope geo-space habitat design. In fusing the slope form with the geo-space habitat, many psychological and technological arguments against geo-space living could be eliminated. Moreover, by combining the two styles, many of its users can reconcile the absence of flat-area land use. Some of the more obvious benefits of placing the geo-space habitats within slopes are sunshine and natural light penetration directly into the habitat, easy access between the indoors and outdoors, enhanced quality of life, freedom from claustrophobia, environmental improvements, a pleasing landscape, and fresh air. However, there are technological obstacles and construction difficulties associated with the slope supra-space development that must be adequately addressed in the design of sloped habitats (Table 2.1).

Although geo-space cities call for the introduction of all types of land uses in the subterranean space to complement the supraterranean city, there is a sharp distinction between the placement of residences and other land uses. Habitats are to be built exclusively within the slope, while all other land-use forms can be introduced on flat land, as well as within the slope. However, the slope forms are a significant improvement to the geo-space habitats. In countries like Japan, where urbanization is constantly increasing in the lowland and where mountains cover more than 80 percent of the entire land area, slope habitation is a promising new urban option. Human settlements on hillsides have historically gained much diverse experience, from which we can extrapolate stimulating design concepts for geo-space dwellings within the slope.

Energy efficiency is the most commonly known advantage of geo-space design. In general, the geo-space habitat reduces energy consumption, since it is cool in the summer and warm in the winter, it resists outdoor temperature fluctuations with minimal heat gain and loss, it protects humans in cases of extreme weather conditions, and it provides a comfortable ambient environment year-round.

Slope construction at high altitudes provides fresher air than is found in the lowlands, where the cities are located. The slope habitat is also less vulnerable to air pollution and its associated lung illnesses. Additionally, it is more likely that the air near slope habitats will circulate more than in the lowlands, and therefore supply fresher air.

While slope geo-space habitats are not completely immune to natural hazards and require special protective features, the slope area has its own defenses against natural and artificial hazards. Although, geologically, mountains are more likely to be subject to faults and sensitive to earthquakes, it has been found that space hollowed out of terrain is more shielding than the same space above ground, if it is properly designed.[25] Seismic regions are usually those that are marked by foldbelts, uplifts, faults, or thrust faults. It is essential, therefore, to survey and become familiar with the geological nature of the site before any designing is done.[26] Earthquakes can have vertical or horizontal movements, or a combination of the two. Structures should not be built along fault lines (below or above ground). In both cases, the structures would be at a high risk of collapsing.[27] Structures located within the geo-space would be split and lifted up or down in case of vertical movement, or shifted sideways during horizontal movement. However, structures within a massive rock free of faults and loose gravel will move with the rock mass and retain their original forms. Therefore, the deeper the space within the rocks, the safer it is. It may be difficult, however, to convince people of the validity of this explanation.

Earthquakes can intensify the force of landslides on slopes. The higher the gradient of the slope, the greater the intensity of a landslide and the risks associated with it.[28] In order to protect the geo-space habitat from geological shock waves and to avoid the impact of landslides, urban designers must be familiar with the types of possible landslide processes and implement special protective design measures (Fig. 2.1).[29]

Landslides can change land formations. They occur at gradient angles when the land is composed of loess or gravel, or when the earth becomes mud during intense rainstorms. Earthquake vibration is also dangerous to slopes and can change the entire environment. Slope habitats are also open to erosion or deflation. Retaining walls should be constructed as safeguards against these hazards. In any case, the entrance to the habitat is a critical place in either of those conditions. In short, pedology, soil mechanics, and geology should be examined jointly when considering a site for the habitat.

Geo-space habitats are considered to be at low risk for the spreading of fires.[30] As the habitats are below ground, it is likely that they will have fewer windows, which will limit or prevent fire expansion. It would be safer, however, to have more than one entrance/exit, at opposing ends of the habitat. Smoke inhalation and safe evacuation can be serious problems and need special design measures.[31] Also, if built with nonflammable materials, the geo-space habitat would be more resistant to fire than aboveground habitats. We can expect fire insurance rates of the geo-space habitat to be lower than those above ground.

Flood risks can constitute a serious threat to the house and its inhabitants if preventive measures are not taken. We have already mentioned the landslide issue. Drainage design, retaining walls, and runoff diversion should be introduced to protect the home and its residents. These measures can also safeguard against stormy conditions.

Figure 2.1. (facing page) Examples of possible geo-space damage due to landslides (Golany, Chinese Earth-Sheltered Dwellings: Indigenous Lessons for Modern Urban Design, U. of Hawaii Press, 1992, p. 113).

Contrary to the perceived opinion of the general public, geo-space habitats upgrade the quality of living. Belowground habitats are insulated from outdoor distractions, which provides a tranquil and relaxing environment. Creative people, such as writers, painters, musicians, sculptors, or people who seek contemplation time (such as religious people) will greatly benefit from the serenity. Throughout history, Buddhists in India and China often found their need for quiet satisfied in belowground spaces. It is this author's

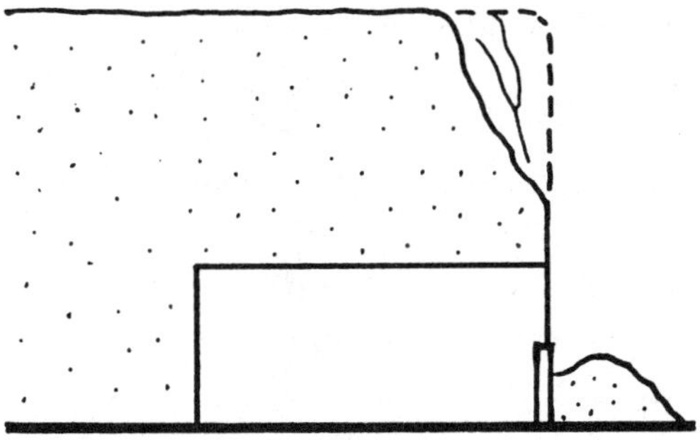

A. UPPER CLIFF LANDSLIDE

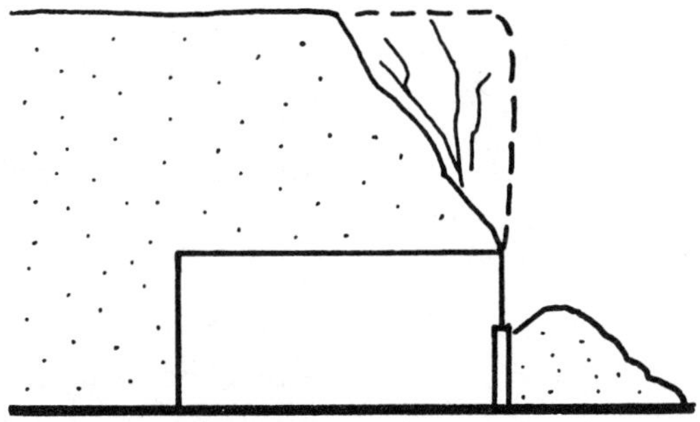

B. CLIFF LANDSLIDE

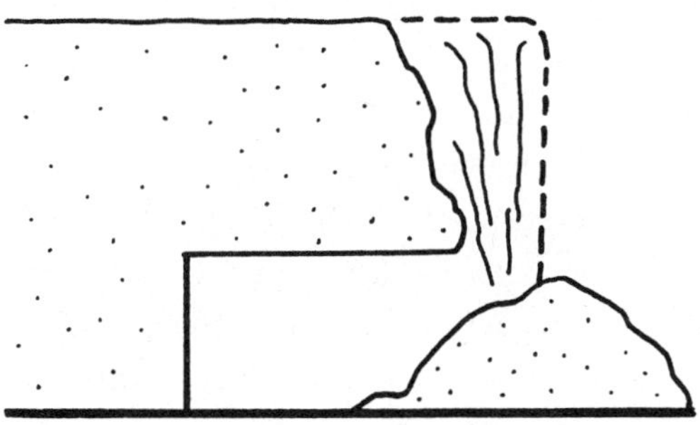

C. CLIFF COLLAPSE

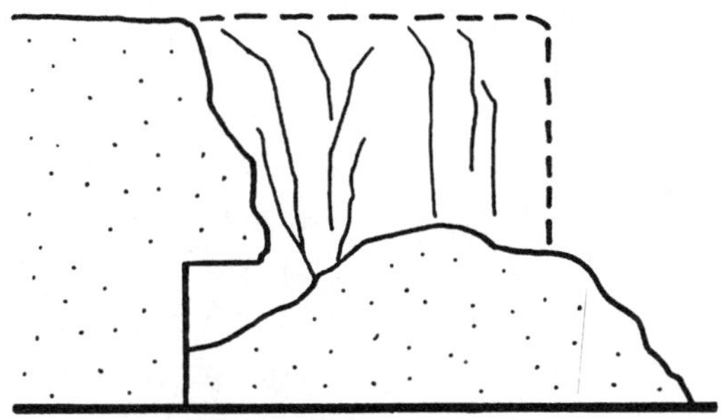

D. CLIFF AND CAVE COLLAPSE

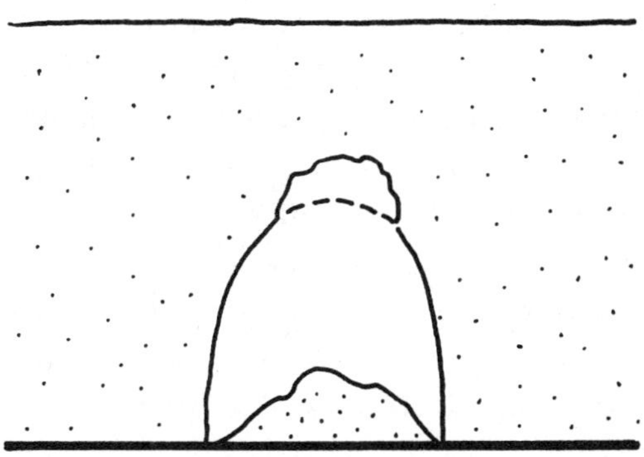

E. CEILING COLLAPSE

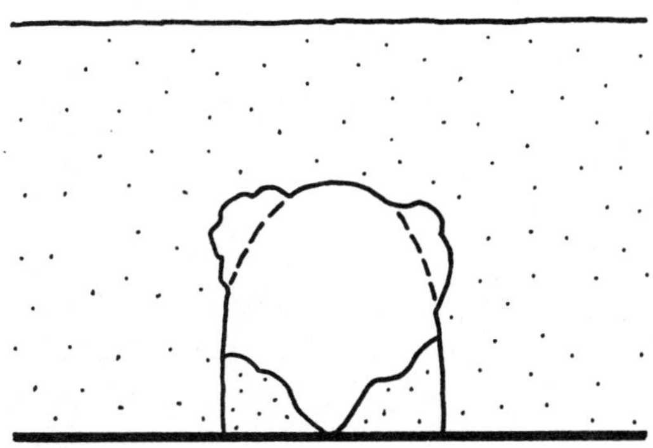

F. WALL COLLAPSE

hypothesis that sleeping below ground provides the right conditions for a deep, uninterrupted, steady sleep that could not be obtained aboveground, where the sleeper would be disturbed by various noises and obstructions throughout the night. Examination of such a hypothesis could change the common supra-space house design and codes to designate that bedrooms be below ground, while other rooms of the house be located above ground. Adding to this, the geo-space habitat is soundproof and does not transfer vibrations, has less noise, and has a lower impact of vehicular traffic than aboveground. The environment also has comfortable and seasonally stable air and an ambient temperature that is cool in the summer and warm in the winter. Habitats built within the slopes are more likely to have lower water tables than flood sites and, therefore, less risk of mold or dampness. However, those with dampness problems could be corrected with modern technology. On the slope, the habitat is closer to nature and to an unobstructed view of the lowland, which adds to the inhabitant's degree of relaxation and tranquility.

The geo-space habitat is protected from the high impact damage of intense storms, such as tornadoes, hurricanes, or electrical storms, and extreme climatic conditions, such as hot-dry or cold-dry weather. In cold climates, the water supply will not freeze and pipes will not burst.

In short, the quality of the geo-space habitat is enhanced by the aesthetic features of the landscape, reduced mental stress, the pleasing environment, and improved health.[32]

To estimate the costs of geo-space, the lifetime durability of the living quarters, the low maintenance and housekeeping expenses associated with this habitat, the possible dual usage of the land, and land preservation should all be factored into the equation. Such considerations will lower the cost.

Aboveground construction benefits from slope development, as we have previously described, but we are now concerned with implementing the advantages of slope usage to various geo-space habitats. Basically, geo-space offers all of the same advantages of slope usage (Table 2.2), and slope and geo-space usage combined introduces a higher quality of life than that of flat areas. Our basic premise is that geo-space

TABLE 2.2.
Environmental Benefits of Slope Geo-Space Habitat

SUNSHINE:

- Option of deep, direct, full sunshine penetration (7 meters and more) into the geo-space habitat on one floor or more when careful positioning, proper orientation, and innovative design are employed

NATURAL LIGHT:

- Possible penetration of natural light during the day deep into the habitat

VENTILATION:

- Increase of wind and air exchange
- Ideal conditions for passive air ventilation through a well-designed air shaft and rotating wind catcher

AIR QUALITY:

- Free from air pollution, smoke, dust, and odor due to active air movement, high elevation, and topography
- Free from inversion and, therefore, lower risk of lung-related illnesses
- Air is more refreshing than in the lowland

NATURAL HAZARDS

Earthquakes:
- Subject to less impact from earthquake disasters than aboveground houses due to its location within the earth
- Requires special design measures against landslide or façade collapse resulting from earthquakes

Fire:
- Low risk of fire spreading to neighboring habitats
- Greater fire resistance than for aboveground houses
- Safe when designed with more than one exit/entrance
- Low insurance rates

Floods:
- Free from floods when proper drainage design measures are considered
- Easy for sewage disposal

Storms:
- Sheltered against tornadoes, hurricanes, and electrical storms

QUALITY OF LIFE:

- Provides a tranquil and relaxing environment
- Provides maximum privacy
- Soundproof and does not transfer vibrations; less noise and less vehicle traffic than the lowland
- Has a low water table and therefore less risk of mold or dampness
- Provides comfortable and stable air temperature: cool in the summer and warm in the winter
- Quiet conditions for good, uninterrupted sleep

(*continued*)

TABLE 2.2. *(Continued)*

- Proximity to the natural environment
- Enhanced and unobstructed view of the surrounding lowland
- Weatherproofed against changing weather conditions and protective against extreme climates
- Provides low housing density by virtue of its geo-space design and geomorphology
- Provides a healthy environment for physical and mental well-being
- Introduces an attractive and pleasing cityscape
- Enhances the aesthetic quality of the landscape contrasted to the uniform landscape of the lowland
- Encourages pedestrian movement, and, consequently, social interaction
- Fuses habitats with the preserved natural environment
- Reduces mental stress
- Offers a stimulating environment conducive to creativity
- Provides an environment that offers faster post-surgery healing
- Provides protection against extreme climate
- Integrates the geo-space with the supra-space

ECONOMICS

- Reduces land prices by using the above ground as an open space; in this way, it compensates for the relatively high cost of construction

living should be associated exclusively with the slopes and not with the flatland. The slopes offer the geo-space habitat environmental benefits that the lowland cannot offer. As mentioned before, like any other design endeavor, the geo-space slope city has both advantages and disadvantages, and these should be viewed as a tradeoff.

Construction

Slope construction may occur in four forms (Fig. 2.2). These are:

1. Aboveground only
2. Semi-belowground
3. Belowground only
4. Combination of 1, 2, and 3

Figure 2.2. (facing page) Some basic examples of slope construction, showing the relationships between building units and different acute gradients.

Development on sloped sites can require higher initial investment than development on flat areas. Such cost increases may result from extensive construction of roads, foundations, and preventive erosion measures. Still, there will be savings from such development, as in the low cost of energy

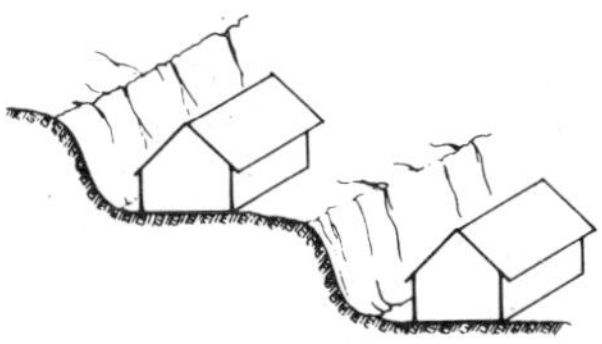

A. SINGLE, ABOVEGROUND UNITS

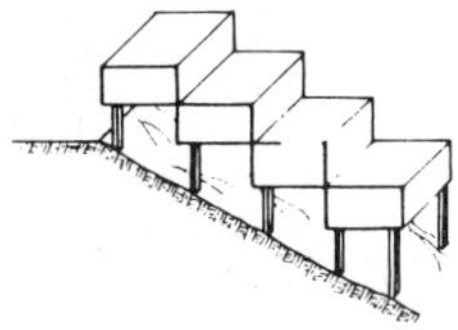

B. SINGLE UNITS, ELEVATED FROM THE GROUND

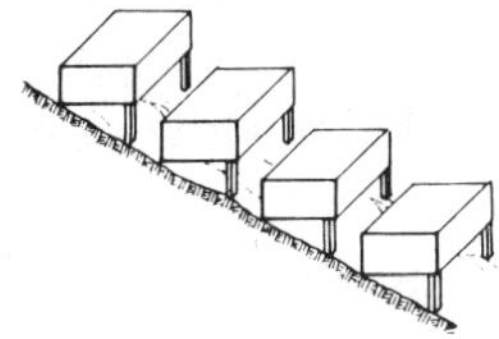

C. SINGLE, ELEVATED, DETACHED UNITS

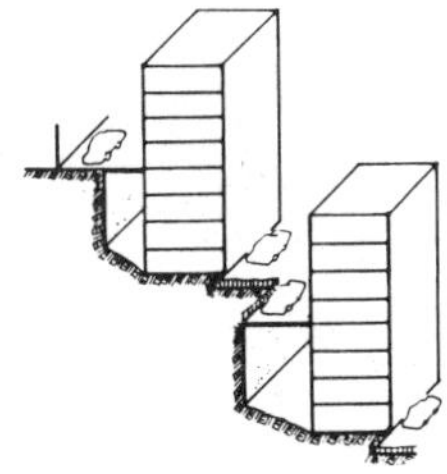

D. MULTI-STORY UNITS, DETACHED FROM THE SLOPE

ABOVE GROUND

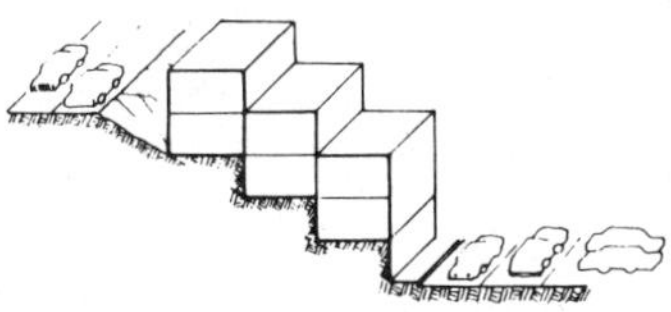

E. DOUBLE, ATTACHED, TERRACED UNITS

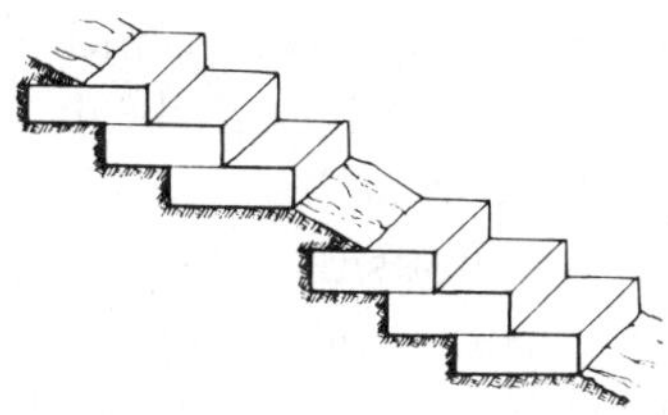

F. SINGLE, ATTACHED, TERRACED UNITS

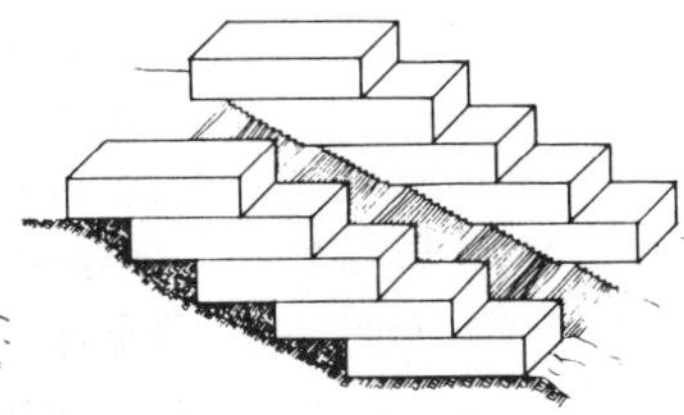

G. SINGLE, CLUSTERED, ATTACHED, TERRACED UNITS

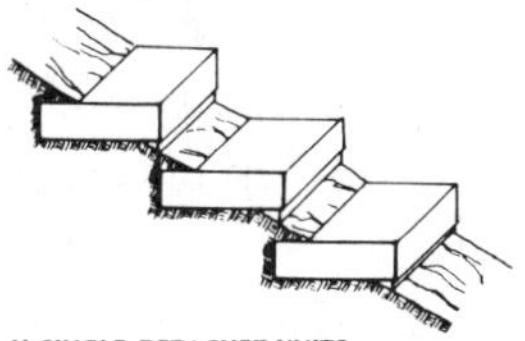

H. SINGLE, DETACHED UNITS

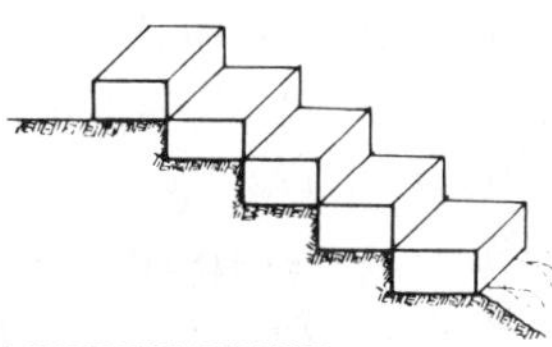

I. SINGLE, TERRACED UNITS

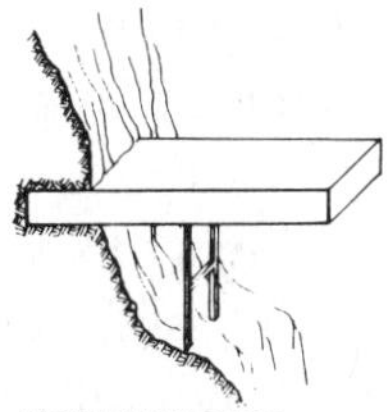

J. SINGLE, AIRY, EARTH-INTEGRATED UNIT

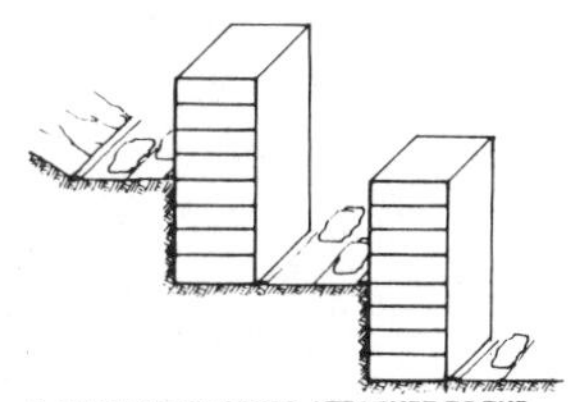

K. MULTI-STORY UNITS, ATTACHED TO THE SLOPE

SEMI-BELOW GROUND

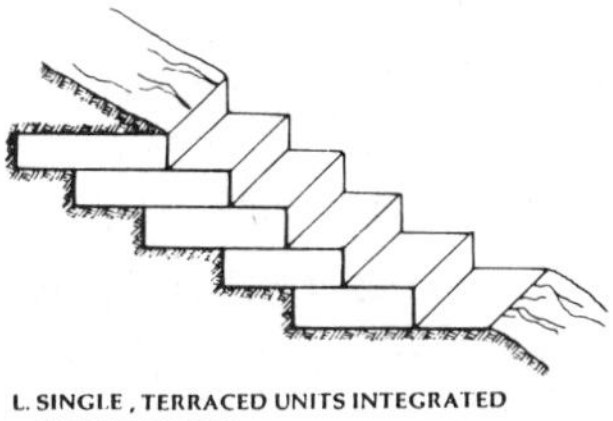

L. SINGLE, TERRACED UNITS INTEGRATED WITH THE SLOPE

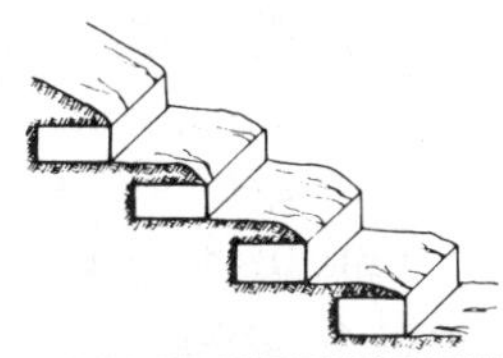

M. SINGLE, DETACHED UNITS INTEGRATED WITH THE SLOPE

N. SINGLE, CLUSTERED, EARTH-INTEGRATED UNITS

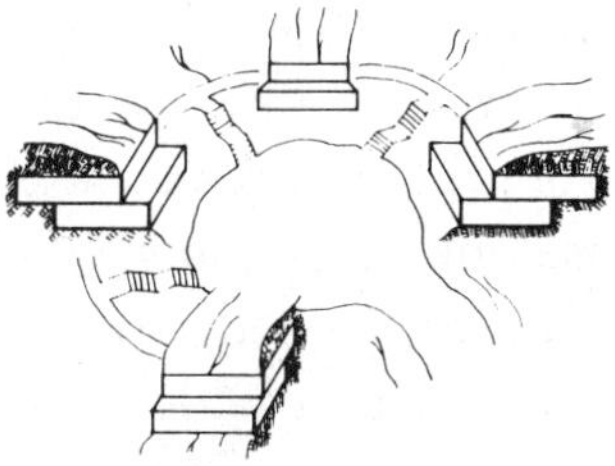

O. CIRCUMFERENTIAL, TERRACED, EARTH-INTEGRATED UNITS

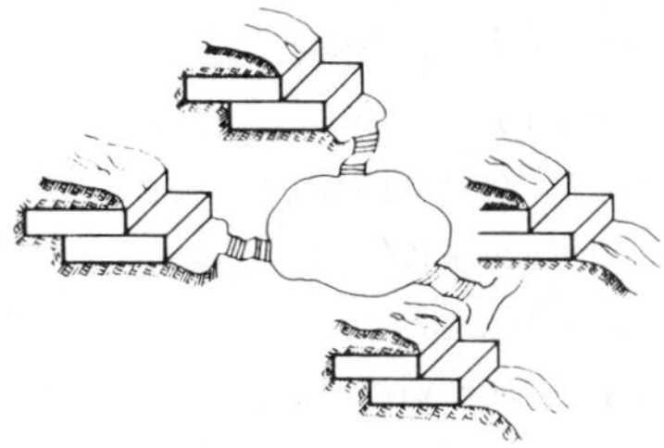

P. CLUSTERED, EARTH-INTEGRATED UNITS

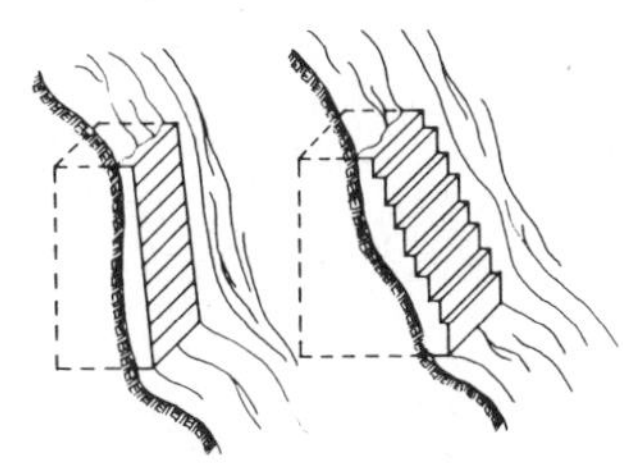

Q. CLIFF-INTEGRATED, MULTI-STORY UNITS

BELOW GROUND

for cooling and heating, significant improvements in health and safety of movement, and the pleasing quality of the environment. In any cost-benefit analysis, we should take into account the cost factors of the life span of the building, and the mental and physical health of the residents. When introducing slope geo-space development, the quality of human life is always a primary consideration.

An aboveground sloped habitat provides privacy in the form of single houses. A variety of compacted types of housing can be built above ground. Over time, a wide variety of sloped houses have been built according to the terrain, including single detached, single attached, terraced, one-floor, or multifloor housing. Recently, with a shortage of space for urban expansion and, after recognizing the significant benefits offered by the slope, there has been more of this type of development. One of the most commonly used is the terraced type, which furnishes wide balconies using the roofs of the lower units (Fig. 2.2*c*). High-rise semi-belowground buildings of six to eight floors can be exempted from elevator installation, since the topography naturally provides access from two different levels (Fig. 2.2*k*). The aboveground terraced units can also produce cottage-type homes of two or three levels for a single family (Fig. 2.2*a*). All in all, the slopes generate a wealth of flexible options, besides the added attraction of the environment itself.

Semi-belowground habitats integrate belowground space with aboveground space. The spaces are of different proportions. Semi-belowground habitats further enrich the sloped aboveground habitat by allowing expansion without increasing the size of the lot. Moreover, semi-belowground permits more complete land use without increasing the price or the height of the building. Semi-belowground habitats can also adopt terraced housing, as can high-rise buildings. Most importantly, the semi-belowground habitat, if designed properly, can receive sunshine, plenty of natural light, a wide view to the open landscape, ventilation, privacy, and quiet.

Belowground or geo-space habitats are fully integrated within the slope, leaving at least one side open to a view and sunshine. Such habitats can be designed singly (detached or attached), in clustered form, as cottages, or as multistory buildings. They allow the possibility of building in high density with a slope gradient of around 85 degrees, or on a cliff. In the latter case, a terraced slope should be constructed with a transportation system accessible at the back of the building via a special tunneling network. Geo-space housing provides a range of benefits resulting from its integration with the earth.[33] Here, too, properly designed, sloped geo-space habitats can enjoy sunshine penetration at more than 7 meters deep, receive natural light, have plenty of ventilation, and provide privacy and relaxation.

Although our concept of the slope geo-space city will have its own identity as an independent city unit, it is also possible to combine the uses of the sloped city with the existing flat city. Slope geo-space and lowland geo-space usage together can ease some of the urban problems that exist in the lowland city.

SITE SELECTION CRITERIA

The most crucial stage in the development of a sloped site with combined geo-space and supra-space land uses is site selection.[34] The method of site selection should use multi-criteria, comprehensive, quantitative synthesis combined with intuitiveness and imagination, flexibility to adopt changes, and analysis. Above all, it should focus on human mental and physical health needs.[35] The major criteria when selecting a site for the geo-space slope city are:

1. Geological structure and soil composition
2. Geomorphology
3. Slope gradient
4. Hydrology, runoff, water table, and erosion
5. Altitude, thermal performance, and relative humidity
6. Orientation
7. Accessibility

Geological Structure and Soil Composition

Many factors must be evaluated and examined extensively throughout the site selection process. Planners should be thoroughly acquainted with the slope site in order to choose the proper location for the geo-space structures. The geological nature of the site has economic implications, as well as an impact on safety and security.

In many cases, a slope, especially one with a sharp cliff, is the product of a major tectonic fault, making any cliff construction highly sensitive to earthquakes and landslides. The advantages that cliffs offer for geo-space development offset the high cost of special design and construction measures required for protection against such disasters. It should be noted, however, that slopes with two crossing faults are more likely to be at the epicenter of a strong earthquake than any point along the fault. Such an area should certainly be avoided as the site for any type of geo-space construction, let alone any new supra-space city.

From a design and construction perspective, there are generally three types of geological ground structures:

1. Stratified layers of limestone;
2. Magna solid rocks of granite or tufa; and
3. Deposits of ground soil mixed with gravel or with alluvial soil.

Stratified layers of limestone are found scattered throughout the world. Limestone ranges from soft to hard; was located, at one time or another, below sea level; and is the product of deposits of sedimentary layers. Over time, these stratified layers emerged above the sea as the result of internal disturbances that either created elastic curves of synclines or anticlines, or shaped the strata in horizontal form. The latter can be of interest to geo-space designers if they appear in alternating soft and hard layers, with the soft layer at least 3 meters thick, and where pockets of space can be created within the soft layer (Fig. 2.3). This type of stratification is found in Kansas City, Missouri, an active center for

Figure 2.3. Geological horizontal stratification consists of alternating layers of soft and hard limestone of thicknesses sufficient to accommodate the necessary floor-to-ceiling height. This is an ideal geological condition for geo-space housing, and enables the employment of the cut-and-use method, as has been done in Kansas City, Missouri, and Douiret, Tunisia.

modern geo-space design. The horizontal configuration is obtained by extracting the soft layers and retaining wide pillars. This leaves the upper hard stratification as a ceiling and the lower hard stratification as a floor.[36] The resulting large spaces are used for storage, refrigeration rooms, or offices. Similar geological stratifications exist in the indigenous, belowground cliff villages of Chenini de Tataouine, Douiret, and Guermessa located at the Matmata plateau in southern Tunisia, bordering the northern Sahara Desert.[37] The natives have adeptly used geological stratified formations in an efficient way.

When development of geo-space takes place within slopes composed of solid rock, the "cut-and-use" method can be adopted. This method can lower the construction expenses, since it requires few or no building materials and consumes limited labor; this is commonly done in the geo-space of the loess soil in northern China. Around 40 million Chinese live in these so-called cave dwellings. In the case of the cut-and-use method, the height and width dimensions of the geo-space are dictated by the natural constraints of limestone stratification. The slope geo-space built within geological stratifications can have diverse land uses because of the potential depth reached with the extraction of the horizontal soft layer from the slope. The depth of the sun's penetration depends on the height of the extracted opening: the higher the opening, the greater the sun and light penetration. This issue of sunshine penetration will be discussed later.

Massive unstratified rock formations are most commonly found in granite or tufa stone. Granite is very hard, contains few or no faults, and requires the use of machinery for cutting. Due to its solid and unstratified nature, granite enables the development of much wider spaces and higher ceilings than limestone. Accordingly, the most effective use of granite is for large-area geo-space, such as public gathering spaces, theaters, museums, storage areas, businesses, or restaurants. Although cutting granite can be expensive, its use could dispense with the need for building materials. This massive granite block dominates the base rock of northern Europe and has been widely employed in Scandinavia, especially in Sweden, where such geo-space is already being used for industry, storage, parking, and military ports.

Tufa rock results from the gradual accumulation of volcanic ash deposits, which settle in wide and relatively deep layers. The stone is relatively soft to cut, and its use would also eliminate the need for building materials or extensive labor. Structurally, tufa can offer wide spaces with high ceilings and is suited to residential and nonresidential uses. Such diverse uses were practiced throughout history in the cave dwellings of Matera, southern Italy, and Cappadocia, central Turkey (Fig. 2.4). An eight-story building was erected on a tufa rock cliff in Cappadocia, as were a large number of beautiful churches in the Byzantine period.

Soil mixed with gravel is the least adaptable to geo-space construction. It is loose and subject to erosion and landslides, especially on slopes, and to increased risks during earthquakes. However, it is still feasible to construct within

Figure 2.4. Slope geo-space habitats between a plateau and a sharp cliff. The habitats are located at the edge of two zones in Ganzu Province, China, which are geomorphologically and productively different.

this soil if precautions are taken in the design, construction, and maintenance of the geo-space. It would be best, however, to excavate only small spaces. The cut-and-use system could not be applied here. There will most likely be a need for an interior concrete structure for support and for exterior retaining walls to resist the impact of landslides. Both protective measures will increase the costs of construction.

Loess soil has been one of the most suitable types of soil for belowground usage throughout history. Most loess soil is free of gravel. It is found in five major regions of the world: southern Germany (from which its name is taken); southwestern Illinois and Missouri; Siberia; southern Israel; and northern China, where it is most widespread. There are three types of loess soils: the glacial type, formed by the retreat of glaciers (as in Germany); the fluvial type, formed by water (Illinois); and the aeolian type, formed by wind (as in China and Israel). Throughout history, the aeolian type has proven to be best suited to geo-space construction. Aeolian loess is free of stone, firm and strong when dry, easy to cut and use, and, as is commonly the case in China, does not need to be supplemented with building materials. Loess soil has its own limitations, however. The maximum known geo-space width developed in aeolian loess is 3 meters. In China it was common to have an average width of 2.5 meters with a similar height dimension, and a minimum cover thickness of 3.5 meters to prevent water penetration or collapse.[38]

Karst is another phenomenon associated with the geological structure of the earth and resulting from underground erosion or geological movements of the earth over time. Karst encompasses natural limestone marked by hollow spaces, all types of horizontal or vertical cavities, tunnels, sinks, abrupt ridges, irregular protuberant rocks, caverns, ravines, and underground rivers or streams. These phenomena are formed mostly by water erosion, as in the case of the Dolomite plateau in northwestern Yugoslavia, from where the name karst is derived. However, a thorough three-dimensional geological survey of such sites would encourage the imaginative geo-space designer to produce an innovative design of belowground spaces. Spaces like this usually have a steady ambient temperature in the winter and summer seasons and are mostly comfortable, if the humidity is not too high. Such thermal performance can be appropriate for selective industry and for land uses that require a controlled temperature and tranquil environment for human beings.

Geomorphology

Geomorphology is the study of the characteristics, origin, and development of the earth's surface features. It is the science that deals with the earth's relief features, using the principles of physiography in the descriptive aspects and the dynamics of the structural geology. However, geomorphology of a site represents a complex of forms that could be related to microclimate (comfort or discomfort), erosion, and soil characteristics in which the geo-space habitat and other land-use design are most concerned. Our interest here is choosing the optimal site of land for the geo-space habitat. For the sake of our discussion, the following analysis will focus on flat versus slope land in all areas. In a zone bordering two geomorphological forms, such as plateau and acute slopes often found in loess soil, geo-space housing will be advantageous (Fig. 2.4).

The comparison between flat and slope site geomorphology reveals significant benefits of the slope geo-space habitat. The introduction of slopes into the geo-space city brings a new and strong dimension into both the geo-space and supra-space cities. This complementary form is applicable to expanding cities that are located near a slope, which have not yet been developed. It is also applicable to new town projects whose sites have not yet been chosen. Cities that are located on a flat site and lack a nearby slope will be unable to make use of belowground habitats.

As was said before, the concept of slope use is not new. Many early towns and villages, especially in the Middle East, the Mediterranean, and many parts of Europe, were extensively developed on slopes. Only in the last few centuries have settlements been developed primarily on flat sites.

Flat sites, geomorphologically, do have advantages in providing easy transportation, relatively low investment for land development, easy access, and comfortable pedestrian movement. The British, who established much of the foundation of urban design codes, introduced the idea that slopes should be preserved for landscape, natural resources, and other recreational needs. Accordingly, cities were to be located on flat land, rather than on slopes. American cities, with few exceptions, have been developed on low land, near rivers, in valleys, or on flat land, making the grid system of streets easier to implement. Although low land—especially near a river—is suitable for agriculture, it is also commonly associated with interrupted drainage, low ventilation, flooding, relatively high reflection of radiation, and air inversion. Flat urban sites located in valleys are characterized by a combination of high inversion and air pollution from public transportation and industry, which is deleterious to the health of the residents. Also, the high water table associated with low land limits the possibility of using space below ground.

Slope sites, on the other hand, geomorphologically provide the optimal location for combined geo-space and supra-space development and overcome most deficiencies usually associated with indigenous belowground space. Practically speaking, the combination of geo-space and supra-space can enrich both types of development. Slopes have a variety of significant advantages for urban housing, which flat land does not offer. Among them are efficient drainage, good ventilation, low risk of air inversion, fresh air, less dust, unobstructed sunshine and natural light penetration, safety from problems associated with heavy urban transportation, security, absence of pollution, proximity to the natural environment and a varied landscape, more privacy, a low water table, healthy and tranquil surroundings, and the availability of a large variety of design options.

The geomorphology of sloped sites is usually not uniform. The overall angle of the slope does not express these gradient variations in its configuration. Topographical contour lines, on the other hand, do represent different natural forms of the geomorphology, which can be used or adjusted as needed. The slope offers a horizontal line for all types of movement along the contour lines; the designer should work with these natural forms rather than creating drastic changes through vertical movement. Therefore, it is advisable to use terraced horizontal and vertical forms alternating along the contour lines. Terraces can support two-way traffic movement, plazas, parking areas, parks, and other wide horizontal spaces. The location of a habitat within a geo-space can be implemented in almost any geomorphological site, if it is accessible. Round, curved, or curvilinear forms can stimulate the imagination of the designer. However, habitat design should take into account the existing natural forms and effect a minimum number of changes to the morphology (Figs. 2.5, 2.6, and 2.7).

Slope Gradient

Slope gradient is an important element of geo-space habitat planning, since the use of such a habitat is designated exclusively for slopes. Our approach would need to examine all degrees of slope gradients, from the conventional 7 degrees

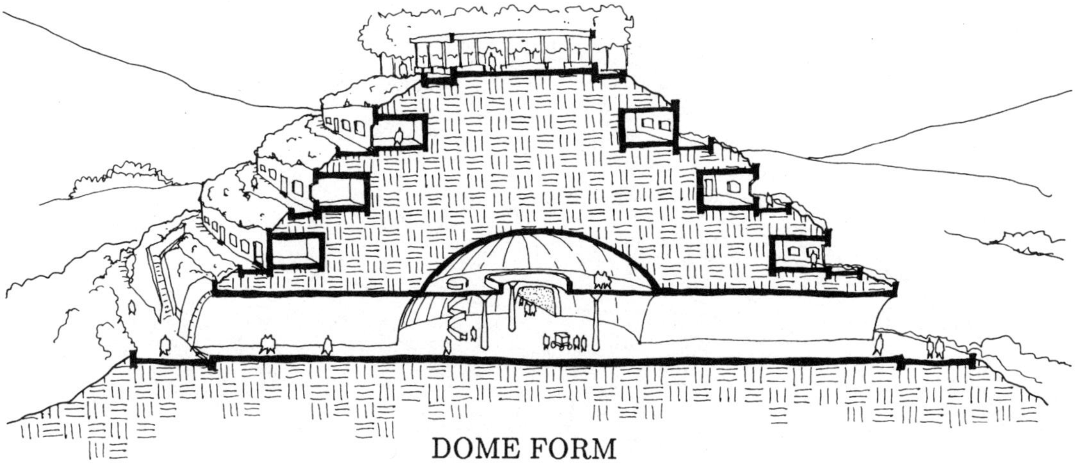

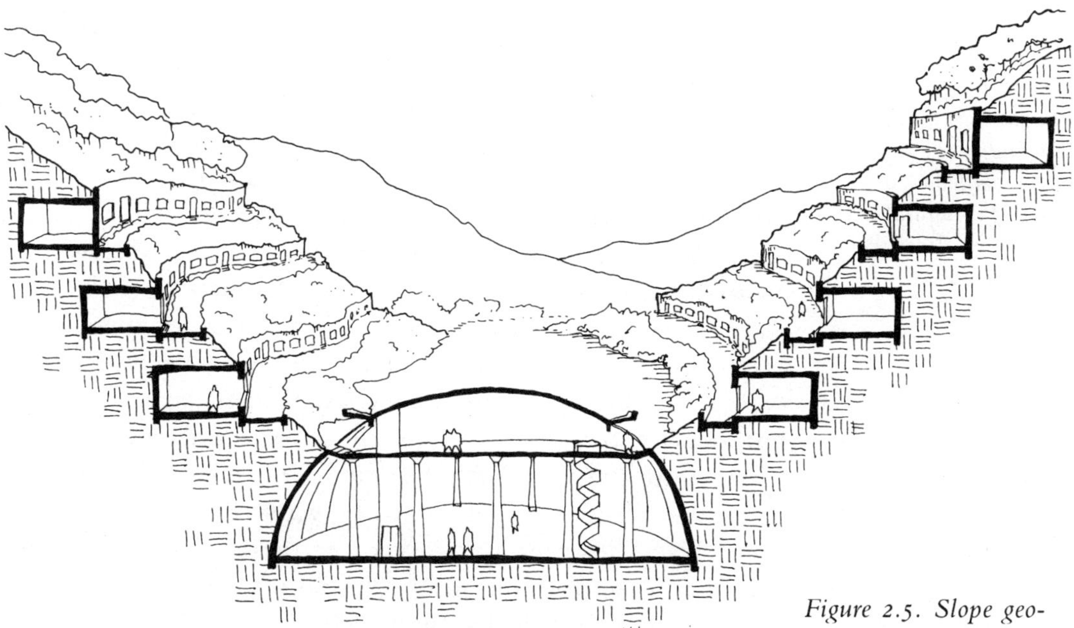

Figure 2.5. Slope geo-space designed within the geomorphology of the land. Cross section showing the adjustment of the slope geo-space habitats and community centers that conform to the geomorphological features of the hill and valley.

to our proposed unconventional 85-degree angle, and explore the feasibility of their development (Fig. 2.8). The issue of slope is related to three important elements: the accessibility of the residents to goods and related items; design and construction; and provision of daily environmental commodities. Each of these factors needs to be treated inno-

Figure 2.6. Overall view of slope geo-space neighborhood clustered around a natural and uniform geomorphologic formation.

vatively by the urban designers of the geo-space city. Until the 19th century, slope-constructed cities and villages were remarkable in their development; they can offer valuable lessons for modern design of geo-space habitats.

Terraced and cliff habitats are the two major methods of slope gradient development. The first is feasible on the lower and more moderate gradients, while the latter is applicable to more acute gradients, or, in some cases, moderate gradients. The acute gradient has advantages and disadvantages. The design and construction are costly, yet it provides a great deal of privacy, features a wide vista of the lowland, is healthy, and has a comfortable ambient environment. The basic rule of slope geo-space is that the steeper the slope, the smaller the residential area. Also, the steeper the slope, the greater the need to increase the pedestrian network and lessen road traffic penetration. Acute slopes intensify the effects of erosion, in comparison to low-gradient slopes, and would require special attention to design.

As mentioned previously, Jerusalem is a mountain city of intertwining ridges, slopes, and ravines. Both the early and recent master plans formulated land-use design principles in

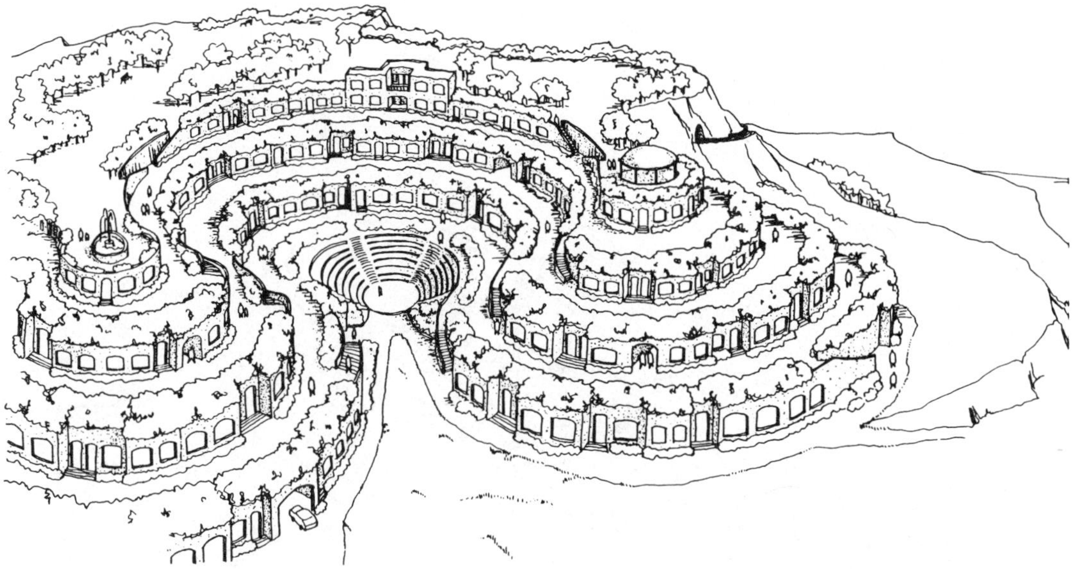

Figure 2.7. Overall view of a slope geo-space neighborhood integrated with diverse geomorphologic forms. The integration provides privacy and a sense of community.

relation to these three levels, and designated the ridge for public open space or buildings, the slopes for residential uses, and the ravine as a green strip. Because of its unique topography, Jerusalem is an ideal city to transform into a geo-space city. All its supra-level transportation could be moved to the geo-space level, converting that freed aboveground space into urban parks. The subject of transportation was discussed in chapter 1 and will be referred to later in this chapter as well.

Hydrology, Runoff, Water Table, and Erosion

Hydrology, runoff, water table, and erosion are interrelated factors associated with the site selection criteria. Hydrology differs from one site to another, depending on the precipitation intensity and the gradient angle of the land.[39] Basically, in hot-dry climates, limited precipitation is characterized by short sporadic bursts of torrential concentrated rain. Consequently, the ratio of runoff is high, and causes intense erosion and flooding.[40] Hot-humid climates are marked by torrential rainstorms, which last for long stretches and strike often. These, too, create erosion, yet the ratio of water seepage into

Figure 2.8. Slope geomorphology of low and acute multiple gradients. The unconventional angle can accommodate different types of habitats and orientations.

the ground is very high. In cool-and-warm climates (such as Pennsylvania and other parts of the United States or England), precipitation lasts almost continually throughout the season, causing a high water table and relatively little runoff. In cold, humid climates, such as Canada and northern Europe, frost and freezing cause soil contraction in the winter. Sometimes, oversaturation leads to soil expansion. All of these environmental characteristics should be taken into account when selecting the site for a geo-space city and, especially, the location of slope geo-space habitats.[41] These features will directly affect habitat construction and the geo-space city lifestyle.

Erosion is an issue of concern to urban areas. All cities act as nonporous bodies, since they are covered by buildings, asphalt roads, and paved alleys, leaving very little, if any, surface soil exposed to soak up water. Rainwater is usually channeled into the sewage system, which, in flat sites, is often unable to absorb large quantities of water, and creates flooding in the streets. This phenomenon is problematic mostly in stressful climates, such as hot-dry or hot-humid regions. If a city is located near a river in the lowland of one of these climatic regions, serious flooding can result in the built-up areas. The development of a city on a slope, however, may intensify slope erosion if preventive design measures are not taken. Flooding is less likely to occur on slopes if water drainage is adequately channeled to saturate the water table.

Altitude, Thermal Performance, and Relative Humidity

In the selection of the optimal slope geo-space site, urban designers must be aware that the elements of altitude, thermal performance, and relative humidity are all interrelated. One basic rule to remember is that the higher the elevation, the lower the temperature, and vice versa. For every increase of 100 meters in elevation, the temperature is reduced by one degree Celsius. Thus, warm and humid regions will have better thermal performance at higher elevations. Moreover, the higher the elevation, the greater the relative humidity. This fact can contribute significantly to site selections in hot-dry climates, and, to some extent, in hot-humid climates when combined with active ventilation. In hot-dry climates alone, humidity mitigates the intense heat by evaporative cooling. These facts relate to energy consumption in the winter and the summer, and to the improvement of the health of the habitat dwellers.

In hot-humid or hot-dry regions, higher elevations offer a significantly healthier atmosphere, in addition to a comfortable ambient temperature. This is partially due to the fact that higher elevations have cleaner air and are usually farther from marshlands and the havens of mosquitoes and other insects. Ambient summer temperature in the slope geo-space is far more comfortable than that in the supraterranean house on an adjacent lowland. Resort towns are also often located at higher elevations to serve as escapes from the trials of urban life.

Slopes also introduce effective ventilation and air circulation, thereby significantly decreasing the effects of car exhaust and other city pollutants. In addition, slopes are free from air inversion and have more comfortable temperatures than the nearby lowlands, which eases living and working conditions and reduces energy consumption for cooling. In central and western Japan, for example, where temperature and humidity are very high in the summer, housing should be located at higher altitudes in order to enjoy lower temperatures and greater ventilation.

Orientation

Orientation is a crucial factor in geo-space site selection. It is often cited as the answer to questions about light, sunshine, ventilation, and weather damage. No matter where the site is located, careful and innovative treatment of a design should attempt to address these questions.

Detailed studies of the site and its climate should be undertaken before construction starts. Data should be gathered on altitude, wind frequency and behavior, orientation, rain patterns, and humidity of a given site in order to select the desired orientation. In general, slopes offer good ventilation and strong winds. Weather on slopes and mountains is dynamic and fluctuates between day and night due to variations in the topography. In general terms, we can state that hot-humid climates should welcome wind movement while hot-dry climates should avoid it, unless a special passive evaporative cooling system is installed.

Slopes oriented toward the southeast, south, or southwest will receive maximum light and sunshine. The same is true of indirect natural light, which penetrates deep into the habitat. In regions with low temperatures and humidity, these southern orientations will be preferred, since they provide sun for part of the day. Northern slopes are less desirable, since they offer no sunshine or limited light penetration into the house. In cold or temperate climates, northern slopes are also subject to ice and longer frost duration. In these regions, a southern slope would be more advantageous. In addition, southern slopes offer more light and longer working days to aboveground industrial sites. Conversely, hot climate regions may prefer slopes facing north, as some ancient cities did. Still, there are some distinct differences among these three orientations, and they should be selected according to the climate of a given site.

Slope design should carefully consider the microclimates of the various parts of a slope and the impact that new development will have on the microclimatic patterns (wind velocity, temperature, and humidity). Terracing, for example, can affect microclimatic patterns just as supra-space buildings do.

Accessibility

Accessibility refers here to passenger services, emergency needs, modern vehicle movement, and pedestrian movement to and within the site. Accessibility is possible on a low gradient, becomes more difficult as the gradient increases, and becomes impossible using conventional modes of transportation as the gradient edges closer to the extreme of 85 degrees. However, two alternatives are recommended to be introduced simultaneously. One option is aboveground technical transportation, such as electric cables, and the second is a completely belowground network that would approach the geo-space neighborhood from the rear. The latter was discussed in chapter 1, in the section on urban transportation (see p. 37). Urban planners have already accumulated much technological experience on underground tunnel construction for vehicles, subways, railways, and parking areas within flat land. There is no technological reason to avoid such construction in hilly or mountainous areas where the geo-space habitats will be located. As for pedestrians, special designs should be outlined to combine relaxation and circulation throughout the aboveground pedestrian network. Slope development usually offers an attractive view of the lowland, depending on the angle of the viewer and the positioning of the network. In any case, a shorter belowground pedestrian network should be provided to link the rear underground vehicle and train networks with the residential areas. Also, special aboveground vehicle movement should be available for use during emergencies.

The removal of sewage waste may be easier on a slope than on a flat area. Garbage disposal can be compressed downward with the help of gravity. However, the water supply may need to be pumped upward if it is not accessible at the higher elevation.

In conclusion, the choice of a site will affect more than just the physical layout of the city. Even in small-scale design and construction (such as that of a single building), the architect makes not only a physical decision, but also, perhaps unconsciously, social, economic, and environmental decisions as well. The impact of the decision becomes much greater when considering the influence of the city's site selection on urban life for hundreds, perhaps thousands, of years (even if the structures themselves may only last a few decades).

Throughout history, site selection was decided in one of two ways—either determined in advance or chosen inadvertently due to complex influential forces. One of these two methods typically dominated during a given era or in a given region. Although preplanned urban design and monumental architecture were explored thoroughly throughout history, especially in ancient Greece and Rome, it was practiced primarily beginning in the Renaissance period and has increasingly been used since then. The evolutionary process continues to be featured among indigenous cultures.

Two other approaches to site selection are still commonly used. One is intuitive or instinctive site selection, which is often used by architects. The other approach, usually employed by an interdisciplinary team that includes politicians, is systematic and considers all possible contributing factors.

Topography is an important consideration in both methods. Many factors affect the choice of a sloped or a flat site. Some of these factors relate to defense, proximity to natural resources, environmental benefits, transportation, and accessibility. Slope sites have been found most often in the Middle East, the Mediterranean region, Asia, and Europe. Flat sites have been commonly chosen in North America during the last few centuries, and are currently the foremost choice throughout the world.

This author emphatically recommends the selection of a

slope site. This selection would allow the flat and rolling topographical forms to be used for agriculture, forestry, lakes, major transportation, or landscaping. Although the selection of a slope site would appear to require a higher initial investment than a lowland site, this cost would be offset by the slope's lower land costs. The slope site would also improve the overall quality of life of the community. By introducing a pleasing environment for all age groups, it would foster greater social interaction in neighborhoods than is currently found within our individualistic urban society. There are comprehensive, long-range benefits to be gained by integrating the concepts of slope site usage and geo-space usage for habitats.

It is important to recognize that there are advantages and disadvantages associated with different site selections. The characteristics of both types need to be reviewed and analyzed with these in mind. In addition to the tangible cost-benefit analysis, which has been previously argued, there is also the invisible "cost" of social benefits, privacy, tranquility, health, well-being, and comfort, which may be gained or lost due to site selection. Even environmental and energy consumption costs are factors in the site selection of a city.

HABITAT DESIGN AND CLIMATIC CONSIDERATIONS

Geo-space would best be used in regions with extreme climatic conditions—for example, hot-dry lands, such as Australia, and cold-dry areas of central Canada or Siberia. Many of these regions are rich in natural resources and often need large-scale housing to accommodate the company employees whose job it is to exploit these resources. In fact, climatically stressful regions are becoming the new urban frontiers. Regions with warm, humid climates would require a special slope geo-space habitat design. In regions of high humidity, weather conditions can be significantly eased by designing spacious rooms with high ceilings, and by installing effective passive ventilation systems. Ventilation can also effectively reduce condensation, which often builds up in the geo-space of such zones. In addition to the general climatic study of the slope geo-space habitat design, altitude, wind, solar radiation, orientation, and sunshine penetration need to be investigated.

Winds

In addition to regional patterns, local winds are influenced by the differentiation in temperature and land configuration. Slopes that line the sides of valleys experience a wind tunnel effect channeled by the valley slopes. Slopes are also subject to downward or upward winds. In the afternoon hours, when the heat in the lower part of the valley becomes intense, downward winds along the slopes are likely to develop. In the case of inversion, when low temperature and air stagnation strike in the lower section of the valley, the slopes are more likely to experience diminished air movement. In places where air turbulence is frequent and temperature is low, a sheltered sloped site is preferable. In supra-space cities, cross ventilation is achieved through the side windows on perpendicular walls. In geo-space, single side windows may not be sufficient, although wind movement will probably have a higher velocity than on the lowland. To create air circulation within the habitat, a controlled air shaft system needs to be constructed (Figs. 2.9 and 2.10).

Housing located at the lower part of the slopes is exposed mainly to high temperatures throughout the day and low temperatures in the early morning. Generally speaking, such houses have poor ventilation and are prone to air stagnation. As a result, they consume a great deal of energy for cooling and heating while the air pollution is a potential health threat. The high slopes are much less vulnerable to pollution (dust or smoke) than the lowland.

Radiation

Radiation in lowlands is usually intensified in a city located in the lower part of a valley. Radiation is received directly from the sunshine and indirectly from reflection off the nearby

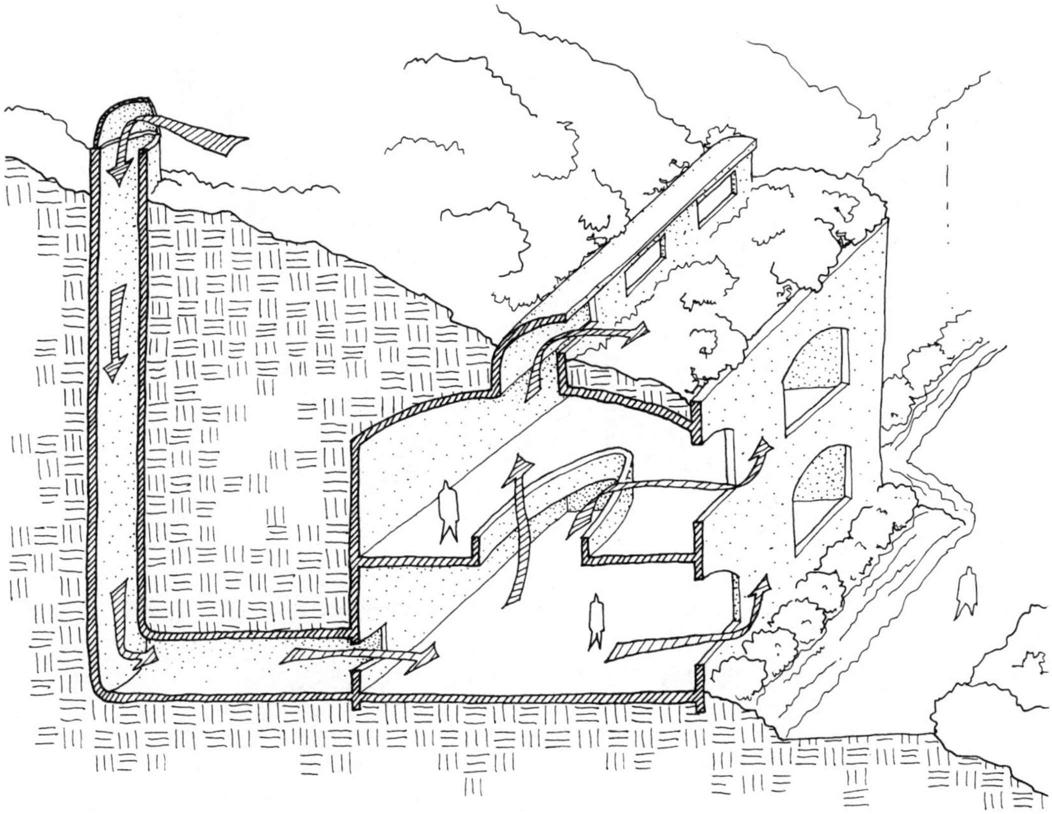

Figure 2.9. Ventilation system design. Passive ventilation is supported by dynamic air movement, which is typical of the slopes.

slopes. Lack of ventilation will increase the intensity of the radiation and produce higher temperatures within the community. Intensified radiation is less likely to occur within habitats located on a sloped site, which enjoys much more air circulation.

Sunshine Penetration

Sunshine penetration is a major challenge for the slope geospace habitat designer to face. This issue has a dual implication, as it affects both the psychological state and the health and well-being of the dwellers. Understandably, most, if not all, people are averse to the idea of belowground usage because of the image they have of a dark, dank, underground space. However, innovative habitat design methods and well-calculated orientation toward the sun can certainly result in natural light penetration throughout the day. Sunlight can

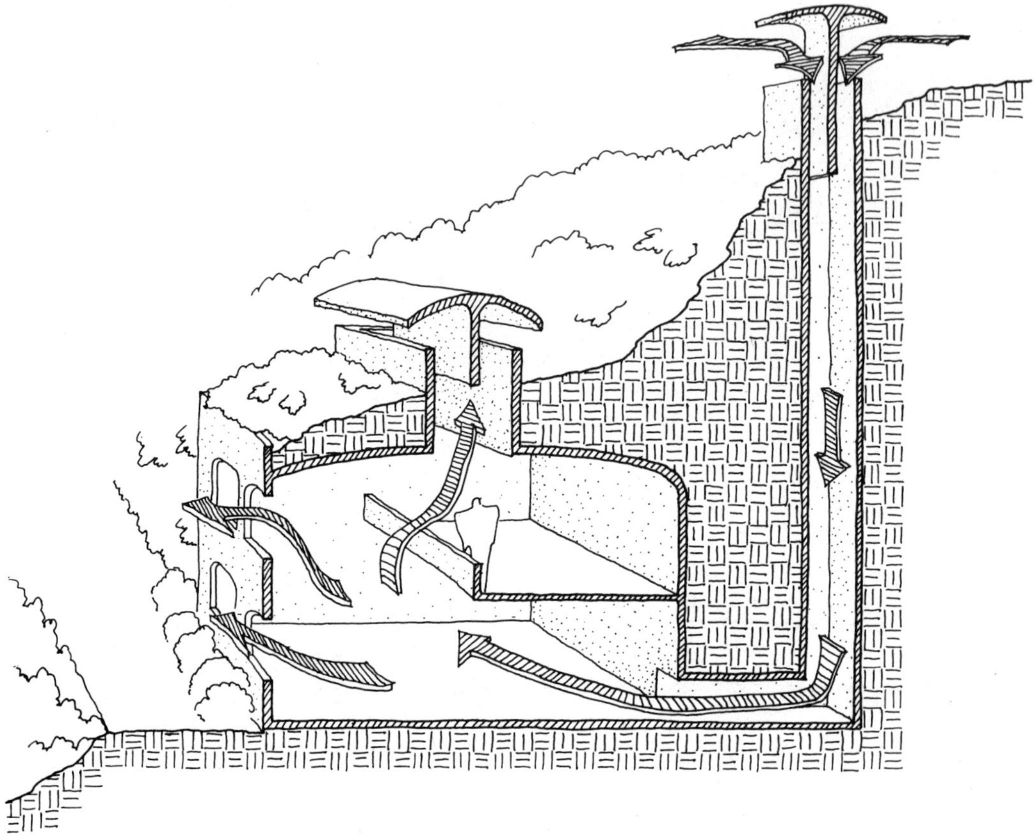

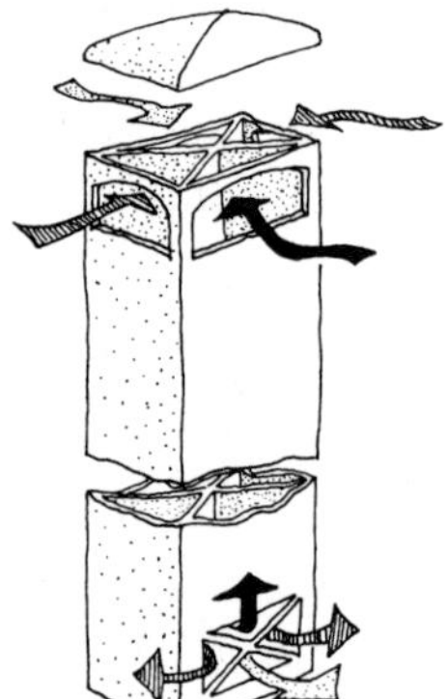

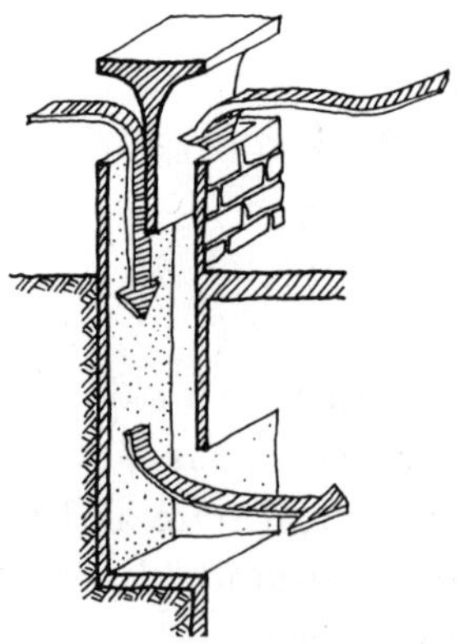

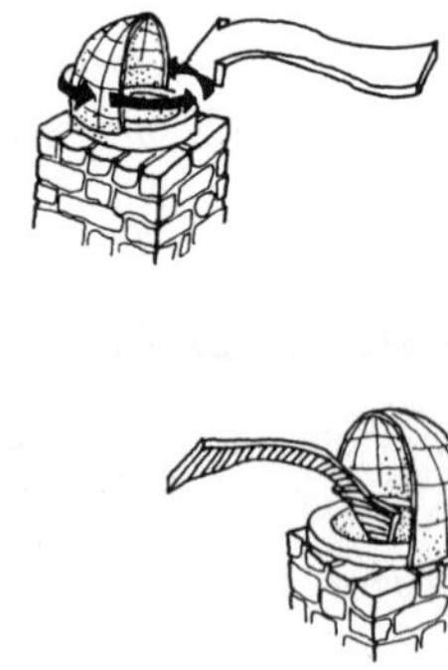

reach deep into the back of rooms to 7 meters in depth and more at latitudes ranging from 24 degrees N to 52 degrees N. Figure 2.11 provides examples of rooms of different floor-to-ceiling heights (3, 4, 5, and 6 meters) and a constant horizontal depth at different latitudes of 7 meters for comparative purposes. (Many habitats in congested supra-space cities do not receive natural light or sunshine at all.) This potentiality has led this author to persistently and exclusively promote the use of the geo-space habitat on slopes and not on flat land (Fig. 2.12). Also, sunshine and light penetration have positive psychological effects on their users, which can reduce or eliminate the public resistance to the use of geo-space.

THERMAL PERFORMANCE OF THE GEO-SPACE HABITAT

Soil thermal behavior has a major impact on geo-space habitat ambient temperature and its function.[42] It is therefore essential to fully understand the thermal behavior of the soil to determine geo-space form and design.[43] It was already explained in chapter 1 that insulation and heat retention are two functions of the soil. Its thermal pattern provides the geo-space habitat with cool, comfortable temperatures in the summer (around 20 to 22 degrees C) and relatively warm ones in the winter (12 degrees C).[44] If insulation alone was required, the habitat could be built above ground and could be enveloped with layers of earth, as are the adobe houses of the southwestern United States. In fact, this insulation can be achieved today with thin walls of high insulation quality, which can minimize heat gain and loss. Many modern American habitats of the 1970s and 1980s used the aboveground, earth-enveloped form to achieve high insulation. In this way, designers overlooked the significant advantages of geo-space soil thermal retention—the natural process that releases cool air in the summer and warmth in the winter. Similarly, thermal benefits can be obtained in an aboveground house enveloped in very thick walls of earth. Such is the case with the Jerusalem indigenous house.

Figure 2.10. (facing page) Example of a proposed passive ventilation circulation system for slope geo-space habitats. Entrances and outlets are to be regulated. Indoor air is usually heated by human activity and electric appliances, which cause the air to be drawn into the floor of the house and the indoor air to escape through the windows or ceiling openings.

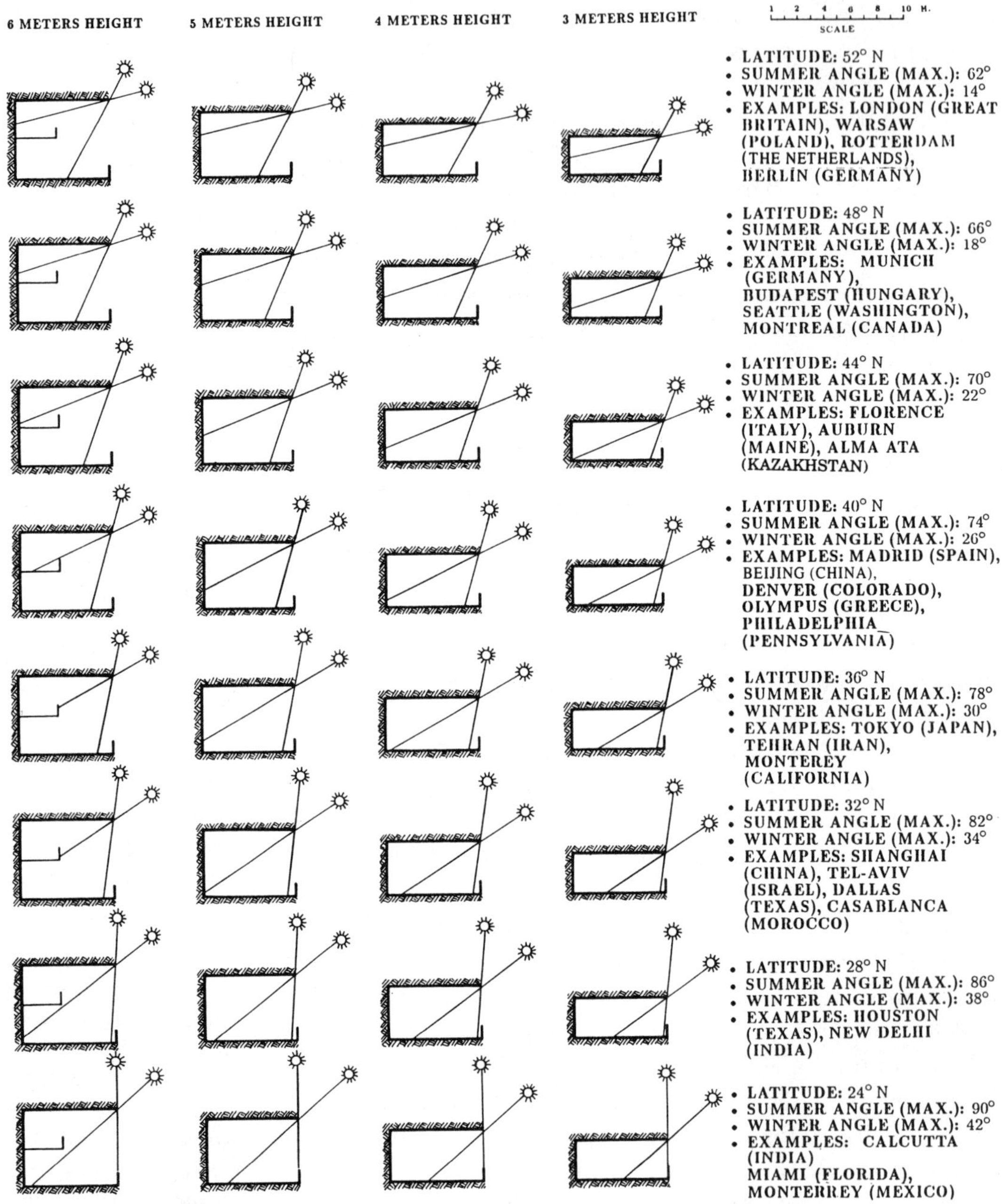

Figure 2.11. Maximum sunshine penetration at mid-summer (June 21) and minimum at mid-winter (December 21) in a slope geo-space habitat of 7 meters in depth. Sunshine penetration progressively increases as ceiling height increases from 3 to 6 meters. Also, sunshine penetration is reduced in summer and is augmented in winter as the latitude increases. City locations are approximate.

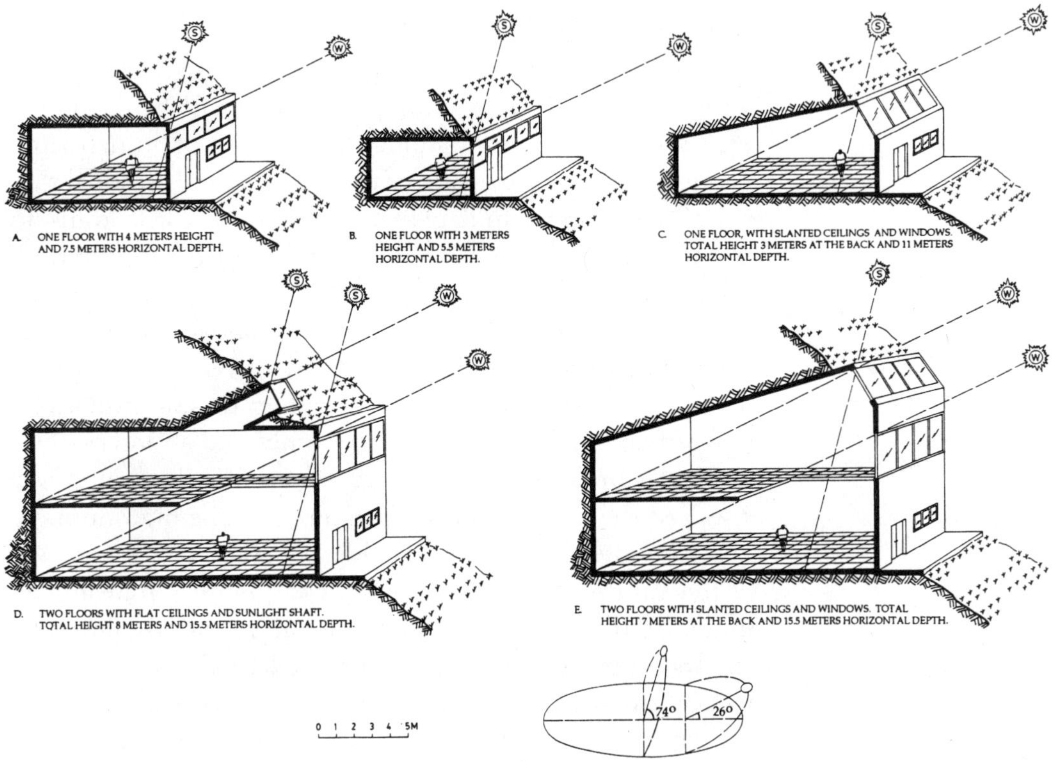

Figure 2.12. Cross section of five slope geo-space habitats. The design allows maximum light and sunshine penetration in 40° N latitude (Madrid, Spain; Beijing, China; Denver, Colorado; Olympus, Greece; and Philadelphia, Pennsylvania).

Jerusalem Indigenous House Example

Despite the advantages of geo-space soil, it is still possible to build a habitat above ground and also benefit from the heat retention process. Throughout the evolution of human settlements, humans were aware of the soil's natural seasonal thermal process and simulated it by building thick walls for aboveground habitats. A good example of this is the traditional house design of Jerusalem.

The Jerusalem house, constructed above ground until around the third decade of the twentieth century, was made of thick walls formed of 1 to 2 meters of soil sandwiched between an exterior wall of large stones and an interior wall of small stones and mortar (see Fig. 1.2*b*). The outer wall absorbed the low winter temperature and the high humidity that penetrated the soil. The colder air reached the inner wall by the summer and established a cool indoor environment. Similarly, the summer's high temperature and low humidity penetrated the outer wall and proceeded to reach the inner wall by the winter. In short, due to the thick wall cover, the interior of the Jerusalem house was cool in the summer and warm in the winter, functioning as an insulator and retainer of temperature. The thickness of the wall determined the period of heat gain and loss and the heat retention capacity. To a great extent, this house design is a simulation of the thermal performance of the geo-space habitat.

Thermal Fluctuation Impact

There is a strong correlation between the depth of soil and temperature fluctuation. The governing rule is that the greater the depth from the surface, the smaller the thermal fluctuation.[45] In general, at a depth of 10 meters, the temperature is seasonally and diurnally stable at around 22 degrees C in the summer and 15 degrees C in the winter. At much greater depths, the temperature increases due to heat emissions from deep within the earth.

Geo-space has its own ambient microclimate independent of the outdoor air temperature fluctuation. This fact is especially significant in regions with harsh climates and that are rich in natural resources. These areas attract large temporary or permanent settlements and will need housing appropriate to the weather conditions.

It is possible, however, to minimize temperature fluctuation between a depth of zero and 10 meters by distorting the normal thermal pattern. This can be achieved by placing insulation between the geo-space ceiling and the soil surface, extending horizontally beyond the geo-space dimensions.[46] In this case, it will take the surface temperature a longer time to travel through the greater mass of soil in order to reach the walls of the geo-space.[47] Another method is to cover the ground surface above the geo-space with buildings or dense trees in order to minimize the heat from the sunshine.

It is also possible to accelerate the soil thermal process by watering the soil surface above the geo-space in the evenings. The temperature received throughout the day at a depth of around 5 cm will then penetrate deeper into the soil. The watering should take place in the evening hours, after the soil surface has received its maximum thermal share for the day and before the soil temperature is diffused into the air. By repeating this process for a few evenings, the surface temperature will reach the geo-space more quickly than it would by the normal process.

Observations of living creatures in their natural habitats indicate that some creatures that live in regions with stressful climates (very hot, such as the Sahara, or very cold, such as Canada or north central Asia) or where there is an extreme diurnal temperature fluctuation have constructed their habitats within geo-space for thermal protection.

Condensation

Condensation is the release of humidity caused by the contraction of air when the temperature is reduced significantly. In geo-space, condensation can take place in the summer when the walls provide cool low temperatures, when the relative humidity of the air is very high, and when air exchange is minimal or nonexistent. Geo-space in rainy summer regions, such as Japan, will most likely be subject to condensation.

To minimize or eliminate condensation, it is necessary to increase ventilation. Passive air movement can be achieved through the construction of vertical air shafts where, due to the temperature differentiation between the outdoor (high) and the indoor (low) air pressure, air movement will take place. If a window or door onto a patio is opened, the outside air will move downward into the geo-space habitat.

Another ambient thermal issue associated with the geo-space is the contrast between indoor and outdoor temperature. As was mentioned previously, indoor winter temperature is higher than outdoor temperature due to soil retention of summer heat and the time lag process. Our research findings in the subterranean dwellings of the Matmata plateau, Tunisia, indicate that this indoor-outdoor temperature differentiation often causes the inhabitants of the dwellings to catch cold.[48]

Thermal Performance and Geo-Space Form

Do regions with different thermal performance require different geometric forms for geo-space design? Or are the geo-space forms that are suited to hot-dry climates, such as in North Africa, also ideal for cold-dry climates, such as in central Canada?

Observations of geo-space thermal performance indicate that when the summer temperature is usually between 20 degrees and 22 degrees C, a cooling system is not needed. In the winter, when the temperature is between 12 degrees and 15 degrees C, some type of heating system is needed. Thus, regions with lengthy summer seasons in hot-dry climates, where cooling would be needed throughout most of the year, will require maximization of the surface space (walls, ceiling, and floor), since the entire surface acts as the cooling system of the geo-space habitat. On the other hand, the major problem with cold-dry climate regions with lengthy winter seasons is that heating consumes large amounts of fuel throughout most of the year. This habitat will require minimization of the surface in order to reduce energy consumption. Our research indicates that the most suitable design form for the hot-dry climate is the trapezoid, and for the cold-dry climate, the circle, when both forms have the same floor size and ceiling height.[49]

Other design forms can alter the thermal performance of a habitat, such as having the geo-space part of a house and its surrounding soil totally covered with buildings.[50] This occurs when the geo-space part is positioned less than 10 meters below the surface of the ground. In such spaces, temperature fluctuation increases as the depth decreases. In warm-dry climates, it is desirable to delay the temperature from reaching the belowground space. A horizontal insulating layer above the ceiling of the building, of a size far greater than the ceiling itself, will cause this time lag and, consequently, keep the geo-space house cooler than normal. Another solution is to continuously cool the ground by covering it with trees whose foliage begins high up on the trunk to support the flow of air below.

In hot-humid climates, where ventilation is always desired, the introduction of effective ventilation systems should reduce the temperature within the geo-space. In hot-dry climates, it is crucial to introduce an evaporative cooling system. The Baghdad house is an example of a habitat that has successfully reduced indoor temperatures. This indigenous house, which evolved throughout centuries in a hot-dry climate, combined all three types of levels: aboveground, semi-belowground, and belowground. Its system of cooling is based on a high water table, passive ventilation, air cooling by evaporation, and belowground placement. All of these elements were integrated to cool the belowground space, as well as the aboveground patio. Moreover, the house introduces four levels of coolness, which provide different degrees of relief to suit the needs of different age groups. These four levels consist of a mezzanine level with semi-belowground space exposure (*tachtabosh*); the semi-belowground space with an air-shaft and water jar evaporative cooling system (*neem*); the subterranean space (*serdab*); and the belowground space, which reaches a body of water provided by the water table (*bir el-tbile*).[51]

Air quality differs between the lowland and the slope. In the lowland, inversion is common, especially in the late hours of the night. Inversion can become harmful when the city produces auto exhaust pollution, and, in hot-dry climates, dust pollution. The slope, on the other hand, has fresher air with fewer dust particles, is free from inversion, has better air circulation, and, therefore, has a lower risk of lung-related illnesses.

SLOPE GEO-SPACE HABITAT DESIGN

All slope geo-space housing units should be designed to maintain a view of the outdoor environment and the lowland parts. Slopes have different forms and degrees of gradient, which determine the type of geo-space housing unit and its grouping patterns. Some generalizations can be drawn that classify the housing patterns in relation to the slope gradient (Fig. 2.13):

Figure 2.13. (facing page) Geo-space habitats introduced into slopes with gradients from 7 percent to 80 percent. While the slope offers many benefits, including ventilation, sunshine, a view of the surrounding environment, a tranquil atmosphere, and privacy, steeper slopes introduce difficulties in construction and accessibility. Note the correlation between slope gradient and habitat lot size.

1. **Low gradient slope:** Due to its low gradient and limited rockiness, this type of slope is usually uniform, with little variation. The low gradient slope is flexible in that it can adopt a variety of housing types with limited changes to the grading. Clustered housing, detached or attached row housing, or single, dispersed units are applicable. This type of slope can also accept a combination of geo-space and supra-space development, and it enables the integration of mixed land uses.
2. **Moderate gradient slope:** The moderate gradient is the most commonly chosen slope because it can accommodate a rich variety of housing forms, and because the cost of construction is lower than that on the acute gradient slope. The moderate gradient can also house mixed land uses. Terraced forms are common, enabling buildings to rise above a single story (Fig. 2.14).
3. **Acute gradient slope:** The acute gradient slope usually incorporates a variety of geomorphological configurations and, as such, can enrich the design alternatives for geo-space housing. Development typically

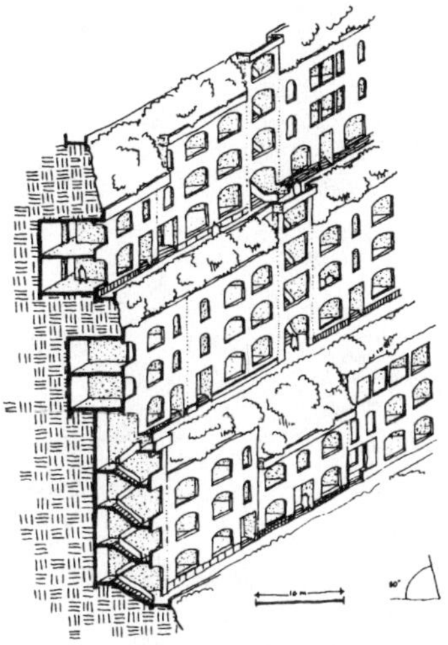

80% GRADIENT SLOPE

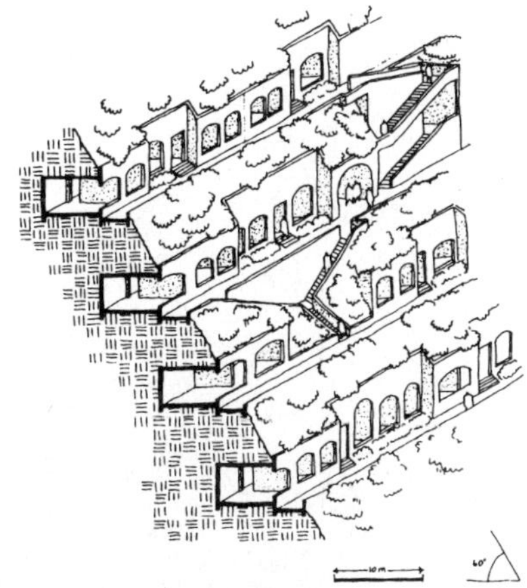

60% GRADIENT SLOPE

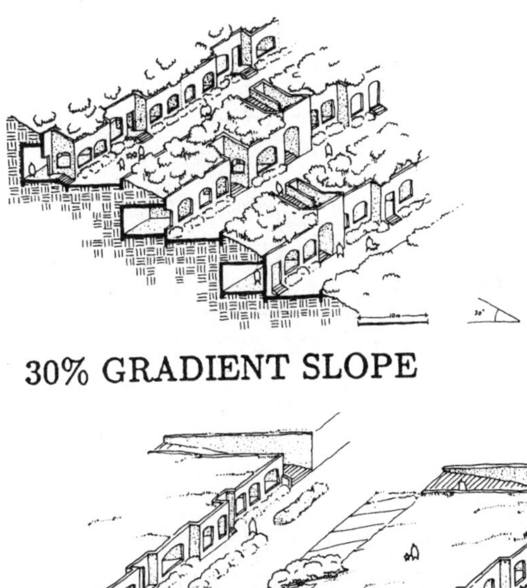

30% GRADIENT SLOPE

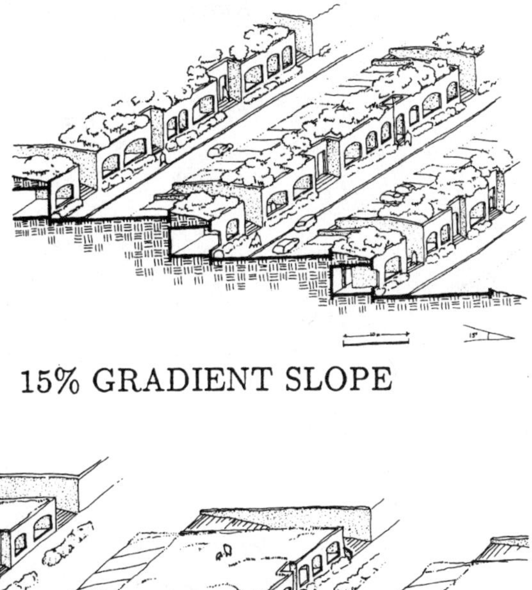

15% GRADIENT SLOPE

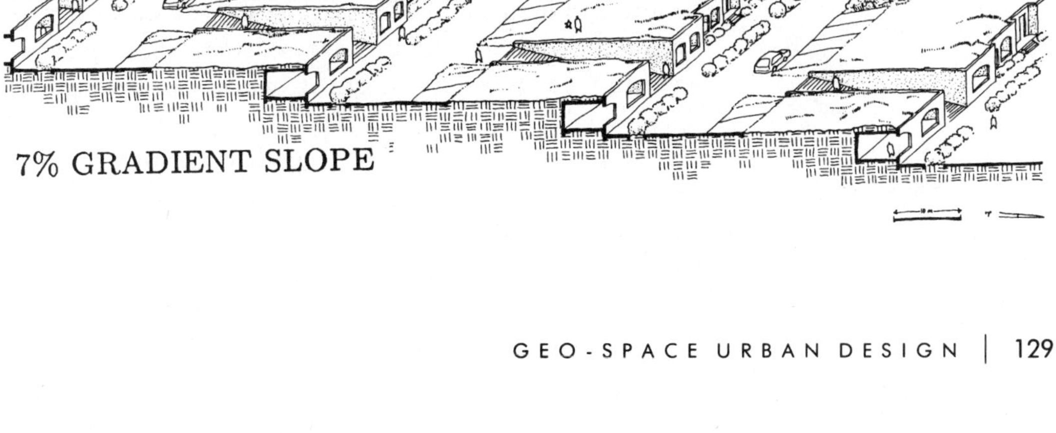

7% GRADIENT SLOPE

Figure 2.14. Integration of subterranean and supraterranean single-family habitats on a terraced slope in a geo-space city.

includes cliff forms and natural terraced forms (Fig. 2.15). It can accommodate cliff geo-space housing or structures, or single detached units. An innovative acute gradient slope design can culminate in a most attractive development integrated with the natural environment, although its construction can be costly.

In the design of the geo-space slope habitat, the following basic guidelines should be adopted by the designer and the developer:

Figure 2.15. A slope geo-space house with two floors and a conventional aboveground façade.

1. Enhance the overall scenery of the geo-space slope city. By virtue of its topography and other natural features, the city has a rich, scenic landscape. The greenery of the slope tends to counter the massiveness of the lowland city.
2. Sketch a complete design of the slope geo-space city with a panoramic and harmonious display of the valley below. This view should represent the fusion of the natural environment of the slopes with the artificial environment of the geo-space façades, supra-space structures, and traffic network.
3. Enhance the slopes by featuring special areas for scenic viewing along traffic routes.
4. Retain a clear skyline of the slope geo-space city so that its appearance shows unity and harmony. The skyline should frame both the slope city and the lowland.

5. Place cultural and entertainment centers that offer evening activities along the slope so the lowland can be viewed at night. These centers should cater to diverse age and gender groups.
6. Introduce an urban environment with a pleasing atmosphere without diminishing the natural environment on the slopes. Thus, the slope geo-space development has the potential to be the pearl of the whole city (including the lowland).
7. Conserve the natural environment of the slope by minimizing the damage from the development of geo-space structures and the terraces.[52]
8. Design façades of the geo-space structures (housing, cultural and entertainment centers, restaurants, religious centers, and the like) to represent their functions. Because of the nature of rocks and soil, the internal architectural form must take the shape of an arch. It is likely that the façade will be shaped similarly. In any case, with the use of a variety of architectural façades, pluralism should be encouraged to form overall aesthetic harmony. Retaining walls should be integrated with the greenery.
9. Install an extensive underground transportation system in the slope geo-space city and develop the supra-space city parts to become a pedestrian city. The front of the city at the face of the slope will have a comprehensive pedestrian network in two directions—horizontally through the terraces and vertically via stairways.
10. Emphasize the natural features of the geomorphology of the given slope. Each slope has its own nuances with its own positive advantages, which should be highlighted. For example, curved sections of the slope could accommodate a small amphitheater between two terraced levels, or a small recess in the topography could hold benches for small gatherings and for viewing the lowland. In short, the outdoor environment of the slope should be attractive and pleasing to all age groups. Emphasis should be placed not only on an overall outdoor appearance, but also on the social

needs of the residents who are living in the confined geo-space environment.

11. Give special attention to landscaping details and lighting design, including alley lights, for nighttime viewing. The view from the interior of the geo-space residence to the lowland should never be obstructed. It is equally important that the ornamental lighting of the geo-space slope city enhance the view from the lowland and the sense of community between the two cities.
12. Aboveground structures in the slope geo-space city should not be too large, nor should they dominate the view.
13. Design a network of outlets in the slope geo-space city for emergency situations. Construction of this network must consider the possibility of landslides, erosion, fires, and earthquakes and should have restricted access to keep it for emergency needs only.

Although geo-space harms the environment less than other types of urban development, it is still important to take into account the sensitive ecosystem of the slope when building geo-space habitats. Each slope has its own unique character, both inside and out. The slope's details can forge a variety of forms and designs.

Finally, sloped topography usually has a low water table in comparison to the lowland or flat land, and therefore has more development potential than those areas. This condition allows for almost unlimited horizontal digging and tunneling, making the development of the slope geo-space city more feasible and practical than other sites.

New Design Principles

With the current state of technology and design capability, it is possible to build any type of subterranean space in almost any slope site and still meet modern standards of living. With an innovative design, we can bring natural light, sunshine, ventilation, and the outdoor environment deep into underground living space. To do so, a few basic principles should be implemented when designing slope geo-space habitats.[53] These five principles are site selection, orientation, accessibility, indoor-outdoor perception, and natural ventilation (Fig. 2.16).

1. **Site selection:** Geo-space habitats for living should be located exclusively on slopes and not on flat sites. Slopes provide a wide variety of configurations, which offer a range of design forms. Slope location will reduce, if not eliminate, the feeling of claustrophobia and offset the loss of other advantages of the flat sites.
2. **Orientation:** The preferred habitat orientation should be toward the south, southeast, or southwest, and not toward the north. When combined with an innovative and careful design, such an orientation will enable sunshine and natural light to penetrate deep into the house. The quality of this lighting will be similar, if not superior, to that of habitats located in flat, congested urban environments.
3. **Accessibility:** Access to the habitat should be by an ascending approach to the geo-space house, as opposed to the traditional descending approach. By mounting stairs to enter the dwelling, some of the initial biases of belowground space should dissipate. This design should diminish the feeling of claustrophobia usually associated with indigenous belowground space.
4. **Indoor-outdoor perception:** The design should allow for direct view between indoor habitat and the outdoor lowland environment. In such a setting, the resident can be part of the surrounding environment and feel the rhythm of the outdoors: the trees, clouds, wind, birds, and the movement of other people. This design makes the natural environment an integral element of the indoor space. Here, too, such design considerations would help to eliminate the feeling of claustrophobia.
5. **Natural ventilation:** The design should provide for the penetration of natural ventilation deep into the habitat with effective passive methods.[54] Slopes usually

Figure 2.16. (facing page) Five basic design principles for slope geo-space habitat design.

RIGHT — WRONG

ONE: SITE SELECTION

SLOPE

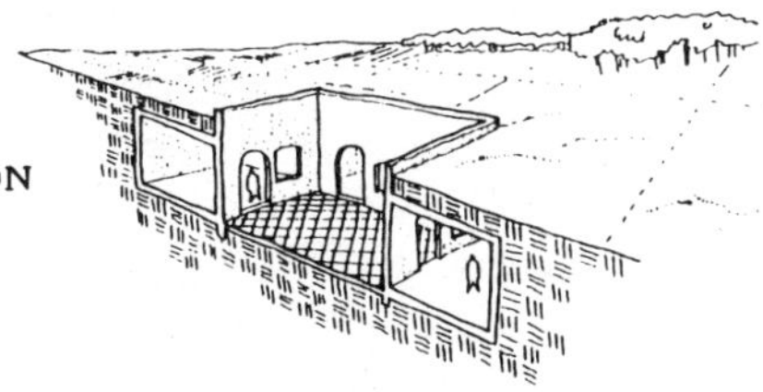

FLATLAND

TWO: ORIENTATION

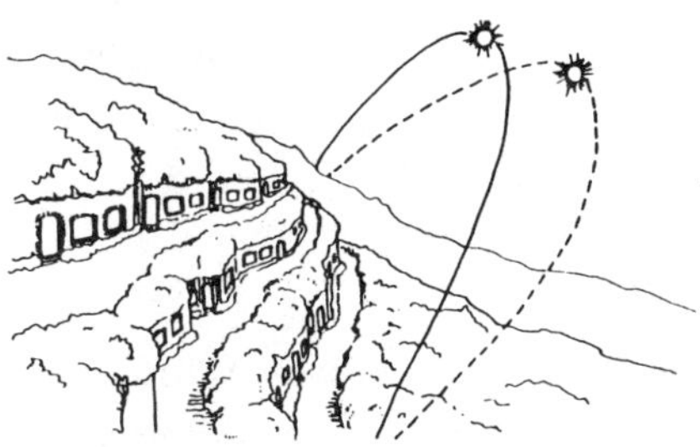

FACING SOUTH, SOUTHEAST OR SOUTHWEST

FACING NORTH

THREE: ACCESSIBILITY

ASCENDING INTO THE HABITAT

DESCENDING

FOUR: INDOOR-OUTDOOR PERCEPTION

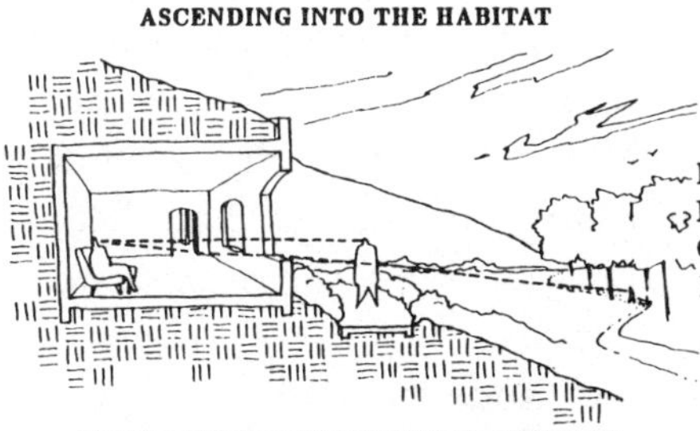

DIRECT INDOOR-OUTDOOR EYE CONTACT TO THE LOWLAND ENVIRONMENT

ABSTRACT VIEW

FIVE: NATURAL EFFECTIVE VENTILATION

WIDE SHAFT WITH ROTATING WIND CATCHER

LACKING PROPER SHAFT

have an active wind movement, which can be harnessed for ventilation purposes with a well-designed air shaft and rotating wind catcher. In a warm-humid summer climate, this principle is essential.

The major deficiency of most of our existing cities is their selection of a flat lowland site. Such a location has proved to be economically expensive, reducing the quality of human health and, in many cases, producing unsightly surroundings. Reevaluation of site selection principles should be also linked with the potential usage of geo-space integration. A flat site is not suitable for dwellings below ground.

Slope Habitat Transportation

Slopes have a variety of topographical features, which can be used effectively for transportation within the residential slope geo-space and can facilitate both vertical and horizontal movement. As opposed to flat land, which enables easy movement because it follows the degree of the topographical contours, slopes face certain obstacles. The primary difficulty is the movement from one horizontal level to another, which can be overcome, or at least eased, by a gradual, but lengthy, configuration of stairways, elevators, escalators, cable cars, climbing wagons, or subways (Fig. 2.17). In the case of acute slopes, geo-space transportation should be built behind the structures as a subterranean network. The urban planner should view the topographical contour as the baseline for transportation and pedestrian movement, and should consider human movement and transportation, emergency access, and evacuation in the city designs.

The subway is an effective mode of mass transportation and a convenient means of linking large-scale active land uses. It can form the approach to the slope geo-space development from a variety of angles, directions, distances, and points of departure as closed or open circulation (Fig. 2.18). In Japan, for example, a new "post-bullet train," an intra-urban regional system 600 km in length, is now being considered for geo-space use. With the adaptation of geo-space usage to slopes, it now is possible to have subways behind

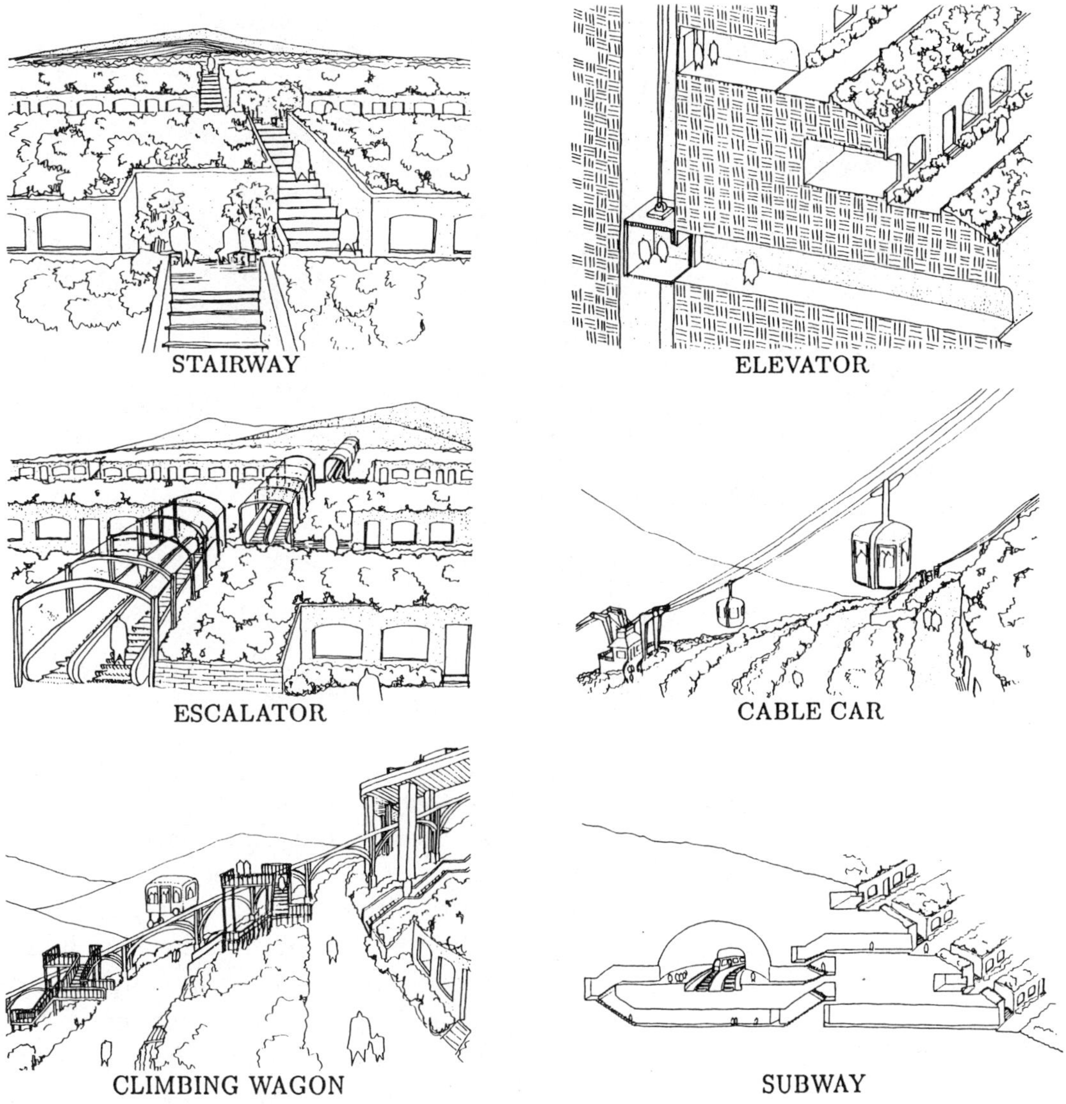

Figure 2.17. Vertical and horizontal traffic movement options along the slope of a geo-space city.

residential neighborhoods linked with other geo-space land uses.

The most difficult problem facing geo-space construction is the design and development of the road network. The geo-space motor vehicle roads and the pedestrian networks can also become a central means of transportation for various slope geo-space land uses (Fig. 2.19).[55] Parking spaces for the neighborhood will also be added, as well as bus stops, sta-

Figure 2.18. Horizontal linear subway train and station serving the residential neighborhood of the slope geo-space city.

tions, and lines. There should still be a limited network of aboveground transportation within the slope geo-space neighborhood for emergency use (Fig. 2.20).

Stairways are the most common method of vertical movement on the slope and have been used for indigenous hillside settlements. The ancient Greeks, who persistently used the grid system in their urban design, built stairways onto their rugged topography and effectively integrated them into the grid system. However, stairways in the slope geo-space city can be quite lengthy and can consume a good deal of human effort. A wide variety of attractive design forms should always be designed to suit the different needs of people (Fig. 2.21). The width of a stairway should accommodate more than one person, and the stairs should be shaded, in a pleasing environment, and offer rest stops along the way. There is almost an endless variety of design possibilities that still provide privacy, intimacy, and an attractive view to the lowland.

Elevators can be used on acute gradient slopes. They can

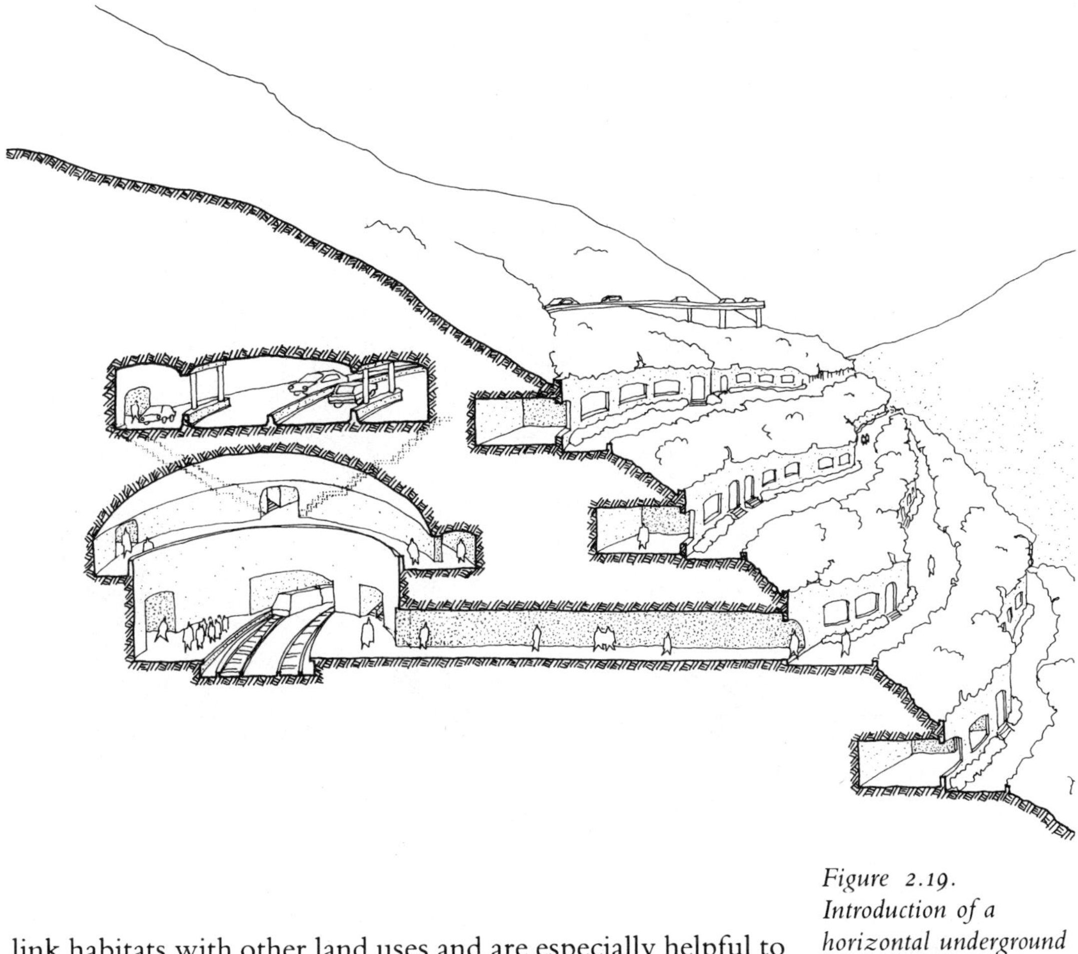

Figure 2.19. Introduction of a horizontal underground transportation network into the slope geo-space city. The network combines automobile and subway transportation.

link habitats with other land uses and are especially helpful to disabled and elderly persons, and to pregnant women. The preferred location for the elevator is outside on the hillside with an open view to the landscape of the lowland. In some cases, the elevator can also be located within the geo-space and can be divided into sections according to the needs of the user (Figs. 2.22 and 2.23).

The escalators that are commonly found in shopping centers can also be used for the slope geo-space city. Escalators are practical for relatively short distances and are serviceable to all age groups, although they may not respond to the needs of disabled users. Escalators can be operated and controlled by the user.

Cable cars are typically found along mountain slopes in ski

Figure 2.20. Example of terraced slope geo-space habitats with limited local aboveground transportation for emergency uses.

Figure 2.21. (facing page) A variety of stairway design forms. The stairways are integrated diagonally, perpendicular to, or parallel with the slopes to provide privacy to each home. The stairways are an alternative access to the slope habitat instead of using the underground motorways located behind the slope geo-space neighborhood.

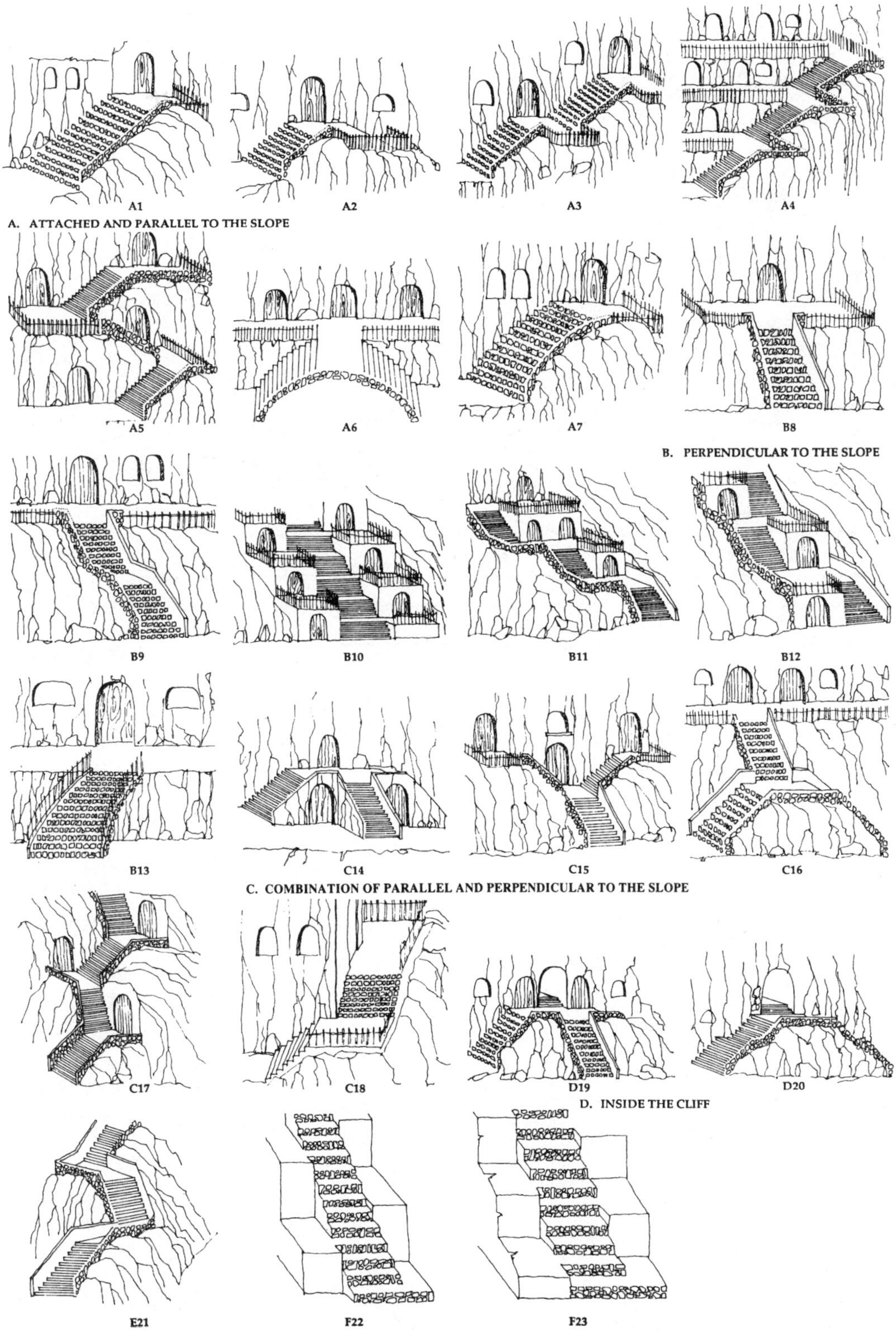
A1
A2
A3
A4
A. ATTACHED AND PARALLEL TO THE SLOPE
A5
A6
A7
B8
B. PERPENDICULAR TO THE SLOPE
B9
B10
B11
B12
B13
C14
C15
C16
C. COMBINATION OF PARALLEL AND PERPENDICULAR TO THE SLOPE
C17
C18
D19
D20
D. INSIDE THE CLIFF
E21
F22
F23

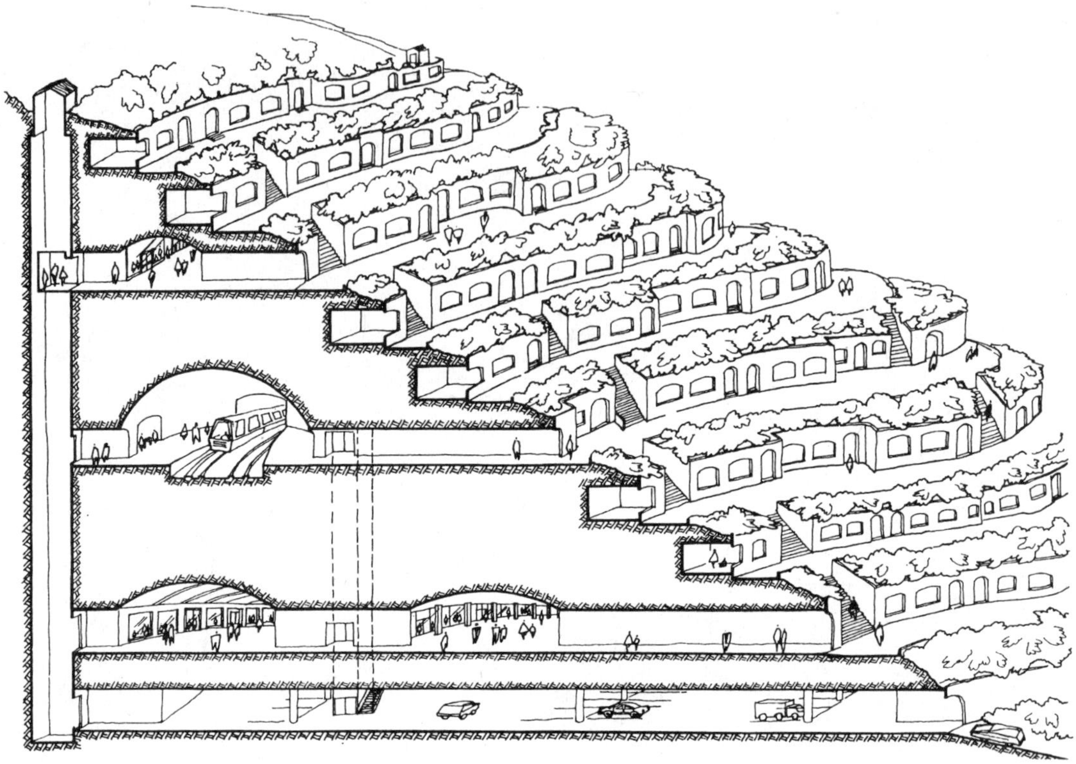

Figure 2.22. Use of elevators in certain sections to ease vertical movement. The elevators can be installed near the shopping district. This method of transportation also meets the needs of handicapped and elderly users.

Figure 2.23. (facing page) Concept of elevator usage in the slope geo-space city neighborhood with an acute gradient of 80 percent. The elevators should be integrated with shops and restaurants. The elevators should also be accessible to the handicapped and the elderly.

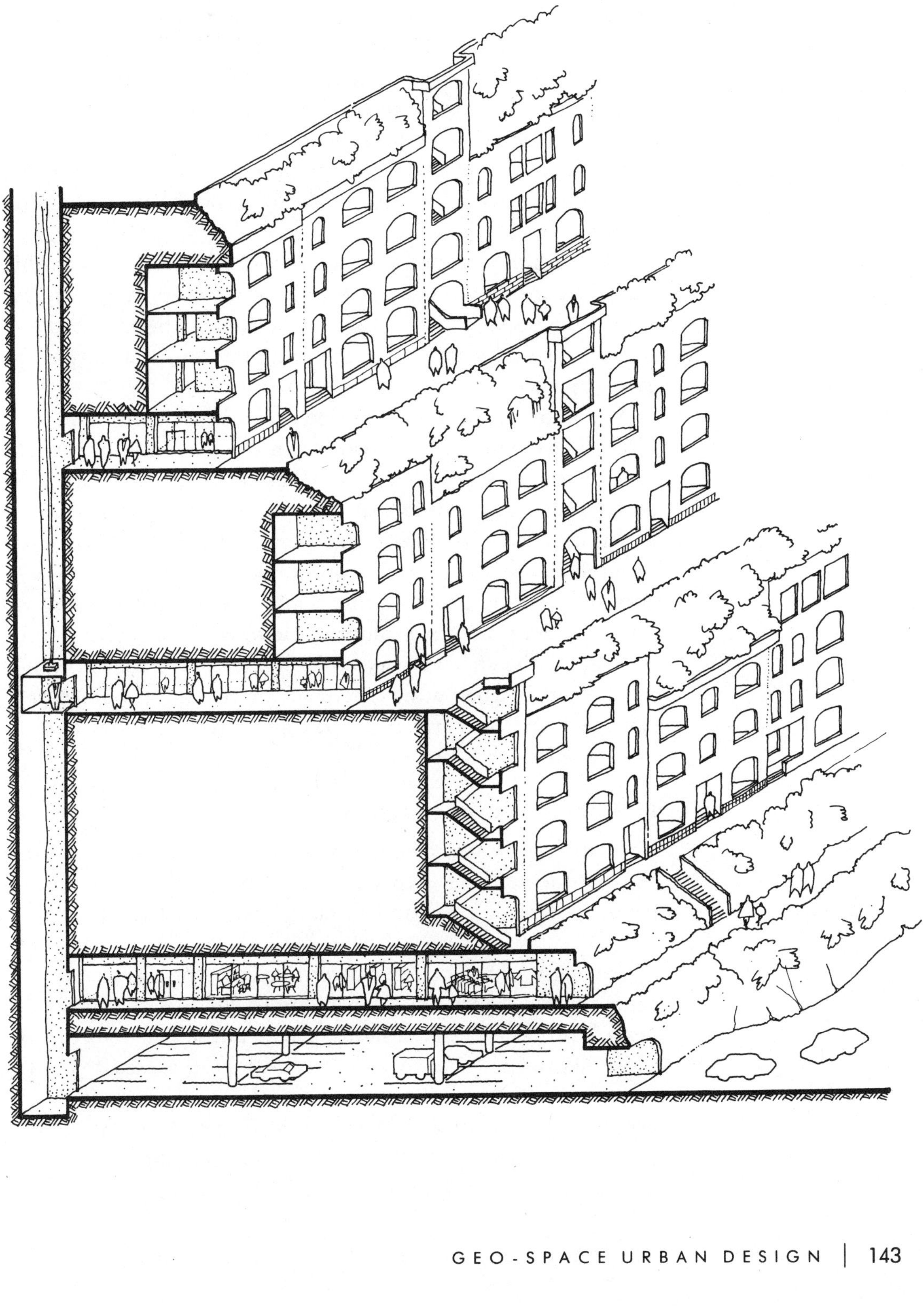

resorts for vertical movement to high elevations. This may be the most adaptable and the least expensive mechanical method of vertical transportation for the slope geo-space city. It also has the benefit of exposing the user to the pleasant view of the surrounding environment. However, the use of the cable car may raise objections from groups such as environmentalists, who believe it greatly disturbs the natural landscape.

The climbing wagon on rails is another means of vertical transportation. While it has the ability to stop frequently at short distances, it may cross the horizontal lines of movement on the slope too often, requiring numerous underpasses or overpasses.

The above-mentioned concepts present belowground solutions that will preserve pedestrian movement and vegetation in the geo-space city, and therefore ensure a safe and pleasing environment free of pollution and noise. To guarantee clear circulation during emergency situations, the aboveground pedestrian network should also be available to accommodate emergency traffic. Stairways along the slope can always be an alternative solution. However, pedestrian traffic by way of the stairway network should enrich the community and the urban atmosphere of the neighborhood, especially when it is furnished with an attractive view to the surrounding landscape.

Habitat Forms and Design

Habitats built within the geo-space will offer people a visually unique aesthetic experience due to the immediate environmental setting of the habitat, as well as its interior design.[56]

The form of the house is determined by the physical environment and, more importantly, by the design that supports maximum penetration of light into the house. Therefore, it is most likely that the designer will adopt the idea that the width of the house (which equals the size of the façade) should be greater than its depth. Thus, the house will take a rectangular form (or similar shape, such as elliptical) rather than a form of equal dimensions, such as a square.

The façade becomes the focal point of the habitat unit and is completely immersed within the geo-space. It will be the only visible part of the house to the outsider. The façade will include doors, windows, balconies, and overhangs. All these features will be designed in harmony aesthetically and functionally to represent the habitat's own unique identity.

The impact of building materials on the geo-space habitat is as an effective visual element. It is probable that many of the habitat façades will not be flush, since the geo-space habitat's natural environment will most likely be of stone. The façade would probably be made from uniformly cut stones of different materials, such as bricks, blocks, cement, adobe, or a combination of these. In any case, this façade will be the dominant visible feature of the house. Aesthetically, it should represent the character of the house. In short, the façade should be of great concern to the architect/designer and to the residents.

As we have suggested before, the floor-to-ceiling height of the interior design should exceed the conventional standards of 2.75 meters or 3 meters in order to achieve the feeling of spaciousness and minimize claustrophobia. In this case, the height and width of the house will be expressed by the façade of the building. Contrary to the traditional supra-space house, the depth of the geo-space habitat will not be noticeable to the outside viewer.

The patio also has an important role in the design of the habitat. An open patio in the front of the façade would establish a space open to the sky and would add a sharp contrast between the interior part of the house and its exterior environment. However, a façade-patio marriage offers all sorts of geometrical forms, including squares, circles, hexagons, and rectangular shapes, and other concave or convex forms (Fig. 2.24). The open patio can be planned for the center of the house and surrounded on all sides with geo-space habitat units.

Interior design of the geo-space habitat will be similar, yet not identical, in its basic design principles to the supra-space house. Indeed, the one-family unit can be made of one, two, three, or more floors, as the client may desire. The greater

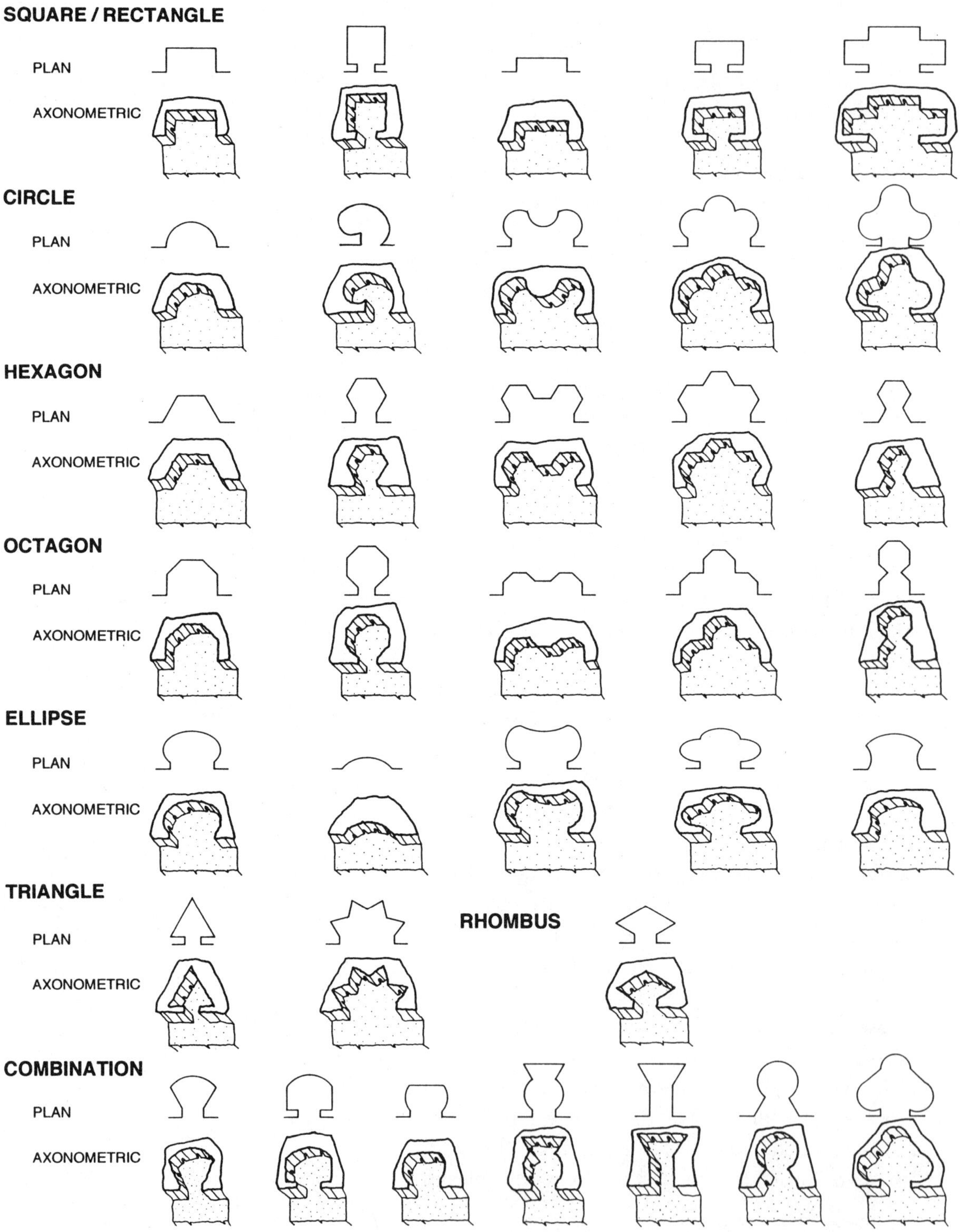
SQUARE / RECTANGLE
PLAN
AXONOMETRIC
CIRCLE
PLAN
AXONOMETRIC
HEXAGON
PLAN
AXONOMETRIC
OCTAGON
PLAN
AXONOMETRIC
ELLIPSE
PLAN
AXONOMETRIC
TRIANGLE
PLAN
AXONOMETRIC
RHOMBUS
COMBINATION
PLAN
AXONOMETRIC

TABLE 2.3a.
Location of Slope Geo-Space Habitat Units

LOCATION OF:	
Bedrooms	-inside
Living room	-front
Family room	-front
Recreation room	-front
Kitchen	-inside
Restroom	-inside
Stairways	-inside
Storage	-inside
Garage	-inside

TABLE 2.3b.
Special Design Features of Slope Geo-Space Habitat Units

DESIGN FEATURES:
• High, curved ceilings (vault, dome)
• Wide spaces for rooms, corridors
• Light and bright colors
• Merging space functions: dining, kitchen
• Minimum dividing walls, when possible
• Wide, generous space at the habitat entrance
• Two floors with common atrium

the number of units, the greater the amount of natural light and sunshine that will penetrate into the housing unit (Table 2.3). If the habitat comprises more than one floor, it would be better if the deep part of some floors be built of glass bricks, which would allow light to bounce deeper into the floor spaces.

SOCIOECONOMIC BENEFITS

The overall value of using the "trinity concept" of slope usage, compact form, and geo-space will be greater than that gained from conventional urban forms. Some of its long-range benefits include economic, social, health and well-being, and transportation factors.

There are also significant social benefits to using this combined design pattern.[57] First, the commuting time between places of work, residences, and shopping would be reduced substantially. Second, urbanity and social interaction would be increased, which would ease the loneliness that is so prevalent in modern cities (Fig. 2.25). Third, children and the elderly would be nearer to the urban setting and more familiar with the city's urban geography, improving the quality of their social lives and their self-reliance. Open spaces, both supraterranean and subterranean, would be abundant and in close proximity to pedestrian outlets, making the city more pedestrian-oriented than conventional cities. This new urban

Figure 2.24. (facing page) Progression of plans and axonometric views of geometric façade forms of slope geo-space habitats. The open patios represent the use of the slope side with a front pedestrian passage.

pattern would make use of land, such as acute slopes, that otherwise would not have been used (Fig. 2.26). The economic savings from this type of design pattern can be used to make significant contributions to the improvement of services and the overall quality of urban life (Fig. 2.27).[58]

To compare the urban economic gains in land use, infrastructure, and urban fixtures, this author researched and compared three separate design patterns for 19.42 hectares (48 acres) of land: conventional sprawl (on a flat site), compact form (on a slope site), and combined form (compact form with subterranean use on a slope site). Although there is an economic gain associated with reduced urban fixtures, it was only after the statistical results were compiled that the magnitude of the gain became apparent. Combining the supraterranean and subterranean models produced greater savings than did the conventional model.[59]

The urban economic return values are diverse and, in general, lower project costs.[60] First and foremost, land prices are reduced by half because of the dual use of the land. Second, the innovative design and construction would also reduce the overall expenditures because slope site, compact form, and subterranean space usage are all integrated.[61] Third, operation and maintenance expenditures should be lower, offering relief to the taxpayer and/or diverting the gain to the improvement of the social quality of life and the environment. Finally, there are environmental advantages to be gained.[62] The environment is protected and preserved because of the minimal number of changes to the natural surroundings, landscape, air quality, and tranquility.[63]

CONCLUSION

A steady increase in geo-space usage—and, with it, the accelerated use of geo-space habitats—is anticipated to continue in the future. In countries where land is scarce for urban expansion, slope geo-space habitats will be adopted out of necessity. In other countries, geo-space design will be practiced for climatic reasons. In still others, particularly those countries

Figure 2.25. (facing page) View of an urban slope geo-space habitat complex.

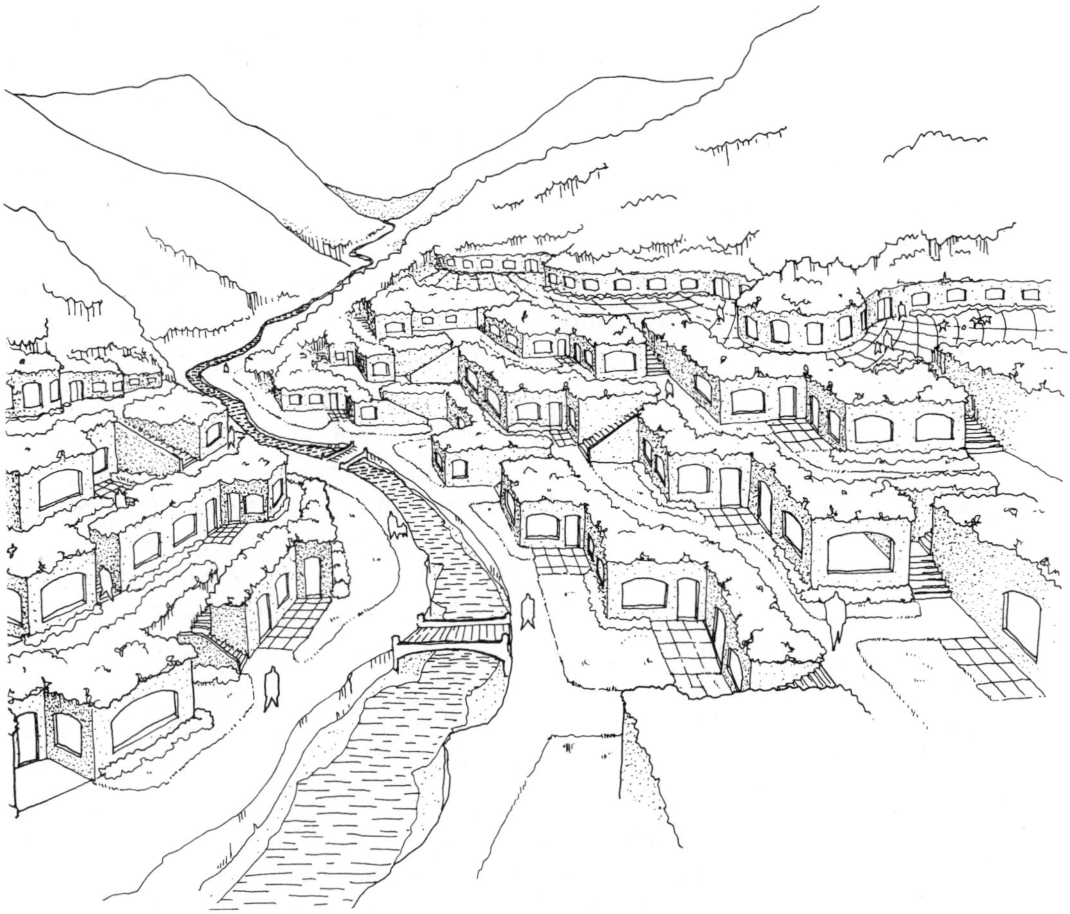

Figure 2.27. Slope geo-space and earth-integrated urban neighborhood built along acute slopes on both sides of a river.

with advanced technology, slope geo-space habitats will be endorsed because of the wide range of benefits they offer. This author believes that a breakthrough will occur when slope geo-space habitats are more widely used by the middle class, which will improve the image of geo-space in the eyes of the general public.

Although slope geo-space construction requires a higher initial investment than does construction on a flat site, this cost would be offset by various short- and long-range benefits of the slope site, including improvements to the overall quality of life.

There has already been significant technological success in the construction of belowground space, but we are still at the

Figure 2.26. (facing page) Geo-space habitats on acute slopes. If not developed, this land would not have otherwise been used.

infant stages in terms of human perceptions and acclimatization to geo-space living. It is this author's assumption that the nature of geo-space would have a far-reaching, positive influence on human beings, although, to be sure, these effects are not yet fully known.[64] In short, the nature of geo-space can be narrowed into three elements:

- Diurnal stability and synchronization of temperature and humidity
- Deep quietness, tranquility, and freedom from disturbances
- A low degree of pollution of all types

In prioritizing the research on geo-space use, we must first treat the effects of geo-space on humans. In the near future, research should be undertaken by a collaborative team to address the variety of present issues and debates. The following subjects should be priorities for further study:

1. It is this author's assumption that sleeping in a geo-space environment affects the physiological, psychological, and mental well-being and tranquility of individuals of various age groups. Research in this area should be used as a guide for the design of future geo-space living areas. In cities built on flat land, aboveground houses can have belowground sections in the form of basements (semi-belowground) or cellars (fully belowground). If sleeping in belowground spaces proves to be healthier than in aboveground spaces, bedrooms should be located in the belowground level. The advantages of belowground space usage should be emphasized to help improve its image.
2. The effect of the ambient temperature conditions of geo-space on the rate of healing and recuperation from external wounds after surgery should be researched.
3. The impact of geo-space on individual and group creativity should be studied. It is my assumption that per-

sons engaged in writing, music composition, painting, sculpture, or other activities that require quiet contemplation and reflection will be more productive in the geo-space than in the supra-space.

4. Longevity and geo-space habitation may be related because of the quality of the geo-space environment. The high concentration of cave dwellings in northern China would make an ideal laboratory for such research.
5. Improved work performance and productivity may occur within the geo-space environment. It is my assumption that geo-space, compared with aboveground spaces, would affect people positively by increasing production. If such an assumption is proven to be correct, there will be an additional economic implication for geo-space usage.
6. Bias and claustrophobia, and the methods of coping with these psychological issues, are a matter of research priority. Also, other research findings can be used to improve the negative perception of geo-space.
7. Research should be conducted on possible design methods for effective natural ventilation systems, especially for warm and humid climates, in order to improve indoor air quality and human comfort.
8. Slopes offer optimal conditions for geo-space living using any geomorphological forms. There has been significant historical design innovation and practice with aboveground hillside housing, but not with slope geo-space housing, requiring research of the development of various design patterns.

Construction of the slope geo-space habitat has its own price. Other than the price defined by the cost-benefit analysis approach often used by economists, it is necessary to consider the price paid for the social quality of life, for healthy conditions, for mental well-being, and for the level of comfort of a supra-space urban flatland site. The slope geo-space habitat offers a compromise between gains and losses. It offers new hope for upgrading our quality of living. The need to adapt earth-sheltered space to the city is based not only on the shortage of land, but also on improving the quality of life.

There is a strong bias against living below ground. Poor design implementation of belowground habitats in the past has strengthened that bias. We believe that the problems associated with geo-space habitation can be solved to adjust to our present living standards. The major limitations of belowground habitation are psychological factors. We offer the following design rules to cope with such problems:

1. Geo-space habitats should be located exclusively on slopes and in terraced form (see Figs. 2.12, 2.13, and 2.14).
2. Geo-spaces should be oriented toward the south, southeast, or southwest to enable direct, deep penetration of sunshine and the maximum amount of natural light.
3. Ascending, not descending, access to the habitat should be adopted (see Fig. 2.15).
4. Direct eye contact between the indoor and the outdoor environment should be provided so that the inhabitants can experience the daily rhythm of their natural surroundings.
5. The outdoor environment should be brought into the geo-space habitat.

The negative image of the geo-space habitat is expected to be eased when such habitats are more often used by middle and upper socioeconomic groups. An increasing understanding of the potential of good modern design to afford comfortable living conditions, even if they are below ground, should also improve the image of the geo-space habitat.

Finally, the use of geo-space for habitation yields more favorable results in extreme climates than in temperate ones; however, every climate type introduces its own problems to geo-space design, which must be resolved separately. The major problem, for example, in rainy-temperate climate zones is the need to counter the high humidity rates with an effective ventilation system.

Although the optimal development of the trinity city concept can be achieved in new towns, it would be better to integrate this concept into an existing city, since the habitats are somewhat isolated on a slope. The response of the general public may be negative as a result of bias, misinformation, inexperience, lack of a holistic view, and short-range planning. To counter this negativity, there is a need for increased public awareness, established public policy, and coordinated design enforcement.

Public awareness of the trinity city concept can be increased through study, research, dissemination of knowledge through all forms of media, discussions, and displays of practical models. The formation of public policy should follow the increase in public awareness and be established as part of the legislation, where appropriate. New zoning ordinances combined with incentives are necessary to achieve enforcement and to protect the public interest. It will be essential to enlist the opinions and knowledge of designers, policy makers, and educators throughout the design and implementation process. This requires the effort and contribution of almost every sector of the public and community.

The trinity city concept introduced here is based on the study of historical vernacular urban design and its adaptability to our modern needs. Although it is based on an ancient concept, its role in modern times stems from its practicality. This innovative concept requires the contributions of aggressive designers, social scientists, economists, and politicians. It is the author's belief that the project is a tradeoff of cost benefits in the social and economic sense, and that for every social venture there is an economic price.

NOTES

1. George Rand, "Thinking Ecologically About Indoor Environments," *Underground Space,* Vol. 6, No. 2 (Sept./Oct. 1981): pp. 105–08.
2. Belinda L. Collins, "Review of the Psychological Reaction to Windows," in *Underground Utilization: A Reference Manual of Selected Works,*

Vol. IV: Human Response and Social Acceptance of Underground Space, ed. Truman Stauffer (Kansas City, MO: A Geographic Publication, Department of Geosciences, University of Missouri, 1978): pp. 532–40.

3. Tadashi Hane, Keiko Muro, and Hidekaszu Sawada, "Psychological Factors Involved in Establishing Comfortable Underground Environments," *Urban Underground Utilization '91, Proceedings* (Tokyo: Urban Underground Space Center of Japan, 1991): pp. 480–87.
4. Elizabeth Wunderlich, "Psychology and Underground Development," in *Underground Utilization: A Reference Manual of Selected Works, Vol. IV: Human Response and Social Acceptance of Underground Space,* ed. Truman Stauffer (Kansas City, MO: A Geographic Publication, Department of Geosciences, University of Missouri, 1978): pp. 526–29.
5. Paul B. Paulus, "On the Psychology of Earth-Covered Buildings," *Underground Space,* Vol. 1, No. 2 (July/Aug. 1976): pp. 127–30.
6. J. B. Hughey and R. L. Tye, "Psychological Reactions to Working Underground: A Study of Attitudes, Beliefs, and Evaluations," *Underground Space,* Vol. 8, No. 5–6 (1984): pp. 381–91.
7. Junji Nishi, Fujio Kamo, and Kunihiko Ozawa, "Rational Use of Urban Underground Space for Surface and Subsurface Activities in Japan," *Tunneling and Underground Space Technology,* Vol. 5, No. 1/2 (1990): pp. 23–32.
8. Masato Sato, Dr. Masao Inui, and Junichiro Sawaki, "Psychological and Behavioral Effects of Visually Closed Office Spaces," *Urban Underground Utilization '91, Proceedings* (Tokyo: Urban Underground Space Center of Japan, 1991): pp. 493–99.
9. Yuji Wada and Hinako Sakugawa, "Psychological Effects of Working Underground," *Tunneling and Underground Space Technology,* Vol. 5, No. 1/2 (1990): pp. 33–37.
10. Gideon S. Golany, *Earth-Sheltered Dwellings in Tunisia, Ancient Lessons for Modern Design* (Newark, DE: University of Delaware Press, 1988): p. 144.
11. Gideon S. Golany, *Chinese Earth-Sheltered Dwellings, Indigenous Lessons for Modern Urban Design* (Honolulu, HI: University of Hawaii Press, 1992): pp. 141–43.
12. Barbara Routh, "A Study of Human Response to Attitude Statements About Use and Occupancy of Underground Space: A Preliminary Report," in *Underground Utilization: A Reference Manual of Selected Works, Vol. IV: Human Response and Social Acceptance of Underground Space,* ed. Truman Stauffer (Kansas City, MO: A Geographic Publication, Department of Geosciences, University of Missouri, 1978): pp. 541–45.
13. Yuji Wada and Hinsko Sakugawa, "Psychological Effects of Working Underground," *Tunneling and Underground Space Technology,* Vol. 5, No. 1/2 (1990): pp. 33–37.
14. Gideon S. Golany, *Earth-Sheltered Habitat, History, Architecture, and Urban Design* (New York: Van Nostrand Reinhold, 1983): p. 20.
15. Gideon S. Golany, "Can We Live in Geo-Space?" *Urban Underground*

Utilization '91, Proceedings (Tokyo: Urban Underground Space Center of Japan, 1991): pp. 18–23.

16. T. E. Wirth, "Landscape Architecture Above Buildings," *Underground Space,* Vol. 1, No. 4 (1977): pp. 333–46.
17. Sydney A. Baggs, "The Dugout of an Outback Opal Mining Town in Australia," in *Underground Utilization: A Reference Manual of Selected Works, Vol. IV: Human Response and Social Acceptance of Underground Space,* ed. Truman Stauffer (Kansas City, MO: A Geographic Publication, Department of Geosciences, University of Missouri, 1978): pp. 573–99.
18. Ray Sterling, "Earth-Sheltered, Energy Independent (ESEI) Building Design Project," in *Underground Utilization: A Reference Manual of Selected Works, Vol. IV: Human Response and Social Acceptance of Underground Space,* ed. Truman Stauffer (Kansas City, MO: A Geographic Publication, Department of Geosciences, University of Missouri, 1978): pp. 569–72.
19. Peter K. S. Pun, "Design Policies for Mixed Land Use Development: A Case of Hong Kong," *Regional Development Dialogue,* Vol. 12, No. 2 (Summer 1991): p. 44.
20. Yoshinobu Ashihara, "The Aesthetics of Cities Built on Hillside Areas," *Hillside Cities,* report and summary of the proceedings of the International Conference on Hillside Cities: Focus on the Revitalization of Local Cities (Nagoya, Japan: United Nations Center for Regional Development, 1989): pp. 69–105.
21. Hindenobu Jinnai, "A Comparison of Japan Space in Hillside Environment: Japan and Italy," *Regional Development Dialogue,* Vol. 12, No. 2 (Summer 1991): pp. 37–38.
22. Id., p. 38.
23. Mararu Mizuta, "The Conservation and Development of Hillside Environments: the Nagasaki Example," *Regional Development Dialogue,* Vol. 12, No. 2 (1991): pp. 143–44.
24. L. F. Goldberg, "A Comparative Experimental Evaluation of Five Earth-Sheltered Houses," *Underground Space,* Vol. 8, No. 2 (1984): pp. 36–43.
25. Kazuhiko Urano and Katsumi Matsubara, "Seismic Design Method of Shield Tunnels with Axial Prestressing by Means of Rubber and PC Bar," *Urban Underground Utilization '91, Proceedings* (Tokyo: Urban Underground Space Center of Japan, 1991): pp. 412–18.
26. Arnold N. Lowing, "Earthquake Considerations for Earth-Sheltered Residential Buildings," *Underground Space,* Vol. 8, No. 3 (1984): pp. 169–73.
27. R. L. Loofbourow, "Effects of Seismic Movement on Underground Space, with Special Reference to Kansas City, Missouri," *Underground Space,* Vol. 9, No. 4 (1985): pp. 225–29.
28. Etsuro Saito, Hiroshi Ikemi, Hiroyuki Nakano, and Masahiro Nakamura, "Estimation of Anxiety in Underground Urban Space During Earthquakes," *Urban Underground Utilization '91, Proceedings*

(Tokyo: Urban Underground Space Center of Japan, 1991): pp. 512–18.

29. Golany (1992): pp. 111–14.

30. John Degenkolb, "Fire Protection for Underground Buildings," *Underground Space,* Vol. 6, No. 2 (Sept./Oct. 1981): pp. 93–95.

31. Yoshihei Takeuchi, "Safety Precautions Against Fire in Urban Subsurface Space," *Urban Underground Utilization '91, Proceedings* (Tokyo: Urban Underground Space Center of Japan, 1991): pp. 564–67.

32. Kenneth B. Labs, "Planning for Underground Housing," in *Underground Utilization: A Reference Manual of Selected Works, Vol. IV: Human Response and Social Acceptance of Underground Space,* ed. Truman Stauffer (Kansas City, MO: A Geographic Publication, Department of Geosciences, University of Missouri, 1978): pp. 553–61.

33. Mitsumasa Okamura, "On the Construction Technology for Underground Structures with Large Section and Length in Metropolitan Area," *Urban Underground Utilization '91, Proceedings* (Tokyo: Urban Underground Space Center of Japan, 1991): pp. 387–90.

34. Pierre Duffaut, "Site Reservation Policies for Large Underground Openings," *Underground Space,* Vol. 3, No. 4 (Jan./Feb. 1979): pp. 187–94.

35. Gideon S. Golany, *New Town Planning: Principles and Practice* (New York: John Wiley & Sons, 1976): pp. 60–93.

36. Truman Stauffer, ed., *Underground Utilization: A Reference Manual of Selected Works,* Vols. I–VIII (Kansas City, MO: A Geographic Publication, Department of Geosciences, University of Missouri, 1978).

37. Golany (1988): pp. 72–76.

38. Golany (1992): pp. 14–28 and pp. 115–18.

39. Brent Anderson, "Waterproofing Materials and Techniques for Cut-and-Cover Structure," *Underground Space,* Vol. 8, No. 2 (1984): pp. 99–110.

40. Satoshi Imamura, Shuji Tanaka, and Kuniaki Sato, "Technology Assessment Technique on Groundwater and Land Environment Conservation Concerning Underground Space Development in Urban Area," *Urban Underground Utilization '91, Proceedings* (Tokyo: Urban Underground Space Center of Japan, 1991): pp. 429–34.

41. Kuniaki Sato, "Underground Space Use and Environment Protection Technology in Japan," *Urban Underground Utilization '91, Proceedings* (Tokyo: Urban Underground Space Center of Japan, 1991): pp. 439–47.

42. Yoshiyuki Shimoda and Minoru Mizuno, "Utilization of Thermal Characteristics of the Ground at Underground Space," *Urban Underground Utilization '91, Proceedings* (Tokyo: Urban Underground Space Center of Japan, 1991): pp. 608–15.

43. Gideon S. Golany, "Soil Thermal Performance and Geo-Space Design," *Japan Society of Civil Engineers, Geotechnical Engineering, Proceedings of JSCE,* No. 445/III-18 (1992–3): pp. 1–8 (invited paper).

44. Gideon S. Golany, *Design and Thermal Performance, Below-Ground Dwellings in China* (Newark, DE: University of Delaware Press, 1990): pp. 44–125.

45. Hiroshi Yoshino, Muneshige Nakatomo, Shinichi Matsumoto, and Tsutomu Sakanishi, "Five-Year Measurement for Thermal Performance of a Semi-Underground Test House," *Urban Underground Utilization '91, Proceedings* (Tokyo: Urban Underground Space Center of Japan, 1991): pp. 521–28.

46. Thomas Bligh, "A Comparison of Energy Consumption in Earth Covered vs. Non-Earth Covered Buildings," *Alternatives in Energy Conservation: The Use of Earth Covered Buildings,* proceedings of a conference, Fort Worth, TX: July 9–12, 1975 (Washington, DC: National Science Foundation, 1975): p. 92.

47. Golany (1983): pp. 63–68.

48. Golany (1988): pp. 92–108.

49. Golany (1983): pp. 105–08.

50. S. J. Raff, "Ground Temperature Control," *Underground Space,* Vol. 3, No. 1 (July/Aug. 1978): pp. 35–44.

51. Gideon S. Golany, *Baghdad Indigenous House Design,* in review by the press, Chapter IV.

52. Bruce Hamilton Green, "Your Underground House," in *Underground Utilization: A Reference Manual of Selected Works, Vol. IV: Human Response and Social Acceptance of Underground Space,* ed. Truman Stauffer (Kansas City, MO: A Geographic Publication, Department of Geosciences, University of Missouri, 1978): pp. 600–07.

53. Bruce Hamilton Green, "Basic Principles of Underground House Construction," in *Underground Utilization: A Reference Manual of Selected Works, Vol. IV: Human Response and Social Acceptance of Underground Space,* ed. Truman Stauffer (Kansas City, MO: A Geographic Publication, Department of Geosciences, University of Missouri, 1978): pp. 608–13.

54. R. Thompson, "Air Quality Maintenance in Underground Buildings," *Underground Space,* Vol. 1, No. 4 (1977): pp. 355–64.

55. Donald Reis, "Public-Private Cooperation in Developing an Underground Pedestrian System," *Underground Space,* Vol. 6, No. 6 (May/June 1982): pp. 337–42.

56. T. E. Wirth, "Landscape Architecture Above Buildings," *Underground Space,* Vol. 1, No. 4 (Aug. 1977): pp. 333–46.

57. Eugene L. Foster, "Social Assessment—A Means of Evaluating the Social and Economic Interaction Between Society and Underground Technology," in *Underground Utilization: A Reference Manual of Selected Works, Vol. IV: Human Response and Social Acceptance of Underground Space,* ed. Truman Stauffer (Kansas City, MO: A Geographic Publication, Department of Geosciences, University of Missouri, 1978): pp. 523–25.

58. Pere Riera and Joan Pasqual, "The Importance of Urban Underground

Land Value in Project Evaluation: A Case Study of Barcelona's Utility Tunnel," *Tunneling and Underground Space Technology,* Vol. 7, No. 3 (1992): pp. 243–50.

59. Golany (1992): pp. 152–53.
60. Richard M. Michaels, "Social and Economic Feasibility Analysis for Uses of Underground Space," in *Underground Utilization: A Reference Manual of Selected Works, Vol. IV: Human Response and Social Acceptance of Underground Space,* ed. Truman Stauffer (Kansas City, MO: A Geographic Publication, Department of Geosciences, University of Missouri, 1978): pp. 627–33.
61. John D. Rockaway and N. B. Aughenbaugh, "Go Underground for Low Cost Housing," in *Underground Utilization: A Reference Manual of Selected Works, Vol. IV: Human Response and Social Acceptance of Underground Space,* ed. Truman Stauffer (Kansas City, MO: A Geographic Publication, Department of Geosciences, University of Missouri, 1978): pp. 614–18.
62. Dr. Eugene L. Foster, "An Engineer's View of Underground Construction/Mining and the Environment," in *Underground Utilization: A Reference Manual of Selected Works, Vol. IV: Human Response and Social Acceptance of Underground Space,* ed. Truman Stauffer (Kansas City, MO: A Geographic Publication, Department of Geosciences, University of Missouri, 1978): pp. 621–26.
63. Richard F. Dempewolff, "Your Next House Could Have a Grass Roof," in *Underground Utilization: A Reference Manual of Selected Works, Vol. IV: Human Response and Social Acceptance of Underground Space,* ed. Truman Stauffer (Kansas City, MO: A Geographic Publication, Department of Geosciences, University of Missouri, 1978): pp. 564–68. And see Malcolm B. Wells, "To Build Without Destroying the Earth," in *Underground Utilization: A Reference Manual of Selected Works, Vol. IV: Human Response and Social Acceptance of Underground Space,* ed. Truman Stauffer (Kansas City, MO: A Geographic Publication, Department of Geosciences, University of Missouri, 1978): pp. 546–52.
64. Richard D. Harza, "National Commitment to Better Urban Life Assures a Rapid Growth in Underground Construction," in *Underground Utilization: A Reference Manual of Selected Works, Vol. IV: Human Response and Social Acceptance of Underground Space,* ed. Truman Stauffer (Kansas City, MO: A Geographic Publication, Department of Geosciences, University of Missouri, 1978): pp. 619–20.

3

JAPANESE GEO-SPACE SHOPPING CENTERS

T. Ojima

Japan is a small country that has little land where people can live and work. This shortage of space has made it necessary to fully utilize the land, so the Japanese actively have pursued building underground. Their main purpose in developing belowground space is to supplement the existing sites aboveground (Fig. 3.1).

Underground utilization for shopping centers is peculiar to Japan, which has developed the largest number of underground shopping centers in the world. In such a crowded country, the following purposes are served by its underground shopping centers:

- They separate people and cars by developing separate passages underground for pedestrians and automobiles.
- They enable the development of subways.
- They free street space above ground for the expansion of roads and for development of open squares in front of railway stations.

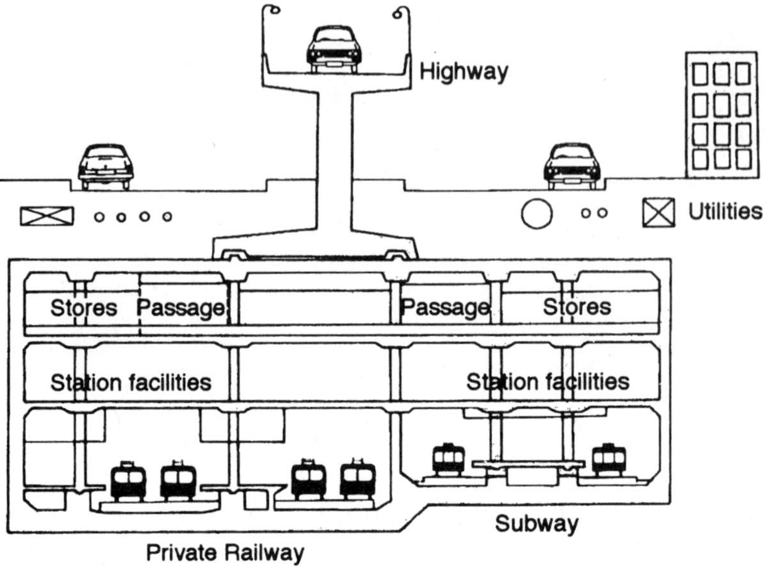

Figure 3.1. Space utilization of underground shopping centers (Osaka Underground Shopping Center Co., Ltd., Introduction of Osaka Underground Shopping Center Co., Ltd., *Osaka).*

- They utilize the basements of buildings.
- They introduce a unified development with above-ground facilities and with parking lots.

The concept of an underground shopping center began as an underground passage constructed for pedestrians to cross under the street. In its early stages, the passage did not have any stores or station facilities. After a short time, however, advertisements were attractively introduced and affixed to both sides of the passage walls. This was the first step in introducing the commercial function into the underground passage, which, up to this time, had been used purely for pedestrian traffic. At the planning stage, show windows were designed for the advertisements by drilling spaces about 1 meter deep into the walls. Initially, the advertisements were placed for indirect commercial use, and products were not exchanged. Eventually, however, display racks were added to the surface on both sides of the wall and salesmen stood next to them to sell their products. By adding the exchange function, the monotonous, dark spirit of the underground passage was lifted, and it became prosperous and lively like the passages aboveground. This transformation was a gradual process that eventually evolved into the complex underground shopping center as it is known today.

One reason why Japan has developed more underground shopping centers than Europe and the United States is its high dependency on railroads for transportation. In many large cities after the 1930s, railway transportation moved underground in the form of subways with a platform constructed between the level of the subway and the surface of the ground to be used as a gate or ticket counter. As the front of a railway station became a congested meeting ground where many people boarded or left the trains, it became necessary to develop space for the passengers to wait. At the same time, this situation was viewed as an excellent opportunity to entice people to gaze at and purchase products, and this space also came to be used as a commercial area.

Thus, the terminal building became a complex of its station and commercial facilities. The basement of this building was then extended to the underground square of the station as a convenience for passengers. The entrances to this complex accessed the surface, and were planned as an addition to the aboveground floors. Thus, the railroad station complexes, like the underground passages, led to the evolution of underground shopping sites.

Street vendors were another factor that contributed to the development of underground shopping centers. After World War II, many of these vendors occupied space illegally and were difficult to evict. By arranging the station square above ground, the street vendors were forced to move underground.

Finally, the increase in automobile traffic after 1960 created a shortage of parking lots. Land prices rose, as little space remained for constructing new parking lots above ground. To resolve this problem, underground shopping centers were planned in conjunction with the parking lots, and both were constructed together. This compensated for the high costs of land for the aboveground parking lots.

HISTORY

Although there are various opinions as to which underground shopping center was the first, the one most recognized is the Ueno Shopping Center, constructed around the Ueno-Ginza subway station as early as 1930 (Table 3.1). This shopping center not only had commercial shops and restaurants, but also barber shops and hotels. The concept of underground shopping centers did not become popular, however, until after World War II, when they were constructed more frequently. The opportunity and desire for building them increased with the expansion of roads, the separation of pedestrians and cars, the arrangement of the station square, and the underground placement of parking lots. Underground shopping centers became the means to acquire space without the cost of aboveground land acquisition. The high cost of underground construction was reduced by planning other facilities adjoining the shopping centers.

Since the 1960s, underground shopping centers have been widely constructed around the cities. They have been Japan's effective method of dealing with traffic congestion by segregating pedestrians and automobiles (Fig. 3.2). This period, furthermore, was one of high economic growth.

With the number and size of underground shopping centers increasing, safety became a concern. The importance of safety was reinforced when a gas explosion in 1980 in an underground center killed 15 people and injured 222. This disaster caused many people to come to think that the underground shopping center was a potentially dangerous place and that its design plan should be reevaluated. The government created some regulations or standards by requiring the formulation of a protection system for disasters. This slowed the construction of underground centers and, from that time until today, fewer underground shopping centers have been built. However, today there are about 76 underground shopping centers, with a total floor area of 900,000 square meters (Fig. 3.3).

It has become quite obvious that, to solve its problems of congested land use, Japan should develop the multilevel utili-

TABLE 3.1.
Historical Development of Japanese Underground Shopping Centers

Year	Event
1927	The earliest subway Ginza line (Ueno-Ginza)
1930	Underground shopping center in Ueno
1932	Sudamachi store (small shopping center in Kanda station)
1952	Miharabashi Underground Shopping Center built by filling in the Ginza moat (accommodating street stall)
1955	Asakusa Underground Shopping Center
1957	Parking Lot Act instated
	Ikebukuro Underground Shopping Center (developing parking lot)
	Shibuya Underground Shopping Center (accommodating street stall)
	Osaka Nanba Underground Center (the earliest full-scale underground shopping center)
1959	Definition of underground shopping center
1963–1971	Construction boom of underground shopping centers
1972	Fire in Sennichimae Department Store
1973	Notification from four ministries (the Ministry of Construction, Transportation, the Fire Defense Board, and the Police Agency) of revised safety standards
	Conference of Underground Shopping Centers
1974	Implementation of Notification
1980	Gas explosion accident in Shizuoka Golden Shopping Center
	Notification from five ministries (adding the Board of Energy and Resources to the abovementioned four ministries) of revised safety standards
	Gas preservation safety restrictions imposed
1986	Establishment of Kawasaki Azelea
1992	Establishment of Kobe Harborland
Present Time	Kyoto Oike Underground Shopping Center and Osaka Diamond are under construction

Underground Space Center of Japan, *Designing Underground City,* Daiichi Houki, Tokyo, 1991, pp. 25–29.

zation of aboveground and belowground urban space. This application of underground space utilization is one of the important means of reviving the urban functions. But numerous developmental problems still exist, such as safety, health, and heavy finances. In fact, although many underground shopping centers have been built, master plans for underground utilization in the area have not been adequately designed. From this, it can be seen that while each under-

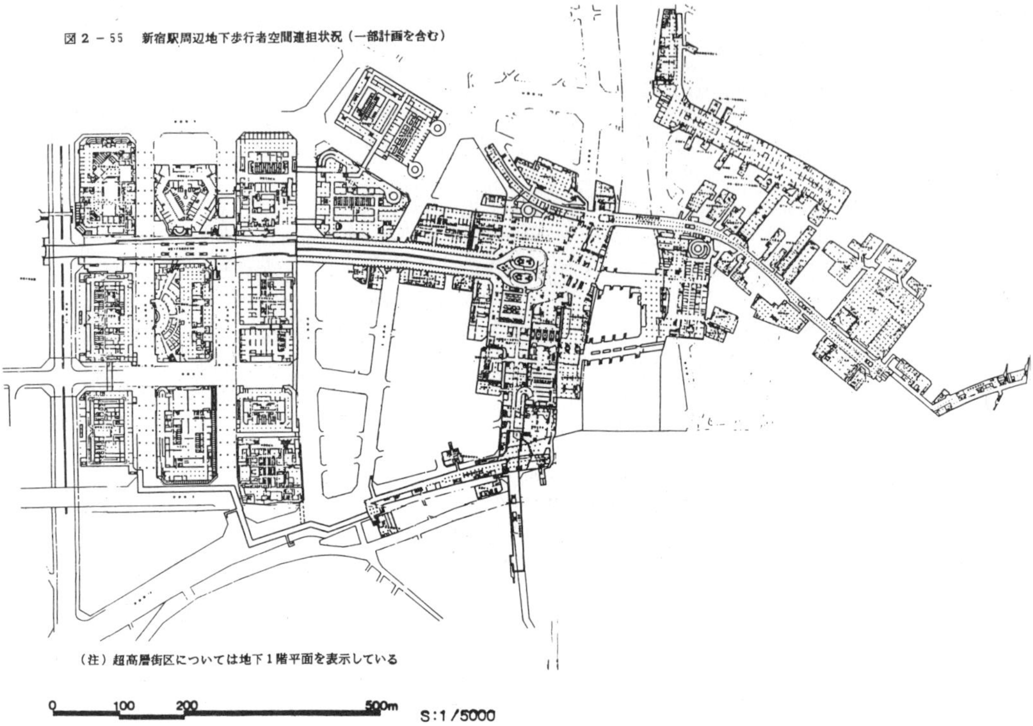

Figure 3.2. Expansion of underground shopping centers (Japan Environmental Systems Co., Ltd.).

ground shopping center may not have a serious problem, the whole pedestrian network often has. This decision to utilize geo-space, which is one of the valuable, yet unused, spaces for mankind, requires very careful planning based on a long-term vision of its use.

LEGISLATION

The underground shopping center, which started from a simple underground passage for pedestrian traffic, grew out of its commercial appeal. With this new dimension as a commercial facility, it became clear that there were problems of safety and health, and the method of planning now became critical. Before 1948, when the Japanese Building Standard Law was established, there were no regulations for underground shopping centers. But, in 1959, the underground shopping center came to be defined clearly as an under-

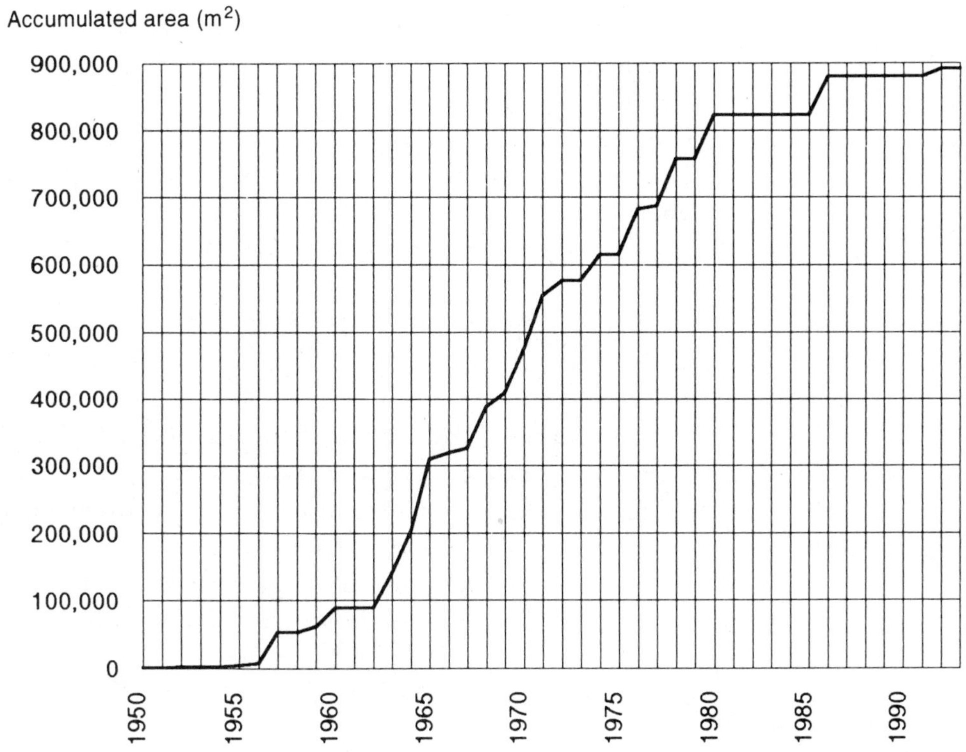

Figure 3.3. Expansion of underground shopping centers (Research Center of Land Development Technology, Survey of Existing Technology of Underground Space Utilization, *Tokyo, 1989, p. 36).*

ground facility that combined an underground passage for public use with stores, offices, and other facilities facing the passage. The Japanese Building Standard Law, the City Planning Law, the Road Law, the Fire Fighting Law, and so on all attempt to deal with the problems of health and safety.

Although there has never been a large fire in the underground shopping center, an accident had already occurred in a complex commercial building aboveground. As a result of this, fire protection and restrictions for using the underground shopping centers were discussed seriously in the National Diet. In 1973, an official safety statement was presented for discussion by four ministries—the Ministry of Construction, the Ministry of Transportation, the Fire Defense Board, and the Police Agency. Then, in 1980, the gas explosion in an underground center forced the original four ministries, with the Ministry of Natural Resources and Energy Board added as a countermeasure for gas preservation, to present another statement (Table 3.2). These regulations thereafter greatly influenced the future development of the underground centers.

The message of the notification was that development of underground shopping centers was prohibited in principle. Furthermore, to develop new underground centers or expand existing ones, the basic idea was as follows:

- The shopping center should not be an obstacle for land-use planning and development of public facilities in the area.
- Development of underground public passages and public parking lots must take priority, and the shopping centers must be constructed with these adjacent facilities so as to utilize minimal area.
- Connections between underground shopping centers and surrounding buildings are prohibited in principle

TABLE 3.2.
Notification from the Five Ministries

1. Underground shopping centers are prohibited, in principle, to be established or extended.
2. Underground shopping centers can be established or extended only when a public underpassage or underground parking lot is urgently needed, and also when an underground shopping center is necessary.
3. The area used must be decided by the city plan and agreed that a multilevel arrangement is necessary. It must be at the road or square in front of the station where the space is already arranged.
4. The shopping center must be on one floor only.
5. The design configuration must be a simple shape.
6. In the case of establishing a parking lot joined to the shopping center, the area of the shopping center cannot exceed the area of the parking lot.
7. The total area for the stores cannot exceed the area of the underground public passage.
8. Access to the aboveground building is prohibited in principle.
9. Enterprises in the underground shopping center must be a state, local public body, or a corporation, and must invest more than one-third of the total cost.

Japan Project Industry Conference, "Current State of Underground City," *Japan Project Industry Conference,* Tokyo, 1986, pp. 144–53.

from the viewpoint of preventing the spread of disasters.

- Not only must the equipment conform to the Japanese Building Standard Law and Fire Fighting Law, but provisions must also be made for acquiring equipment necessary for disaster protection, first aid, sanitation, and dealing with traffic.
- The enterprise must be manageable and operated suitably for the future.

Thus, underground shopping centers may be built only when developing underground public parking lots, in a parking development area, in a station square where automobile traffic is increasing, or when developing public passages to connect main railway stations and bus terminals in commercial areas above ground. In addition, the public passage also must meet certain criteria:

- Public passages must be designated as city planning projects along the road or in the square that has already been developed and where there is a need to develop a multistage passage.
- Public passages must be necessary to make the aboveground traffic flow safer and smoother.
- Underground public passages may be built when it is not proper, from the viewpoint of local traffic conditions, to develop one leading from the road or over the road.
- Public passages should be built only when it is necessary to develop a connection from an underground parking lot to an underground shopping center, and where the facility management and use would improve as a result of the passageway.

New projects for underground shopping centers must meet the zoned use for city planning; in particular, the projects must tie in with the existing facilities of roads, railways, utilities, and urban redevelopment projects. They must further be designed so as not to be an obstacle to the long-term planning of the infrastructural facilities, so that each facility can bring its services into full play.

Also, the design of the project must meet the strict guidelines set by the government. To accomplish this, a conference body was established at the local government level, along with the central connection conference, for the underground shopping centers to bring uniformity to them. From this conference, a notification stipulated that less than one-quarter of the total floor area was to be planned for shops and a parking lot. Here, the underground shopping center would be regarded as a means to develop a public parking lot along with a public passage. Thus, a small shopping area would have to bear the small income from the large public area, and also the cost of the disaster protection system. The concentrated population in Tokyo and the rising price of land, however, have intensified the demand for the development of additional underground shopping centers, so recently these regulatory laws have been eased.

FORM, DESIGN, AND FUNCTION

Site Planning

In Japan, 19 cities have underground shopping centers. Most of these cities are large, since it is difficult to construct underground shopping centers in small cities. Also, most of the centers are constructed in major terminal areas, with the land above used as roads, station squares, or parks. The roads, in most cases, are city and prefecture roads. Many of the centers have strong ties with the railway and subway, and are constructed under both the station squares and the roads linked with them. Some underground shopping centers are constructed under parks, as under Nagoya Central Park and Sapporo Aurola Town Park. These parks are arranged as large median strips between two roads, with the underground shopping centers built under the parks (Fig. 3.4).

The open-and-cut method of construction is used to build most of the underground centers, with their structures being reinforced by concrete frames to resist vibration. The num-

Figure 3.4. Underground shopping centers under the park (Public Corporation of Sapporo City Development, Pamphlet of Sapporo Underground Shopping Center, *Sapporo).*

ber of floors in the center for shops, restaurants, and the like, is limited to one, as specified in the notification. Additional floors below the first level are used for parking. The lowest floor is used for management facilities and control. The average depth of the B1, or first, floor is 6 meters, and the deepest floor is 8 meters (Fig. 3.5). These depths are determined from the allotted space for pipes to the utilities laid below the road. The height from the floor to the ceiling is at least 3 meters.

Floor Planning

Floor designs can be classified into linear or rectangular types, or a combination of both types. The general type is linear, with most of these being used under roads (Fig. 3.6). In the case of wide roads, there are two passageways. A

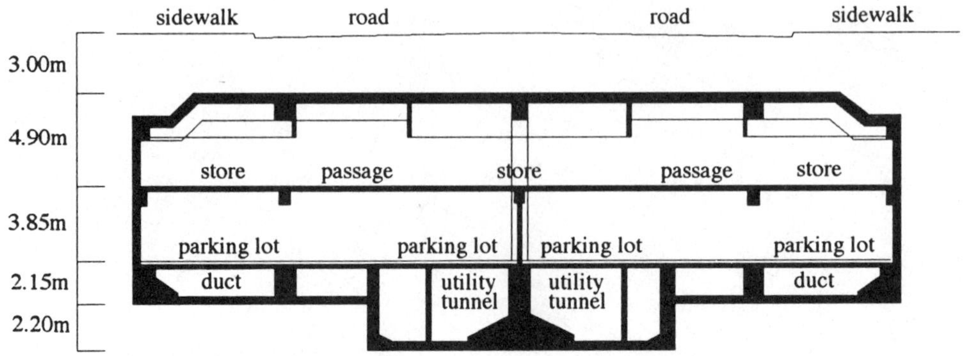

Figure 3.5. Cross section of an underground shopping center (Shinjuku Underground Parking Lot Corporation, Introduction of Corporation, *Tokyo).*

rectangular type is constructed under station squares and parks (Fig. 3.7). Combined linear and rectangular types are constructed under both the station square and the road (Fig. 3.8).

By necessity, linear floor plans are necessarily simple in form, while rectangular types are mostly larger-scale, with more passages. The linear type used under crossings has two passages that meet at right angles and must give a clear sense of direction at the crossing. Rectangular types with many passages need different designs to give each passage a special identity with its square for the purpose of orienting the users' direction.

In the notification, design of the public passage is described as follows:

- The whole plan of the underground shopping center and its passages must provide a clear orientation for convenience to the visitors and for their urgent evacuation in case of emergency.
- The width of the public passage must be a minimum of 6 meters, as set by the guideline

$$W = \frac{P}{1{,}600} + F$$

Figure 3.6. (facing page) Linear forms of underground shopping centers.

with P = maximum number of users per hour for the past 20 years, used as a guideline for planning future development.

NISIBORI ROSA

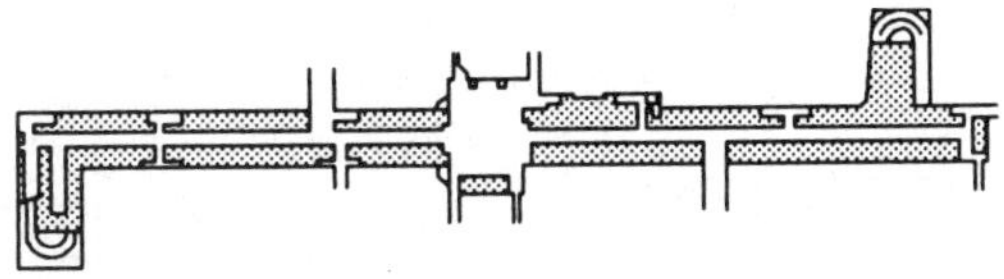

TENJIN CHIKAGAI

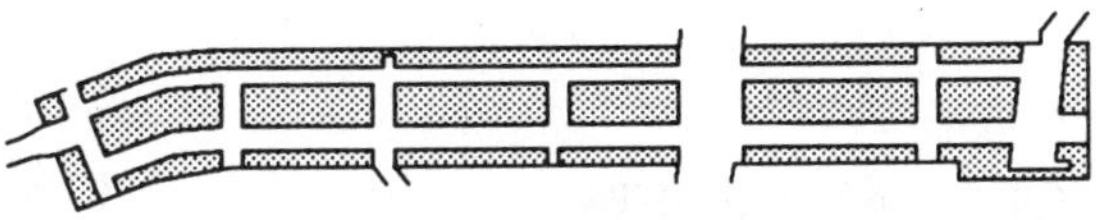

SAKAECHIKA

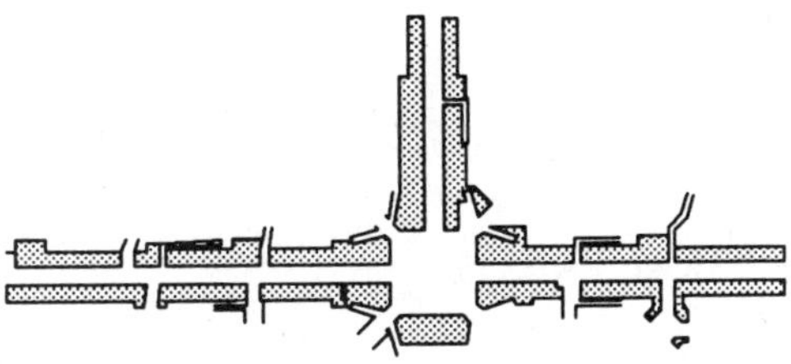

SHINJUKU SUBNADE

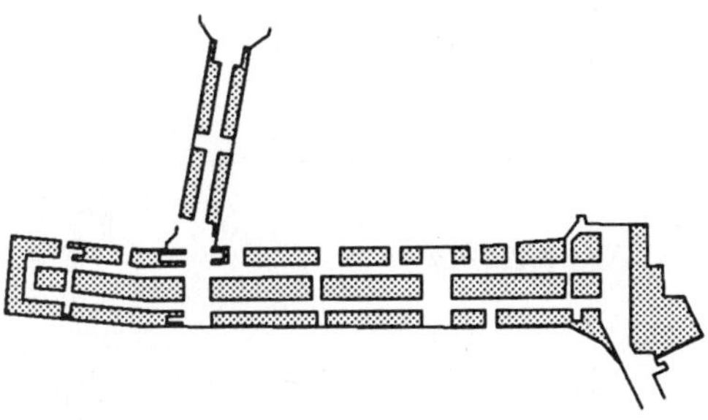

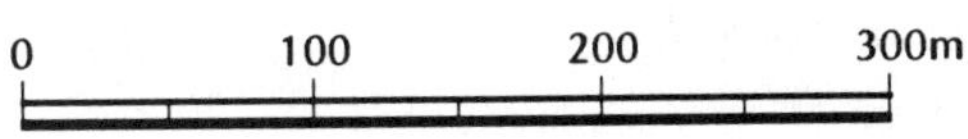

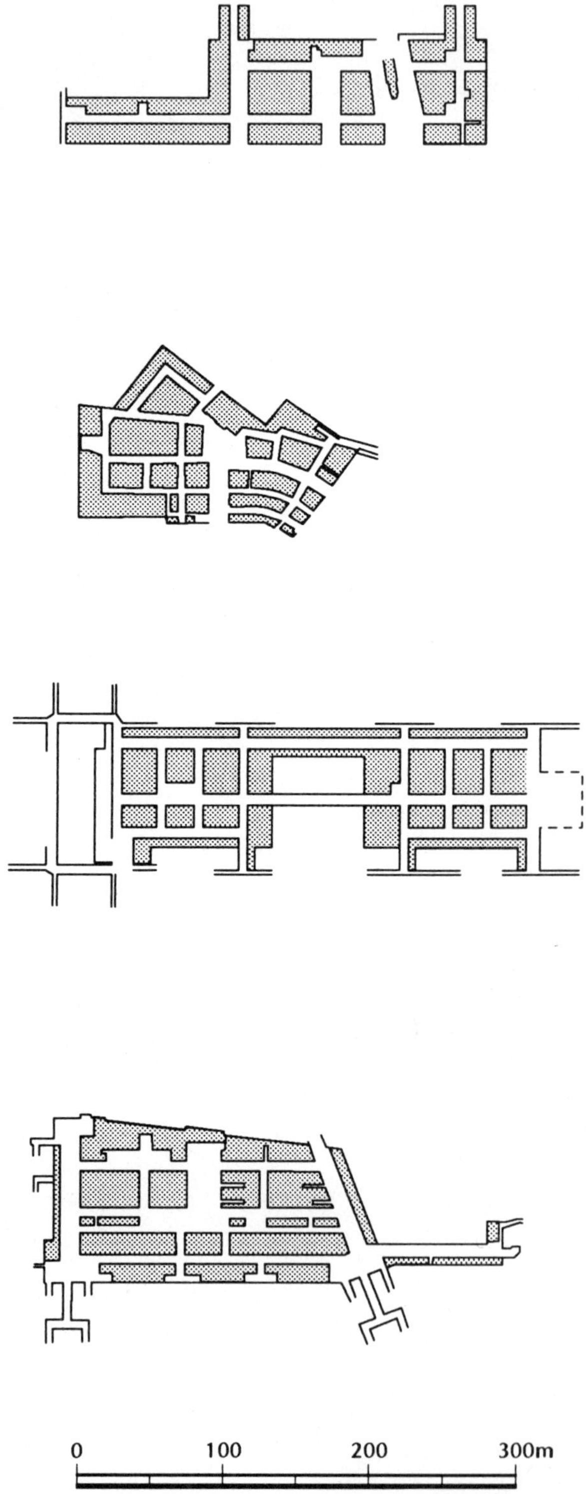

Figure 3.7. Rectangular forms of underground shopping centers.

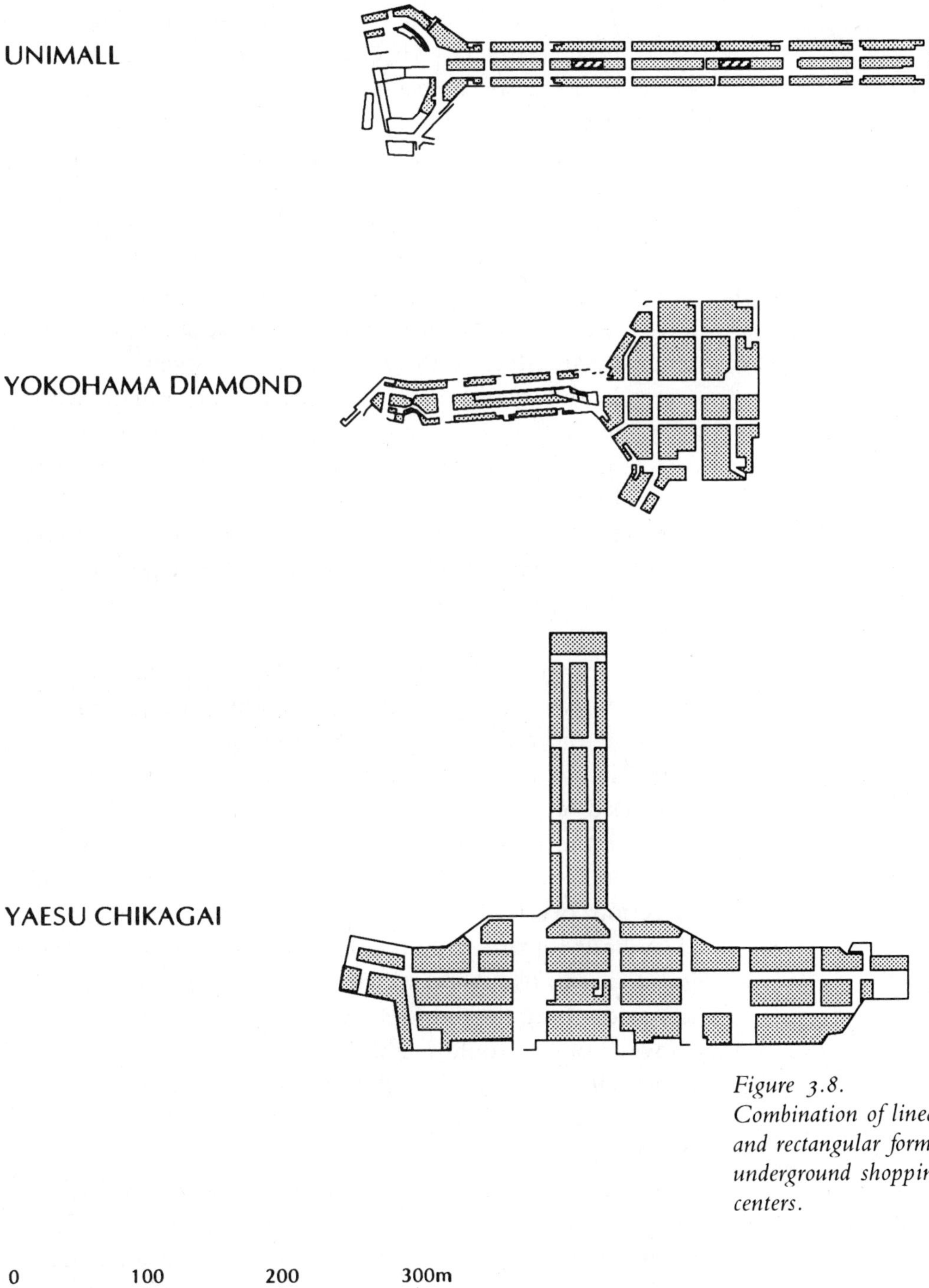

Figure 3.8. Combination of linear and rectangular forms of underground shopping centers.

F = space which exceeds pedestrian traffic passage (e.g., strollers, wheelchairs, people standing, etc.). F = 2 meters for passages with stores, F = 1 meter for passages without stores.

- The effective width of the stairs to the aboveground must be more than 1.5 meters.
- There must be 3 meters at the exit of the stairs to the sidewalk for clearance.
- As a design form, a square must be located at least 50 meters from every point within the shopping center to be an effective means of disaster prevention.
- In the square, there must be no more than two stairs leading aboveground, and there must be an open ceiling for ventilating smoke and for penetration of natural light.
- Fountains, ponds, and the like would be obstacles for emergency evacuation and must not be built in a public passage or square.

Underground shopping centers are composed mainly of passageways and stores. Although stores are joined to each other in the old underground shopping centers (Fig. 3.9), stores in the new underground shopping centers are clearly separated to prevent a fire from spreading (Fig. 3.10).

Since the underground shopping center is also separated from the familiar landmarks above ground and one's sense of direction becomes confused belowground, there must be clear signs for orienting the user. The square is a particularly good area to use as a distinctive, familiar landmark. In addition, designing the public square with an open ceiling leading above ground would enable escape in case of disaster. The open ceiling would also allow smoke to pass.

Scale

While high-rise buildings are large in scale vertically, underground shopping centers are horizontally large-scale. And where there are public passages and parking lots, the floor scale becomes huge.

Of the 76 underground shopping centers in Japan, half of

Figure 3.9. Passage in an old underground shopping center (photo by Shuichi Miura).

them are less than 10,000 square meters each and 30 percent are even smaller than 1,000 square meters. The few especially large underground shopping centers of more than 50,000 square meters are Yaesu (73,413 square meters), Kawasaki Azelea (56,916 square meters), and Nagoya Central Park (55,222 square meters). All of them are constructed in conjunction with city-planned parking lots. There are small shopping centers as well, but they are decreasing in number, and only large centers are now being constructed. Large centers became more common after 1960.

Twenty-two of the 76 underground shopping centers have parking lots, depending on the city plan. Many of the underground shopping centers in Tokyo, Nagoya, and Yokohama are constructed with parking lots, but in Osaka, Kobe, and Kyoto, the centers are constructed separately.

Stores and Restaurants

Since most traffic is located in the terminal area in front of the station, it would initially seem important to introduce com-

Figure 3.10. Passage in a new underground shopping center (photo by Shuichi Miura).

mercial products in this area. But less than 20 percent of the passengers passing through the terminal area are motivated to actually make a purchase.

Stores are built in the underground center itself, and, excluding the parking lot area, occupy about 50 percent of the shop space. The rentable floor area is about 75 percent of the total area, causing the financial management of the center to be difficult. However, some shopping centers have many stores: Nijinomachi Underground Shopping Center has 313 stores, Whity Umeda has 268, and Yaesu has 210 stores.

The underground shopping center may be planned as a city. As such, it must contain various types of stores and restaurants. But, in a closed space, there must also be consideration for the possibility of fire occurring, particularly in the restaurants. Previously, nearly 30 percent of the shops were restaurants; now, this number is decreasing. Some underground shopping centers take the safety precaution of separating the stores and the restaurants into distinct zones. This is feasible from the aspect of the notification, which specifies that the duct systems for stores and restaurants be built separately to prevent fire from spreading.

The average area for restaurants is 400 square meters, while the average area of stores is only 140 square meters. With the exception of a few shops being more expensive, the monthly rental price is about $80 per square meter, about 1.5 times greater than the rental cost in shopping centers above ground. A deposit of about $6,000 per square meter is required initially; this, too, is about 1.7 times greater than that required for centers above ground. These high rental prices and deposits discourage local small stores from renting in the underground centers, which leaves the renting of shops to large independent entrepreneurs. The high rental cost is one of the objections to underground shopping centers.

Generally, the underground shops open by 10 A.M. and close at 8 or 9 P.M., while the passage areas are generally open most of the day, but not for more than 20 hours. Thus, in comparing the sales of underground shopping centers with the general shopping centers above ground, one finds that the underground shops have 2.2 times more sales, and 1.7 times greater sales than shops in terminal stations.

Connection with Surrounding Facilities

Connecting the basements of surrounding buildings to the underground centers is important from the point of view of making a convenient pedestrian network. But this connection increases the likelihood that a fire could spread through the entire underground center. Also, when there is no slope adjustment between the levels of the buildings and the passageways, there may be a difference in the floor levels, making it necessary to build a few stairs. This complicates the connection of passages. However, it is possible to avert some problems by the notification, which restricts the connections to surrounding facilities by specifying that the following conditions be satisfied:

- The basements of connecting buildings must be separated by fireproof dividers.
- Stairs leading directly from the basements of the connecting buildings to the surface, as well as smoke ventilation equipment, must be provided.

- An open ceiling must be built, with smoke ventilation equipment and also with stairs leading directly to the surface.
- Connections must be simple in form.

And finally, people walking above ground must not use the pedestrian space designated for the underground users. By carefully planning the connections with surrounding facilities above ground, the pedestrian space for both types of users can be arranged for an orderly traffic flow.

Entrepreneurs

Shops furnish an important source of income to offset construction costs for the underground center. Equally important is providing pleasure to the monotonous passage and prosperity to management. Because underground shopping centers use the space under the public land of roads and parks, the facilities must have the characteristics of the aboveground public facilities, with sufficient passage between them. Also, as there are peculiar characteristics for safety and the environment under ground, it is dangerous to make decisions solely based on the commercial value of businesses. Thus, most enterprises are jointly public and private. No private enterprises have taken on the financial risk of the underground center since the notification was issued (Table 3.3).

There are few financial resources for construction available from the government. The primary financial sources come from self-funding, about 10 percent; financing from banks, about 50 percent; and approximately 30 percent from deposits from the stores in the shopping center. Bank financing has no advantage over the other forms of financing.

HEALTH

Environmental Characteristics

The general characteristics of underground space are presented in the literature. Effective utilization of the surface of the earth and protection of the view are cited as advantages,

TABLE 3.3.
Investment Source of Some Underground Shopping Centers' Development

UNDERGROUND SHOPPING CENTER	INVESTMENT SOURCES: PUBLIC OR PRIVATE	YEAR OF INITIATION	YEAR OF OPENING	CAPITAL INVESTMENT (MILLION YEN)
Osaka Umeda	50:50	1956	1963	80
Nagoya Sun Road	Private	1953	1957	80
Yokohama Diamond	Private	1963	1964	400
Yaesu	Private	1958	1965	1,000
Shinjuku Subnade	Private	1968	1973	3,600
Nagoya Unimall	Private	1967	1970	600
Sapporo Aurola and Pole Town	49.8:50.2	1969	1971	520
Fukuoka Tenjin	12:88	1972	1976	3,000
Niigata Nisibori Rosa	47:53	1972	1976	580
Nagoya Central Park	Private	1973	1978	1,000
Kyoto Polta	60:40	1977	1980	1,000
Yokohama Polta	68:32	1971	1980	150
Kawasaki Azelea	61.9:38.1	1958	1986	5,000

"Current State of Underground City," *Japan Project Industry Conference,* Tokyo, 1986, p. 9.

while construction costs and people's negative attitudes of using the underground are some of the primary disadvantages. As for environment, in many cases the literature also mentions the steady temperature and humidity as a positive characteristic. This means that the enclosure created by underground space is hardly influenced by outdoor weather conditions. Rather, the indoor environment is determined by its own interior surroundings, and, being separated from the outside world, it maintains a relatively steady thermal condition. Conversely, the inside thermal condition does not influence the outside space environment directly.

A variety of activity performed in space generates heat. In low-temperature regions, the heat that accumulates in winter in belowground space reduces the need for energy consumption to produce warmth. In high-temperature and high-humidity areas like Japan, the use of air conditioning in summer increases energy consumption. For this reason, the thermal conditions produced in underground city facilities should be considered in light of the unspecified number of users who generate heat by their activities (Table 3.4).

TABLE 3.4.
Environmental Characteristics of Underground Shopping Centers Compared with Aboveground Environment

	UNDERGROUND SHOPPING CENTERS	ABOVEGROUND COMMON OFFICE BUILDINGS	ROADS
Time of use	Definite time every day	Business hours	No time limit
Connection with outside	Semi-closed	Closed	Open
Heat gain and pollution sources	• A large unspecified number of people • Restaurants • Fixtures	• A fixed number of people • Fixtures	• A large unspecified number of people • Fixtures • Vehicles
Management	Private, public	Private	Public

Toshio Ojima and Shuichi Miura, "Study on Thermal Environment and Energy Consumption in Underground Shopping Centers," *Journal of Architecture, Planning and Environmental Engineering* 430, Tokyo, 1991, pp. 13–22.

Thermal Performance

Public roads generally are built in an outside space and shopping centers in buildings, in an inside space. However, the underground shopping center brings these two different kinds of spaces (inside and outside) into one underground space (Fig. 3.11). Because of the underground shopping center's unique setting, it is important to clarify its thermal characteristics.

The thermal environment of the underground shopping centers is influenced by the time of day, the centers' connection to the outside, heat gain and pollution sources, and the type of utilization (Table 3.4). Huge areas form open blocks of space larger in scale than those of aboveground buildings. The passages are placed between the blocks of space (Fig. 3.12). The many people who walk or shop in these passages and enclosed spaces generate heat and pollution that should be ventilated—a problem not as great for spaces above ground. Because of the closed space, pollution cannot be cleared by natural circulation of air. The environmental conditions must be controlled by equipment.

As mentioned above, many aspects of this underground enclosed space are different from the aboveground road and shopping centers. The following sections discuss a scientific

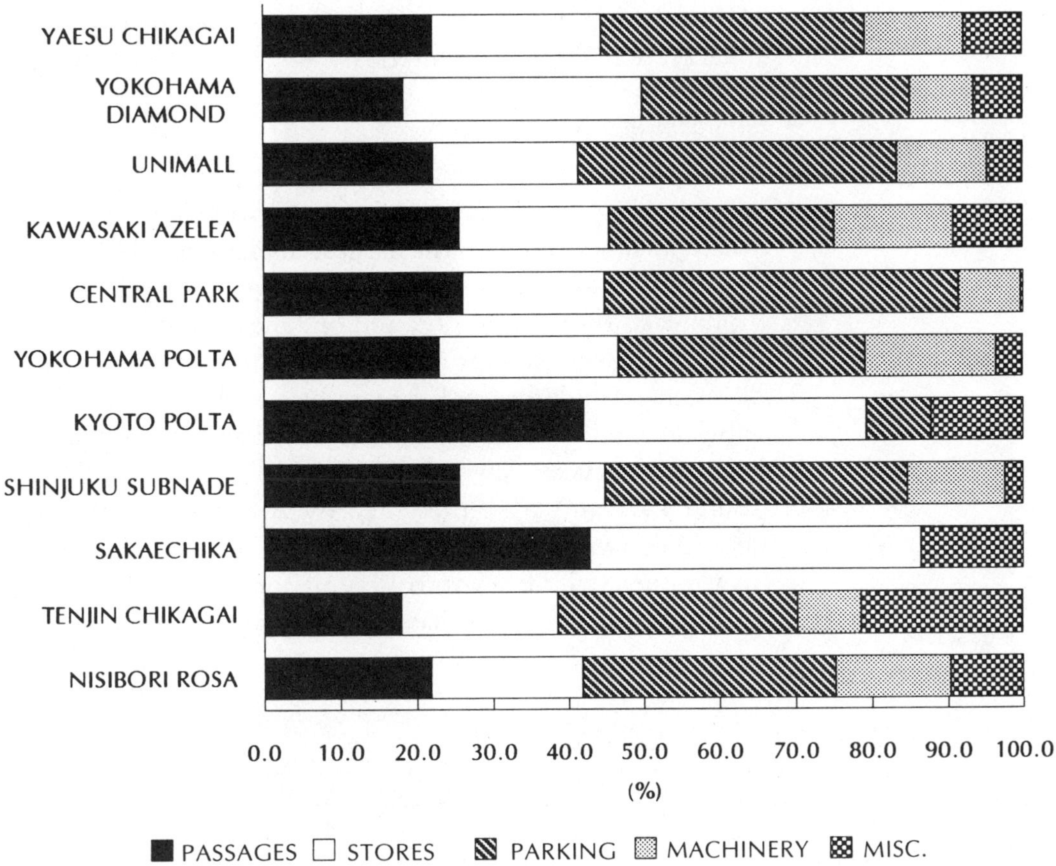

Figure 3.11. Land-use area of underground shopping center.

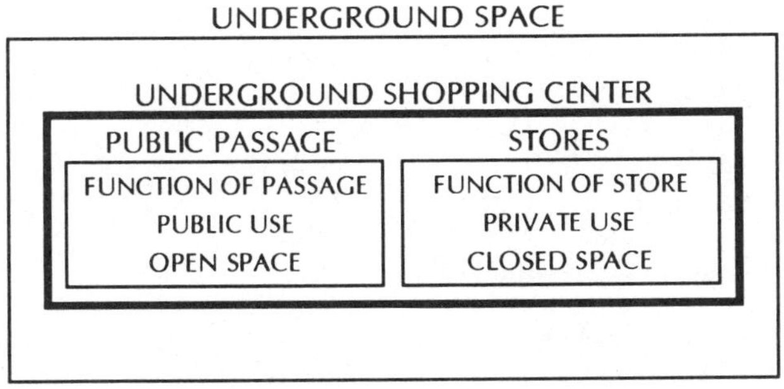

Figure 3.12. Functions of an underground shopping center (Toshio Ojima and Shuichi Miura, "Study on Thermal Environment and Energy Consumpion in Underground Shopping Centers," Journal of Architecture, Planning and Environmental Engineering *430, Tokyo, 1991, pp. 13–22).*

study of the pattern of the environment and energy use in underground shopping centers.

Environmental Measurements

Environmental measurements were carried out at four underground shopping centers in the Tokyo metropolitan area, which included one of the largest underground shopping centers in Japan. In this study, five to seven measuring points within each shopping center and one measuring point above ground were selected for each underground shopping center. All of these underground centers have only one floor level below ground, at a depth of between five and eight meters from the surface. Two of the centers were constructed around 1970 and are represented by A and B in Table 3.5. The other two centers, C and D, were constructed in the 1980s under strict fire control regulations. They differ from the first two centers by having many wide exits and wide

TABLE 3.5.
Temperature of the Four Underground Shopping Centers

			UNDERGROUND SHOPPING CENTER			
			A	B	C	D
August	Outdoor	Mean	29.8	29.7	28.8	27.2
		Max.	33.3	32.6	31.6	33.5
		Min.	26.1	26.0	26.7	23.0
	Indoor	Mean	25.2	26.3	26.7	25.5
		Max.	27.6	30.8	29.0	28.8
		Min.	22.0	23.9	24.0	22.1
October	Outdoor	Mean	16.3	17.4	17.2	22.4
		Max.	21.1	21.2	20.2	27.4
		Min.	11.4	14.0	14.8	17.6
	Indoor	Mean	21.4	21.2	20.1	23.4
		Max.	24.6	24.6	23.6	27.0
		Min.	16.7	16.0	14.0	17.6
January	Outdoor	Mean	9.9	10.4	9.3	10.0
		Max.	13.1	16.8	15.4	13.2
		Min.	7.2	6.4	5.6	7.5
	Indoor	Mean	17.8	17.9	18.7	13.8
		Max.	22.7	24.8	26.0	22.6
		Min.	10.6	11.2	9.6	6.6

Toshio Ojima and Shuichi Miura, "Study on Thermal Environment and Energy Consumption in Underground Shopping Centers," *Journal of Architecture, Planning and Environmental Engineering* 430, Tokyo, 1991, pp. 13–22.

passages to allow people to escape easily during an emergency. Temperature and humidity changes in all the centers were recorded during different months of the year and at different hours of the day.

As stated previously, the commercial zones in the shopping centers generally include restaurants as well as stores. Restaurants comprise 34 percent of the commercial zone in shopping center A, 43 percent in B, 29 percent in C, and 21 percent in D. However, it is known that the restaurant is a potential source for fire. For example, the power capacities of refrigeration machines are between 0.8 USRT/m^2 (2.8 kW/m^2) and 1.0 USRT/m^2 (3.5 kW/m^2), which are about three times that of aboveground office buildings and about twice that of department stores in Tokyo. The capacities of boilers are different in each underground shopping center and are between 130 kcal/m^2·h (0.15 kW/m^2) and 230 kcal/m^2·h (0.27 kW/m^2).

Thermal Environment

The thermal environment was measured with regard to the temperature and humidity in the four shopping centers during the hours that air conditioning was running (between 10 A.M. and 8 P.M.) (Tables 3.5 and 3.6).

In August, the average temperature of the four centers was between 25 degrees C and 27 degrees C. Although these values were not so high, temperature differentials between the points were very significant. The minimum temperature range was 22 degrees C to 24 degrees C and the maximum temperature range reached 27 degrees C to 31 degrees C. This range exceeded the indoor standard temperature of 27 degrees C. The humidity averaged between 50 percent and 70 percent, also with large differences between the points. The minimum humidity was 40 percent to 55 percent and the maximum humidity was 70 percent to 85 percent (Table 3.6).

In October, the average temperature of the four underground shopping centers ranged between 20 degrees C and 23 degrees C. Again, differentials between the points within each site were wide. The minimum temperature was 14 de-

TABLE 3.6.
Relative Humidity of the Four Underground Shopping Centers

			UNDERGROUND SHOPPING CENTER			
			A	B	C	D
August	Outdoor	Mean	63.7	59.5	74.6	71.2
		Max.	80.0	73.0	84.0	82.3
		Min.	49.0	45.0	62.0	54.2
	Indoor	Mean	55.9	55.5	68.2	65.1
		Max.	73.0	70.0	81.7	85.9
		Min.	46.0	44.0	53.0	40.7
October	Outdoor	Mean	38.6	57.9	69.7	73.4
		Max.	79.0	76.0	86.0	96.2
		Min.	26.0	48.0	44.0	38.2
	Indoor	Mean	30.0	46.4	58.8	66.0
		Max.	45.0	63.0	96.0	98.1
		Min.	19.0	32.0	42.0	36.6
January	Outdoor	Mean	40.8	32.6	31.9	47.5
		Max.	60.7	46.7	43.2	70.9
		Min.	26.7	22.9	22.5	31.0
	Indoor	Mean	26.4	22.4	22.9	43.4
		Max.	51.4	42.3	43.1	66.4
		Min.	13.4	10.7	13.9	28.3

Toshio Ojima and Shuichi Miura, "Study on Thermal Environment and Energy Consumption in Underground Shopping Centers," *Journal of Architecture, Planning and Environmental Engineering* 430, Tokyo, 1991, pp. 13–22.

grees C to 18 degrees C and the maximum temperature was 23 degrees C to 27 degrees C. This temperature was warmer than the outside temperature, however, even though the cooling system was used. The underground shopping center was warmer than the outside because of interior heat gain. The average humidity was between 30 percent and 70 percent.

In January, the average temperature of centers A, B, and C was around 18 degrees C, while that of center D was 14 degrees C. Here, the differentials between the points within each site were even greater than in August. The minimum temperature reached 6 degrees C to 12 degrees C, which is very cold, and the maximum temperature was 22 degrees C to 26 degrees C. There were also differences in the average humidity between the sites, with the average of centers A, B, and C registering between 20 percent and 30 percent, while that of center D was around 43 percent. Minimum humidity readings were 10 percent to 30 percent and the maximum humidity was 40 percent to 70 percent.

Temperature fluctuated with time at sites A and D (Fig. 3.13). In August, the temperature at 8 A.M. was especially high (some points were warmer than the outside temperature), since there was no air conditioning and many people were using the underground shopping center to go to work. In center D, the temperature at one point was higher than 30 degrees C at 8 A.M. All temperatures were lowered when the air conditioning was activated, with the greatest temperature differential of around 5 degrees C among the different points. The temperature, of course, rose again after the air conditioning was turned off.

Temperature was recorded at the same hours in January at 8 A.M. before the air conditioning was started and at 10 P.M. after the air conditioning was turned off. The temperature in center A was higher than the outside temperature, but that of center D was almost the same. The thermal storage difference may be attributed to there being more exits in center D. The temperature differentials between the highest and the lowest points at the same hour were large—around 10 degrees C—compared with the highest and lowest point variations in August.

Humidity fluctuations were also recorded in August and January at shopping center D (Fig. 3.14). In August, the outside conditions caused also significant differences between the measured points. In January, the hourly fluctuations and differences between the points were not as great as in August, and the values of all points were near the outdoor humidity.

These results show that there is a large difference in the thermal distribution within one center and among the centers, and that such differences are more significant in winter than in summer. This phenomenon is unavoidable because of the air that enters the underground shopping centers through the exits. A comparison of two underground shopping centers with a different number of exits per floor showed that the thermal influence of the exits on the indoor environment will

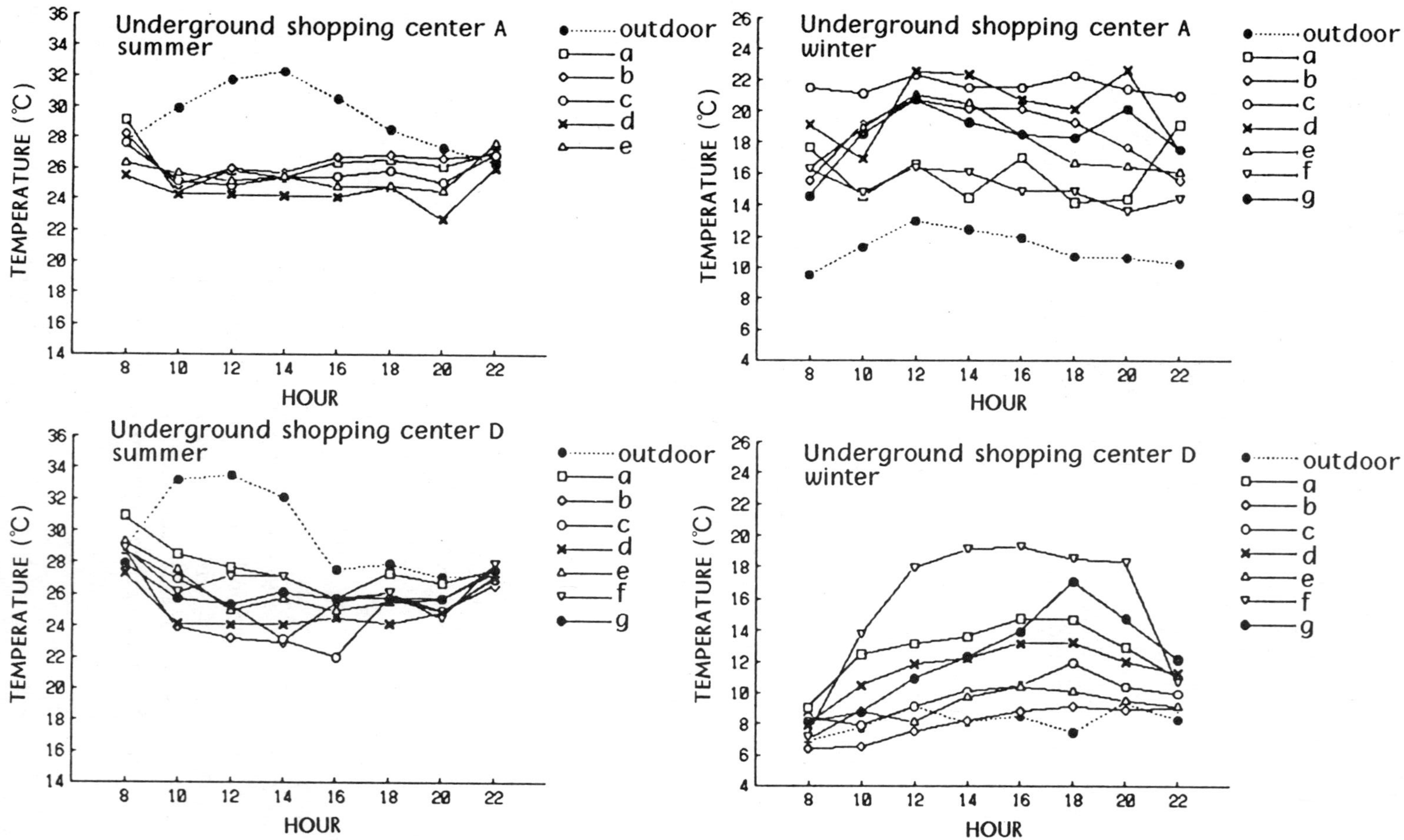

Figure 3.13. Fluctuations of temperature at 7 measurement points in shopping centers A and D (Toshio Ojima and Shuichi Miura, "Study on Thermal Environment and Energy Consumption in Underground Shopping Centers," Journal of Architecture, Planning and Environmental Engineering *430, Tokyo, 1991, pp. 13–22).*

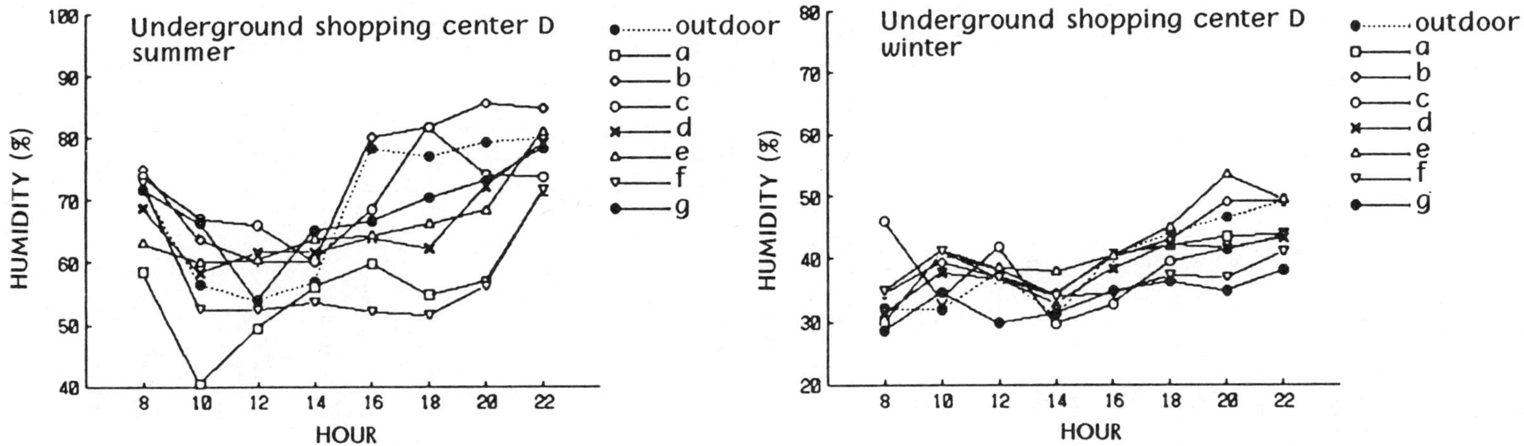

Figure 3.14. Fluctuations of relative humidity over time at 7 measurement points in shopping center D (Toshio Ojima and Shuichi Miura, "Study on Thermal Environment and Energy Consumption in Underground Shopping Centers," Journal of Architecture, Planning and Environmental Engineering *430, Tokyo, 1991, p. 18).*

be very significant. In conclusion, it is difficult to develop a perfect thermal environment for such a large area that would keep the conditions unchanged.

Wind Velocity

Wind velocity was measured in the summer and in the winter in shopping center D. Examining the relationship between the average wind velocity and the average temperature of each measuring point in the center during January revealed that the stronger the wind, the lower the temperature became as the air flowed in from the outside through the exits. However, doors to the exits are prohibited for safety reasons, so the influence of wind on the outside air flowing in would be noticeable, especially in winter, since the temperature in the shopping center is warmer than the outside temperature. Since warm air in the underground center is lighter than the cold outside air, the warm air rises while the cold air above ground flows into the underground shopping center from the exits, which are lower than the ground level. To counter this phenomenon, some underground shopping centers in cold regions are permitted to use doors to close the entrances to the outside. Conversely, during summer, the temperature is cooler in the underground shopping center and, thus, the air there is heavier than the outside air, so the warm outside air does not flow into the entrances. Therefore, in summer, the underground shopping center is in a closed condition, and in winter, it is in an open condition to the outside environment.

Air Pollution

The quality of the air has a direct influence on people's health. The Japanese recognize this fact, and passed a building law that states that the capacity of a ventilation system for indoor areas must be more than 30 $m^3/m^2 \cdot h$ and that an air handling unit must supply more than 10 $m^3/m^2 \cdot h$ of fresh air. Many factors cause air pollution, but in this research, concentrations of carbon monoxide (CO), carbon dioxide (CO_2), and suspended particles were measured in the four underground shopping centers according to the environmental management standards set by the building law (Table 3.7).

TABLE 3.7.
Air Pollution of the Four Underground Shopping Centers

			UNDERGROUND SHOPPING CENTER			
			A	B	C	D
CO (ppm)	August	Mean	2	3	3	2
		Max.	7	9	12	7
		Min.	0	0	0	0
	October	Mean	3	3	3	1
		Max.	5	6	10	4
		Min.	0	0	0	0
	January	Mean	2	2	1	1
		Max.	5	8	7	4
		Min.	0	0	0	0
CO_2 (ppm)	August	Mean	737	564	458	600
		Max.	1,250	1,200	800	1,150
		Min.	500	200	150	400
	October	Mean	593	638	394	345
		Max.	900	1,000	800	600
		Min.	350	450	250	250
	January	Mean	454	480	599	546
		Max.	800	700	950	1,000
		Min.	350	350	350	300
Suspended particles (mg/UG)	August	Mean	0.02	0.07	0.05	0.02
		Max.	0.09	0.22	0.21	0.08
		Min.	0.00	0.01	0.00	0.00
	October	Mean	0.08	0.08	0.04	0.02
		Max.	0.07	0.21	0.14	0.13
		Min.	0.00	0.02	0.00	0.00
	January	Mean	0.02	0.02	0.02	0.03
		Max.	0.09	0.08	0.08	0.11
		Min.	0.00	0.00	0.00	0.00

Toshio Ojima and Shuichi Miura, "Environmental Study of Underground Shopping Center," *Urban Underground Utilization '91,* 4th International Conference on Underground Space and Earth Sheltered Buildings, 1991, pp. 471–77.

Since the main sources of CO are combustion equipment and vehicles, CO concentration levels can be affected by outside air depending on where the vehicles are traveling and on the location of the exits and ventilation sources of the shopping center. But most of the CO concentration levels did not exceed 10 ppm, which, in Japan, is the standard level for indoor rooms.

Carbon dioxide is generated by a mass of people. High CO_2 concentrations were found in terminal areas where there were many passengers. In summer, the concentration levels of CO_2 in centers A, B, and D had a maximum value exceeding 1,000 ppm. However, this, too, meets the standard of Japanese indoor rooms.

Smoking is the main source of suspended particles. Only in smoking areas were the suspended particles found to exceed 0.15 mg/m^3, again the standard required of indoor rooms. However, smoking in underground shopping centers is prohibited, as a rule, except in designated smoking areas.

EFFICIENCY

Construction Costs

As mentioned earlier, construction costs in underground structures generally are very high, especially those of the underground shopping centers once stricter safety regulations were enacted (Table 3.8). Pre- and post-notification cost comparisons were made of the estimates for typical underground shopping centers and their respective revised actual costs that incorporated special building safety requirements.

As a comparison, the construction cost per square meter of an underground shopping center is three times greater than the construction cost of aboveground buildings, such as offices (Fig. 3.15). Although underground development is initially more expensive, the construction costs are offset by higher aboveground land prices for traditional construction.

These high construction costs remain a challenge to management. Many underground centers fall into debt for more than 10 years. The cost of underground shopping centers is very high, but it must be evaluated not only for the profit it generates, but also by the effects of its development on the surrounding area and of the safety of pedestrians.

Energy Consumption

Energy consumption was measured to evaluate the environmental load of belowground buildings. Data were obtained from 32 underground shopping centers, with an average

TABLE 3.8.
Comparative Construction Costs and Total Floor Area of Underground Shopping Centers

NAME OF UNDERGROUND SHOPPING CENTER	TOTAL FLOOR AREA (M^2)	CONSTRUCTION COST (MILLION YEN)	REVISED CONSTRUCTION COST* (MILLION YEN)	CONSTRUCTION COST PER AREA (THOUSAND YEN/M^2)	LAND PRICE (THOUSAND YEN/M^2)	CONSTRUCTION COST/ LAND PRICE
Sapporo Aurola and Pole Town	45,157	6,900	21,770	482		
Yaesu	68,468	11,600	50,000	730		
Shinjuku Subnade	38,564	16,400	57,750	1,497		
Yokohama Polta	29,133	36,300	46,720	1,194	1,730	0.54
Yokohama Diamond	38,816	3,800	16,590	428		
Kawasaki Azelea	56,916	44,500	58,630	1,030	2,790	0.28
Niigata Nisibori Rosa	17,359	6,800	11,850	682	1,020	0.38
Nagoya Central Park	55,702	20,000	31,900	573		
Eska	29,154	3,200	10,090	346		
Nagoya Unimall	24,198	4,200	13,380	553		
Kyoto Polta	26,712	11,000	14,160	530	900	0.46
Osaka Umeda	27,715	5,000	22,040	795		
Abenobashi	9,245	1,800	6,370	689		
Petits Champs-Elysees	3,737	1,600	2,730	731		
Santika Town	17,985	5,800	25,000	139		
Metro Kobe	10,198	500	1,800	176		
Fukuoka Tenjin	35,250	16,000	27,870	791	1,790	0.25
Average	32,018	11,599	24,630	742	1,650	0.38

*After stricter safety regulation notification was enacted.

floor area of 21,782 square meters. The centers were located across Japan from Sapporo in the north to Fukuoka in the south. Here, energy consumption will be expressed by the primary energy unit, where one kWh of electric energy equals 10,258 kJ.

The result of monthly energy consumption in one underground shopping center shows that the maximum energy consumption occurred in August, with the peak cooling load (Fig. 3.16). The next highest value occurred in February, with the peak heating load. The largest ration of annual energy consumption was for lighting, at 43.9 percent, and the second was for power, at 39.7 percent (Fig. 3.17). Cooling took 11.4 percent, and heating 5 percent.

A comparison was also made of annual energy consumption per unit area between underground shopping centers and

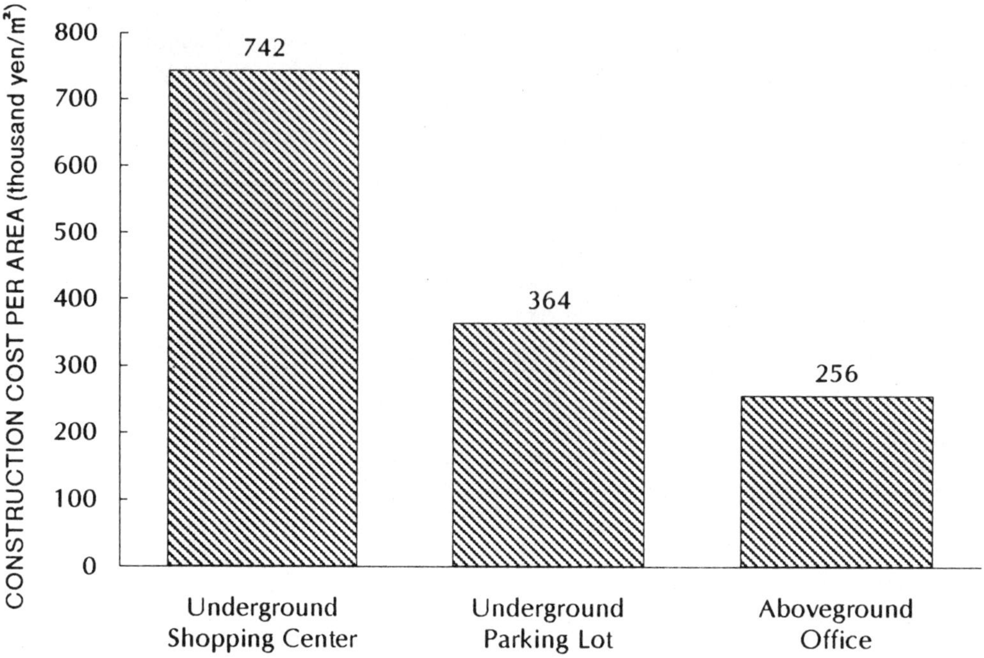

Figure 3.15. Comparison of construction costs for underground shopping center, underground parking lot, and aboveground office building.

aboveground office buildings and department stores in Tokyo (Fig. 3.18). Energy consumption for cooling the underground shopping centers was about 6 times that of the office buildings and 2.5 times that of the department stores, and for heating, about 4.5 times that of office buildings and 2 times that of department stores. This amounted to a total energy consumption of about 4 times that of aboveground office buildings and about 2 times that of department stores.

The large environmental energy load of underground centers for heating and cooling and for clearing the pollution produced by the passageways and stores for long operating hours caused, in part, this heavier energy consumption. Another reason the load was larger is the amount of heat and pollution produced by the pedestrians and stores in the underground centers. The environment of such areas has to be controlled entirely by heating, ventilating, and air conditioning (HVAC) equipment, since the space is generally separated from the controlling effects of the natural environment. Furthermore, the underground shopping center, as a public facil-

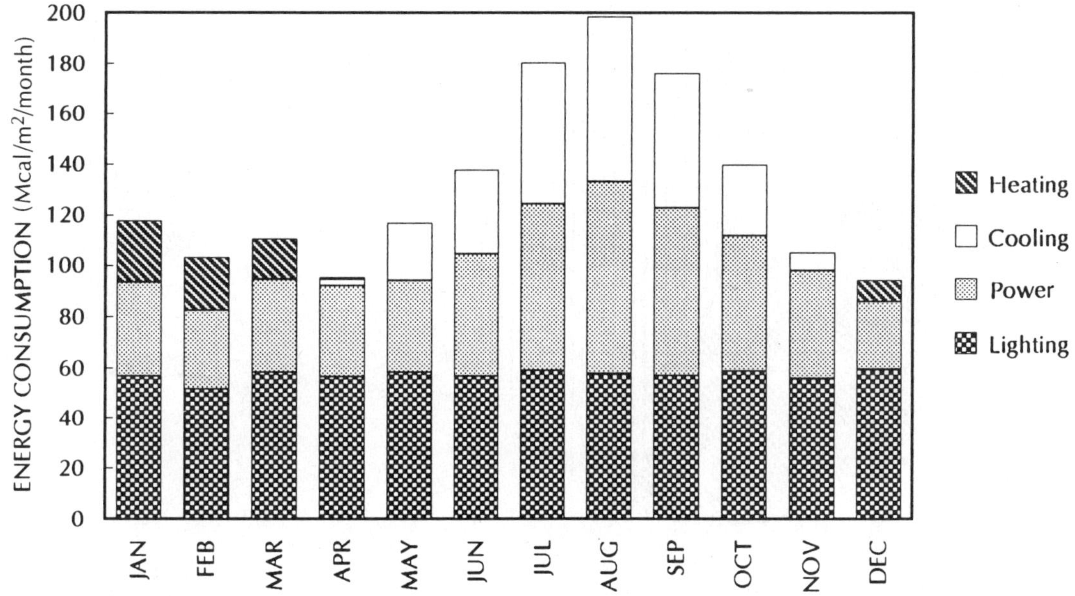

Figure 3.16. Monthly energy consumption in one underground shopping center (Toshio Ojima and Shuichi Miura, "Environmental Study on Underground Shopping Center," Urban Underground Utilization '91, *4th International Conference on Underground Space and Earth-Sheltered Buildings, Tokyo, 1991, pp. 471–77).*

ity, is kept available for people all the time and requires constant environmental control.

SAFETY

Since the underground shopping center is used by many people for long periods of time, safety becomes a critical consideration. As mentioned previously, the safety regulations now being established came about because of disasters that were experienced earlier, both below ground and above ground.

Fire is the most likely possibility, since restaurants use fire for cooking and commodity stores carry many combustible products. Fire can spread among the shops and burn for a long time, accompanied by a great volume of smoke produced by burning materials. As the surroundings of the underground shopping center are blocked off, the smoke would become trapped and soon fill the area, where little oxygen will be supplied, and imperfect combustion will cause more smoke to form.

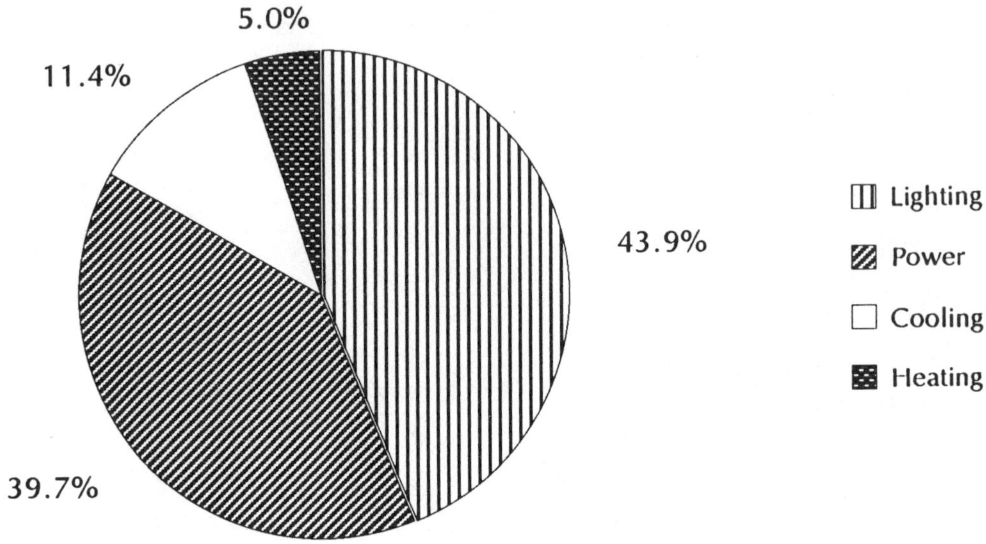

Figure 3.17. Components of energy consumption (Toshio Ojima and Shuichi Miura, "Environmental Study on Underground Shopping Center," Urban Underground Utilization '91, *4th International Conference on Underground Space and Earth-Sheltered Buildings, Tokyo, 1991, pp. 471–77).*

The reason why smoke becomes a particular problem underground is because it flows into and concentrates in the exits, where people are trying to escape. The horizontal speed at which smoke travels is about 1 m/sec, while its vertical speed is about three to five times this. From experimentation, it has been ascertained that crowds of people can escape at a speed of about 1 m/sec, and in the stairwells, their escape rate diminishes to one-quarter of this. In short, the speed of smoke and of people is almost the same. Once enveloped by the smoke, people's visual distance diminishes to 15 or 20 meters, further reducing their escape speed. Someone detained in escaping or unable to use the stairs will suffocate. To compound the problem, one's sense of direction is easily lost in the underground shopping center, where everything looks the same. And, finally, a power failure would totally darken the area, since there are no windows for outside light.

During fires in aboveground buildings, people can escape to the roof and wait for rescue by helicopter, while getting fresh air. The possibility of saving lives this way is much greater than in the underground space, where people may be entirely isolated from both fresh air and the chance of rescue.

The underground shopping center also presents more difficult conditions for fighting fire. Smoke and fire in the en-

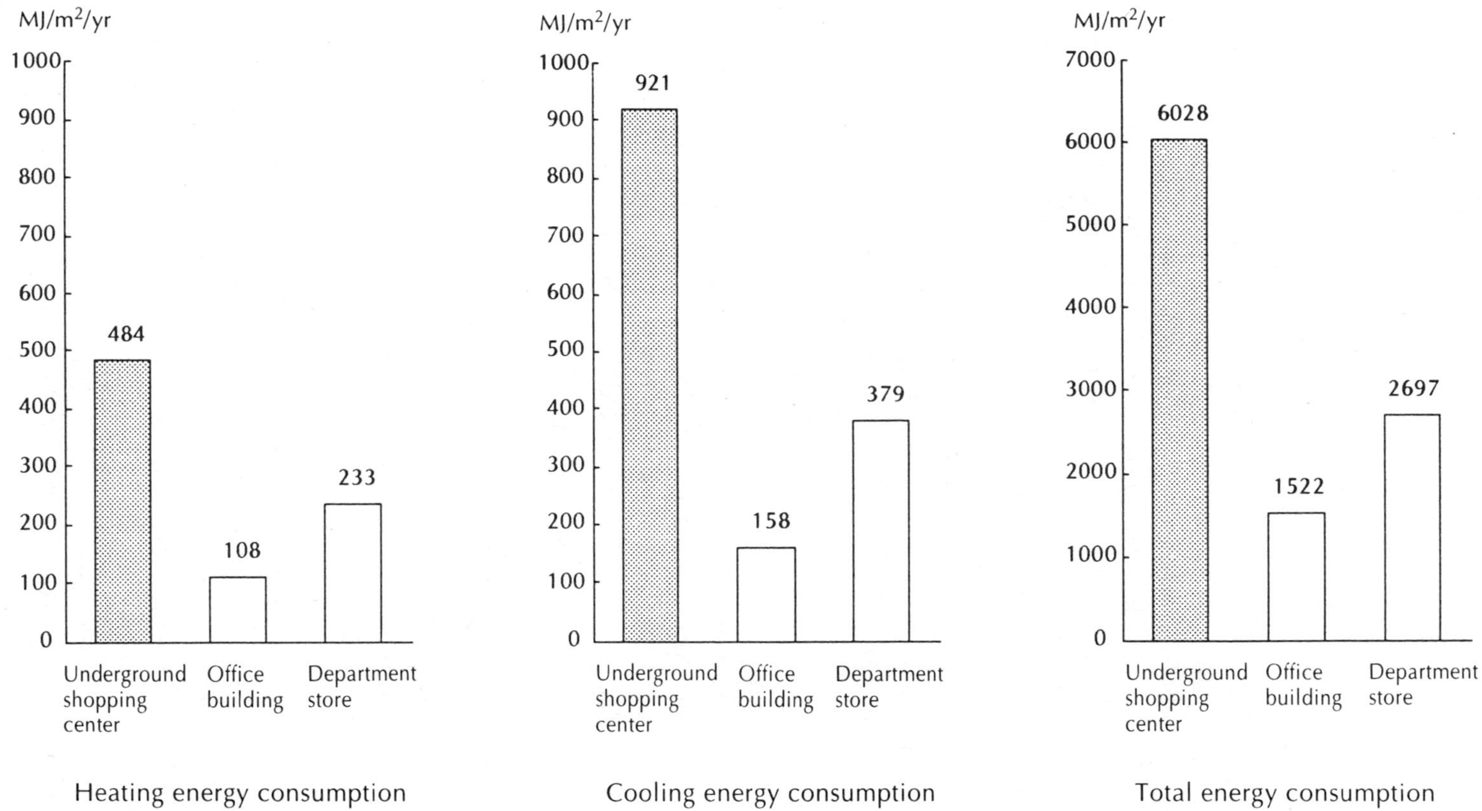

Figure 3.18. Comparison of annual energy consumption of underground shopping centers and aboveground buildings (Toshio Ojima and Shuichi Miura, "Environmental Study on Underground Shopping Center," Urban Underground Utilization '91, *4th International Conference on Underground Space and Earth-Sheltered Buildings, Tokyo, 1991, pp. 471–77).*

trances prevent the firefighting team from entering the area. Once the team enters the area, it is difficult to discover the origin of the fire because of the smoke.

In aboveground office buildings, where people daily use routine passages and locations, it is easy to carry out training for evacuation. But in the underground shopping center, it is very difficult to train sporadic users, who go into many different areas of the center. Without clear knowledge of escape routes, a panic situation increases the severity of the disaster. Earthquakes are another situation where panic can result (Fig. 3.19). With any disaster, people first think of escaping somehow to the above ground, so the entrances become crowded and evacuation disorderly.

Example of a Disaster

A gas explosion occurred in 1980 in a small underground shopping center in Shizuoka city, which turned into a catastrophe. Two explosions occurred. The first explosion was small, but the second explosion, which occurred 30 minutes after the first one, was more massive. Fifteen people died and 222 people were injured. What is significant about this case is that, after the first explosion occurred, no countermeasures could immediately be taken before the next explosion took place, because it was very difficult to ascertain the inner condition of the area since it could not be observed from the outside.

Countermeasures for Disasters

It is possible to design specific countermeasures for disaster prevention in an underground shopping center, particularly for fire (Table 3.9). These measures are as follows:

- Prohibit the use of fire except in restaurants.
- Prohibit smoking except in designated places.
- Use noncombustible interior finishes.
- Maintain firefighting equipment in good condition.
- Thoroughly develop countermeasures to prevent leaking gas.
- Develop countermeasures to slow the spread of fire.

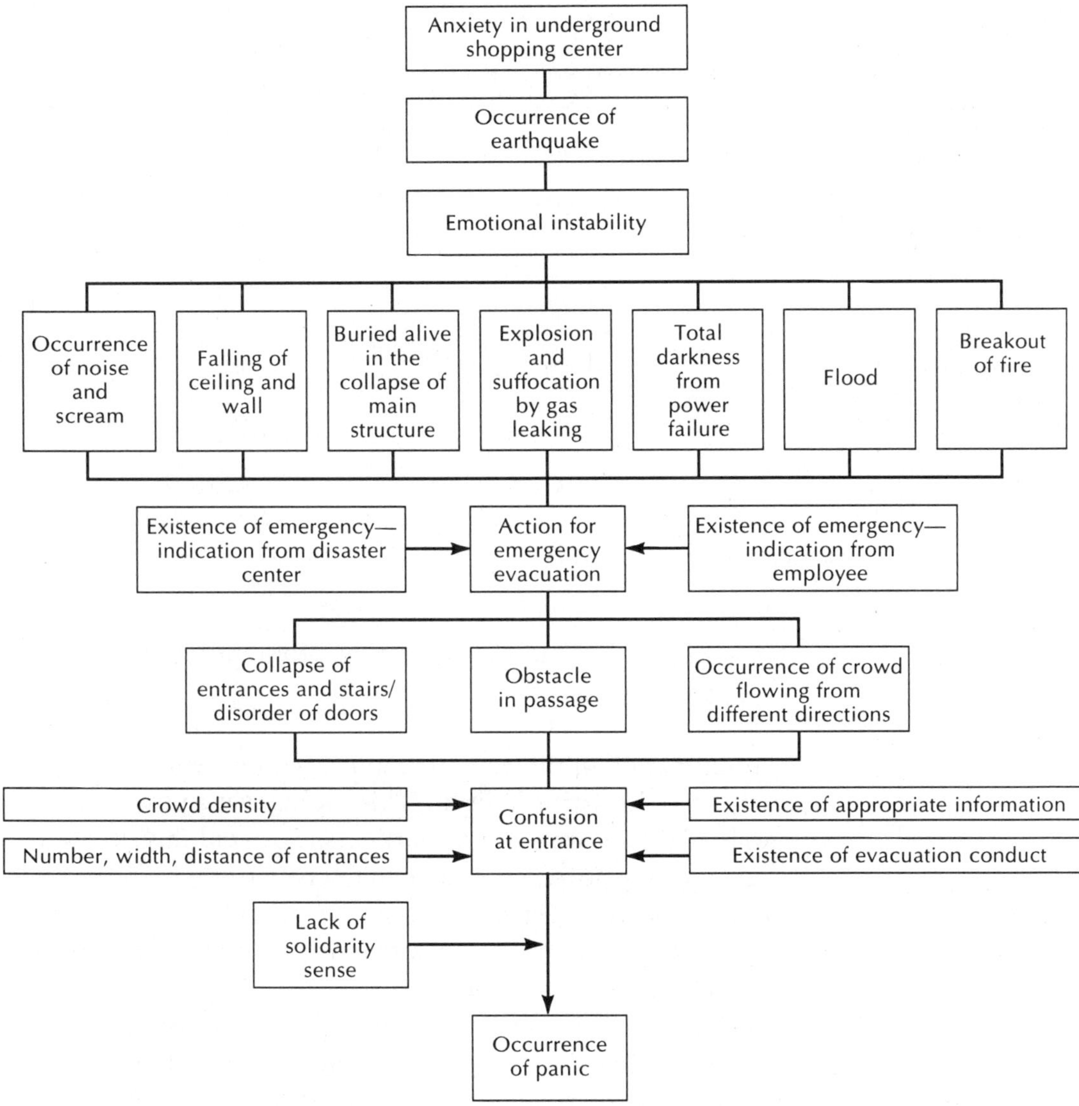

Figure 3.19. Stages of panic during an earthquake (Tokyo Fire Defense Board, Fact-finding Report of Underground Shopping Center, *Tokyo, 1979, p. 5).*

- Discover fires early and begin firefighting in the fires' initial stages.
- Locate fire control facilities for easy accessibility.
- Make thorough, routine examinations of escape routes.
- Ensure that fire protection systems function.
- Strengthen a disaster prevention organization to practice conduct during escape.

TABLE 3.9.
Installations for Disaster Control

	PASSAGE	STORE	STAIR	MAINTENANCE	PARKING
Indoor Fire Hydrant	×	×	×	×	×
Continuous Water-Supplying Pipe	×	×	×	×	×
Sprinkler	×	×		×	
Smoke Exhaust System	×	×			×
Bubble Extinguisher					×
Halogen Fire Extinguishing System				×	
Automatic Fire Alarm	×	×	×	×	×
Leading Lights	×	×	×	×	×
Emergency Receptacles	×	×	×	×	×
Emergency Broadcasting	×	×	×	×	×
Emergency Telephones	×	×	×	×	×
Wireless Communication	×	×	×	×	×
Emergency Light	×	×	×	×	×
Gas Leaking Alarm		×		×	

Kawasaki Underground Shopping Center Co. Ltd., *Countermeasure System for Disaster in Underground Shopping Center Azelea,* Kawasaki.

- Establish a thorough training program to teach emergency escape.

Illumination in the underground shopping center is almost completely by artificial light. This means that a power failure would place the area in total darkness. To avoid this situation, an emergency power source operated by a battery system and an engine-driven generator must be provided. Also, every attempt should be made to introduce natural light into the area as a backup light source.

The process for using these safety countermeasures for fire disaster are sequentially presented in Fig. 3.20. For them to be effective, there must be a widespread installation that would be controlled in a security center (Fig. 3.21).

AMENITIES

Image

Since the climate in summer is hot and humid, as a traditional rule the Japanese house is built on an open plan so air can circulate easily. Most Western countries build houses with

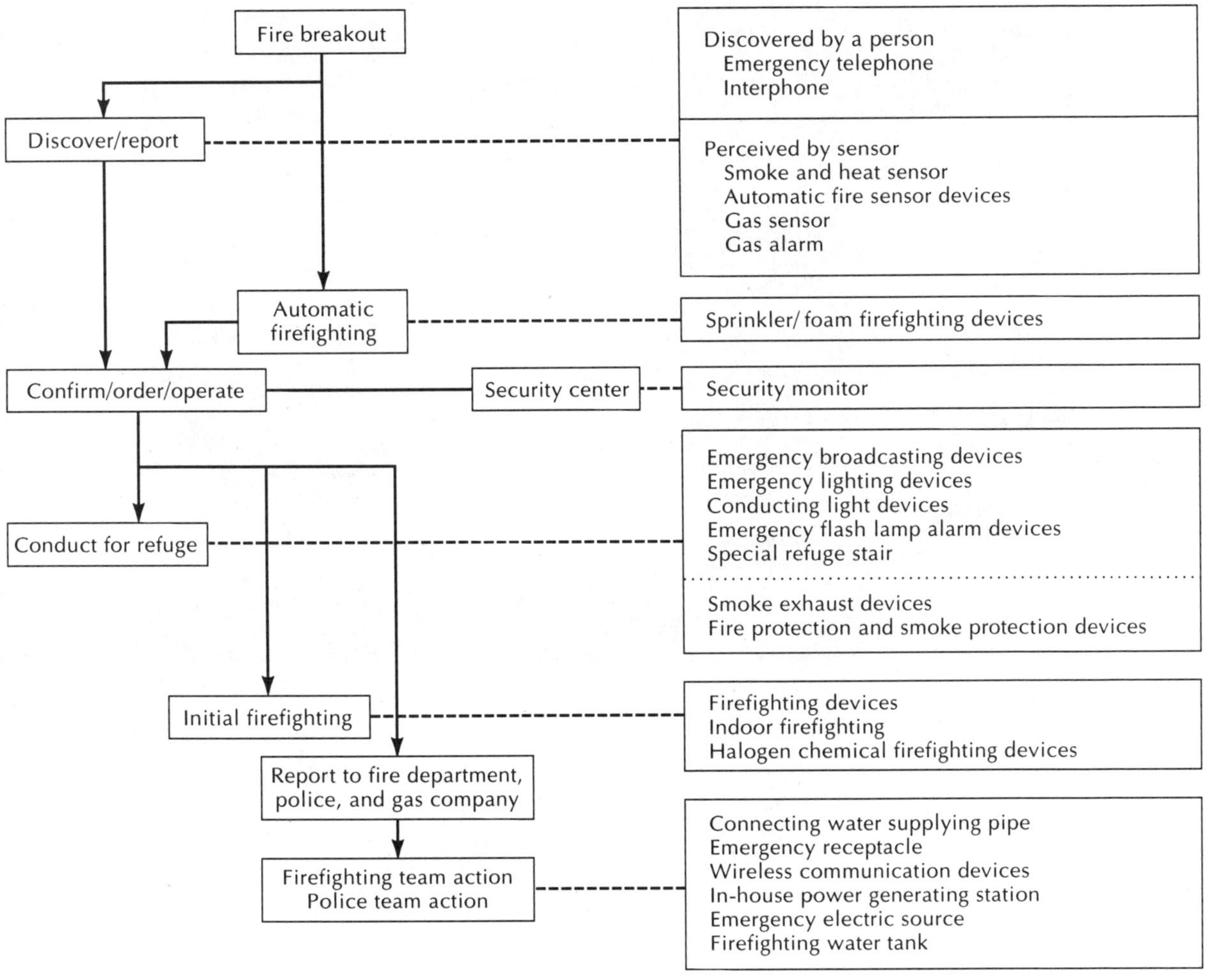

Figure 3.20. Countermeasures for fire disaster (Kawasaki Underground Shopping Center Co. Ltd., Countermeasure System for Disaster in Underground Shopping Center Azelea, *Kawasaki).*

walls of rock and concrete, but the Japanese use a framework of wood, since wood grows abundantly in their country. Because of the openness of their houses and their sensitivity to ventilation, the Japanese hold a negative image of underground space as being dark and humid with little fresh air circulation.

People's attitudes toward the advantages and disadvantages of underground shopping centers were polled in a questionnaire. The major characteristic of underground space—its closeness—appeared to be both an advantage and disadvantage. The advantage cited was the protection it afforded

Figure 3.21. Security control center (photo by Shuichi Miura).

from rain and bad weather, while the lack of sunshine appeared to be its biggest disadvantage. Therefore, in the design for the underground shopping center, the most has to be made of the advantages gained from the enclosed space in order to overcome its inherent disadvantages (Fig. 3.22).

Comfort Design

If the underground shopping center is seen simply as a function of commercial utility, there is little difference in its value compared to the aboveground commercial store. Until recently, the importance of the underground shopping center was attached to its commercial facilities, and not to its comfort. But the overall image of underground space as being dark and enclosed lent little favor to the underground shopping center. Since one of the most important roles of the underground center has been the public space it provides for pedestrians and its substitute for the sidewalk above ground, it should be valued by the same standards as the sidewalk.

Aboveground streets show more variety of design. They

Figure 3.22a–d. Designs to enhance the ambiance in underground shopping centers (photos by Shuichi Miura).

have spaces formed by different widths and materials of road construction, and different colors and forms of buildings along the road. Accordingly, there are unique townscapes on each street, and people can walk pleasantly. On the other hand, the underground shopping center is planned by the minimum of regulatory standards to restrain high construction costs and provide safety. Comfort is not emphasized. Often, its style becomes uniform and monotonous. For example, fountains and ponds are not built for aesthetic purposes because they are considered obstacles to emergency evacuation. The distinctive townscape of streets above ground allow people to recognize their position and direction. Below ground, however, people cannot easily sense their orientation; this causes feelings of uneasiness.

With the strict regulations established after the gas explo-

Figure 3.22b. (Continued)

sion accident of 1980, recent designs of underground shopping centers are at a turning point. Space under ground is becoming more open. Passages are becoming wider and simpler, and many more entrances and squares are being provided. These changes are all aimed at making the underground shopping center a safer place. As the underground shopping center is becoming more open, the aboveground surface, which is also the ceiling of the underground shopping center, is being opened, as many top lights and wide entrances have been opened. It thus seems that the original

Figure 3.22c.
(Continued)

Figure 3.22d.
(Continued)

aboveground surface is disappearing and the level of the underground shopping center is becoming the new ground level. Building for safety has increased construction costs enormously, however, with the burden initially being passed on to the entrepreneur. But on the positive side, these changes are increasing the comfort for the visitor.

Some broad conceptual designs that would further increase the amenity and comfort to shoppers underground are:

- **Use high ceilings and broad spaces.** The sense of closeness that is apt to exist in underground spaces will be reduced by a high ceiling. Also, a broad square where people can rest will become a landmark to help people recognize their location and orientation within the center. Providing a sense of direction can also help the people exit quickly, should such a need arise.
- **Introduce natural light and landscape.** A top light and a ceiling open to the light outdoors will give variable and bright illumination that cannot be obtained by artificial light. The blue sky and the motion of clouds could be recognized. If the landscape above ground can be seen as well, the underground and aboveground will become recognized as one body. Open areas in the ceiling can also be used as natural ducts for smoke ventilation in case of fire. It is difficult to provide top light and an open ceiling, however, because the upper space of the underground shopping center is usually already being used for other purposes. This openness can be provided only in an underground shopping center constructed under a large square or wide road with a center separator, where space is available for building this design.
- **Devise a plan that would lessen monotony and add interest.** Using colors for walls and ceilings can create an upbeat environment. Developing plans to re-create various townscape scenes found above the ground would further please visitors and add variety.

MANAGEMENT

In underground structures, not only are the initial costs of construction and installation high, but the running costs for the facilities' operation are also much more expensive than for general buildings above ground. Moreover, since the underground structure is difficult to extend and repair, the project must be planned from the long-term view of management and use.

There are many differences between managing underground shopping centers and managing aboveground buildings (Fig. 3.23). The special features of the former are the underground location, horizontal expansion of the plan, public use, commercial use, and parking use (Table 3.10). These features create a different set of management problems and issues (Fig. 3.24).

Management of Installations

As mentioned previously, the enclosed area requires special safety measures and disaster facilities. Being different from the aboveground, natural features such as sunlight and wind cannot be received in the underground, so environmental conditions must be produced artificially. Thus, dependence on installations that reproduce aboveground conditions becomes high. In addition, the expectation that public facilities should be open to the public for a long time every day increases the operating hours of these installations.

Air conditioning and ventilation are a necessity, since pedestrians and light produce heat. Air conditioning is used for many hours, even in winter. Depending on the type of business, the amount of ventilation inside differs, as people's reactions also vary. Usually, in an air conditioning system, fresh air from the outside is mixed with returned air from the inside and then circulated again. But the restaurants in underground shopping centers use fresh air from the outside without mixing it with returned air from the inside. In order to avoid air leaks that cause smell and smoke, heating requires an inside pressure lower than that of the outer area.

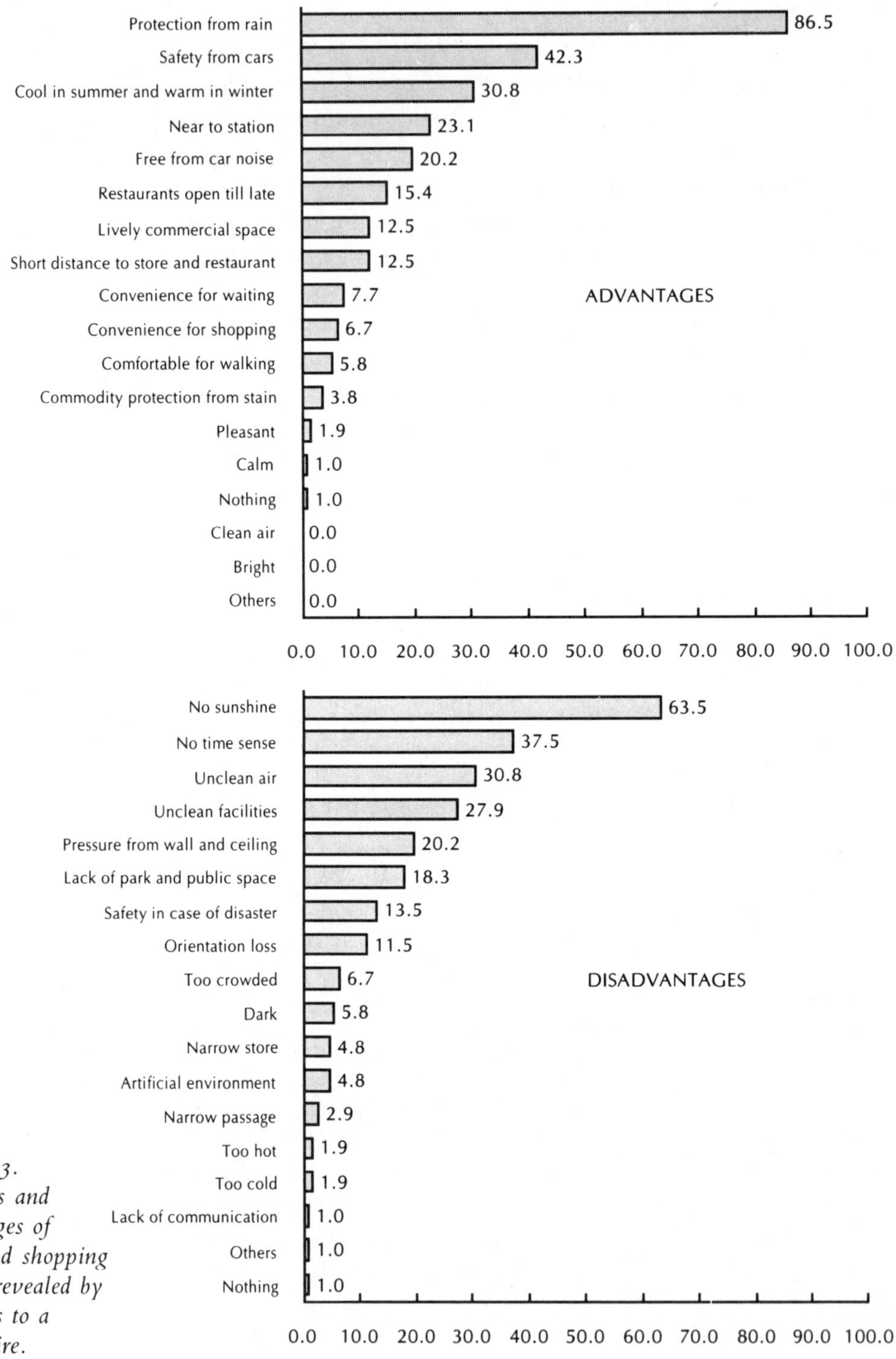

Figure 3.23. Advantages and disadvantages of underground shopping centers as revealed by respondents to a questionnaire.

TABLE 3.10.
Features of Underground Shopping Centers

Underground location	Each underground level is designated for a particular use based on conditions such as optimal pedestrian and motor traffic flow, ventilation, and proximity to aboveground structures.
Horizontal plan	Horizontal space usage is determined by the depth (i.e., number of belowground levels) in any given structure.
Public use	Designs must accommodate 24-hour usage.
Commercial use	Any commercial use (i.e., store function and/or placement) will not inhibit pedestrian traffic flow.
Parking use	In addition to consideration of motor traffic flow, adequate ventilation is provided for all parking areas.

Since the underground space is for common usage, the commercial facilities are also open to the public for long hours. Thus, energy consumption is high and, at the same time, the depreciation of the equipment is accelerated by its heavy use. With a large-scale horizontal area, the pattern of ventilating becomes diversified because each air conditioning system has its own load characteristic. As the duct system becomes longer horizontally, the duct size correspondingly becomes larger. But since the space inside the ceiling is limited, a large duct will not fit into the space well.

Sometimes, air conditioners alone are not effective in all places—near entrances, for instance. Since there are many entrances, and doors cannot be used because of building regulations, the entrance remains open and outside air can flow in. If there is a road near an entrance, gas fumes can easily invade the underground. Further pollution can occur from the underground parking lot. Measures should be taken to prevent this contamination.

Ventilating towers and cooling towers can be used to clear the air, but space to place them is limited. Also, ventilating towers are often set near the aboveground road, where the air is already polluted, so air filters must be changed often. It is difficult to use cooling towers in an underground shopping

- Location underground
  - Reliance on environmental installation
    - Increased installation costs
  - Insurance of safety
    - Increased need for protection against disaster
    - Thorough protection management against disaster
  - Interface between geo-space and city buildings
    - Setting up to surrounding buildings
    - Strain on air filter capacity
    - Inlet air mixed with exhaust
  - Entering and exiting by stairways
    - Occurrence of injuries
    - Counterplan for the disabled
    - Difficulty of cleaning
  - Surrounded by earth
    - Difficulty of structure maintenance
    - Adjustment with other underground cables
  - Close to water table
    - Increase of humidity, mold
    - Danger of flood
- Use for passage
  - Covers large horizontal area
    - Increased need for waste disposal
    - Complication and expansion of air duct and wiring
    - Sewage pipe liable to be clogged on the gentle slope
    - Extension of patroling distance
  - Many stairs
    - Inflow of fresh air and exhaust
    - Invasion of garbage and rain
    - Open and closing of shutters
  - Connection with surrounding facilities
    - Increased need for security and information services
    - Communication system for disaster
  - Use by unspecified number of people
    - Increase in heat gain
    - Unclean public toilets
    - Unclean floors
    - Panic in disaster
    - Vagrants
  - Long hours of use
    - Long operating hours of installations
    - Increase of maintenance time
    - Limitation of inspection time
    - Long hours for police protection
- Use for restaurant
  - Many restaurants
    - Increased need for fresh air
    - Separation of duct system
    - Increase in fire and gas source
    - Production of garbage
    - Outbreak of harmful insects and mice
  - Many managers
    - Various environmental demands
    - Large and complicated organization
- Use for parking lot
  - Many cars
    - Exhaust
    - Parking control system
    - Gasoline
  - Covers large horizontal area
    - Complication and expansion of air duct and wiring
    - Extension of patroling distance

center because they are usually set on the rooftop of a surrounding building, and, in many cases, maintenance becomes inconvenient.

Electrical Appliances

Electricity is used for supplying artificial illumination, since outside light is limited. Many illumination devices of various types are used. Since artificial lighting is used for long periods of time, electrical appliances wear out rapidly. And because of the constant heavy traffic in the passages, the time to replace these appliances is limited.

Most facilities use electricity—for example, restaurants use it in their kitchens because of the potential danger from gas equipment. Thus, electricity is in great demand, so power substitutes are always being sought. If a power failure occurs, emergency lighting, such as leading lights in passages, must be in working order to provide immediate backup illumination. So critical is the need for electrical power backup that providing a private electric generator becomes mandatory. The electric generator uses an internal combustion engine and needs to be given trial tests regularly to ensure its proper functioning at any time.

Plumbing Systems

The plumbing system handles sewage from restaurants, which creates a bad odor and clogs the drainage system. Rest rooms also contribute to clogged drainage systems. The situation is different from that of high-rise buildings, which have a high water pressure and increased energy for pumping and where there are no problems of the kind that exist underground. Underground shops built horizontally on a large scale must pipe sewage a long distance to a drainage tank located on higher ground. It is hard to build drains at a steep angle, and, coupled with long stretches of pipe, the drains easily become clogged.

Measures to prevent floods also play an important role, because underground space is surrounded by spring water. Leaks of water can become a daily occurrence and increase the danger of flooding during a heavy rain.

Figure 3.24. (facing page) Management issues and problems in underground shopping centers.

Fire Protection Systems

Fire protection systems serve a number of purposes. As mentioned previously, it is not easy to fight fire underground. It becomes difficult to point out the fire's exact location from the aboveground vantage point. In addition, there is an added risk of fire spreading when the underground center is built horizontally on a large scale. Strict security measures for gas require the installation of emergency gas circuit breakers. Periodical inspections of the circuit breakers are mandated by the Fire Fighting Law, and a licensed individual, such as a fire protection engineer, performs the task. The breakers usually are not used, but they play an important role in case of an emergency, so their daily maintenance inspection is important. Again, the long hours during which the public passages are used and the heavy traffic on them limit the inspection time. The number of disasters directly caused by an earthquake is small, but an outbreak of fire and a power failure could become secondary disasters.

Underground Structures

Being surrounded by earth, underground structures cause problems regarding structural repairs and preserving the life within the soil. Defects in these structures are not only difficult to find, but also to repair. When the depth of soil above the structure is deep, attention must be given to waterproofing to prevent water leaks and the growth of mold. In the underground space between ceilings and the ground level, utilities need to be protected. This becomes another matter that requires discussion with an agency before construction can begin.

Safety and Security Provisions

Since underground facilities are always open to the public, the possibility of the occurrence of crimes, lost articles, lost children, and personal accidents is high. Taking measures against vagrants is also a problem.

Many stores are potential sources of fire. Clothing stores, for example, contain large quantities of flammable materials, so the possibility of a fire is high. To eliminate confusion during a fire, each store should have a fire prevention manager who could frequently teach disaster prevention measures, such as how to report an emergency, how to guide people to take refuge, and how to extinguish a fire. When a fire prevention manager visits a store, the person must observe if there are any obstacles, such as merchandise or signboards, which would block the passages. It is compulsory to install a disaster-control center where a fire can be reported early, information can be collected, the fire can be extinguished quickly, and people can be directed to refuge sites.

As pointed out in the official notification from the five government offices, an underground shopping center is prohibited from being connected with surrounding buildings. In the earlier days of constructing underground shopping centers, this was not the case. The last shopping center joined to a nearby building was Kawasaki Azelea, with the stipulation that shock absorbers be installed in the connections. Networking with the surrounding buildings is very important. To make an underground shopping center functional during a fire, when people become lost and damage can be extensive, a joint management system for fire protection, both in the center and with its surrounding buildings, must be prepared. A conference for discussing cooperative disaster prevention with both users should be organized to establish a central prevention plan.

Cleaning

Continual attention must be paid to cleaning and trash collection. Floors quickly get dirty from heavy use and from the many entrances where dust and trash from the wind and rain blow in from the outside. A strong industrial cleaner should be used on the floors to remove grit. Garbage cans must also be provided in every passage to prevent littering. Some types of shops, such as restaurants, also add to the amount of garbage. Public rest rooms also require attention. Trash collectors have to haul trash long distances in a horizontally built center. Overall, the underground center requires a full-time staff to keep the shopping center clean.

Special Services

Many types of services must be provided for many types of users since, in many cases, the underground shopping center is the nodal point for traffic, serving not only people in the facilities, but also users of aboveground bus stops, railroad stations, and surrounding buildings. Special services, such as elevators and ramps for wheelchairs for physically handicapped persons, must be installed.

CASE STUDY: KAWASAKI AZELEA

Kawasaki Azelea Underground Shopping Center was completed in 1981, taking 56 months to construct. It was built under the square in front of a railway station and has two floors, with a total area of 56,916 square meters. Its initial capital investment was about $490 million, with $200 million donated from Kawasaki city and $100 million given by Kanagawa Prefecture. It is operated by Kawasaki Underground Shopping Center Corporation. This center is the second largest in Japan, and was the first large center completed after the strict regulations for underground shopping centers were established by the notification.

The purpose of the development was to:

- Separate pedestrians from the car traffic to resolve the congested traffic system above ground.
- Ease the lack of public parking in the area lot by developing a parking lot under ground.
- Develop a comfortable underground shopping center that would symbolize the future of Kawasaki city.
- Entice visitors and purchasing power, which had been flowing out to other cities, by building a core commercial area.

The shopping center has changed the image of Kawasaki city; it now attracts 173,000 people daily.

Figure 3.25. (facing page) Plans of the Kawasaki Azelea Underground Shopping Center (Kawasaki Underground Shopping Center Co. Ltd., Underground Shopping Center Azelea, *Kawasaki).*

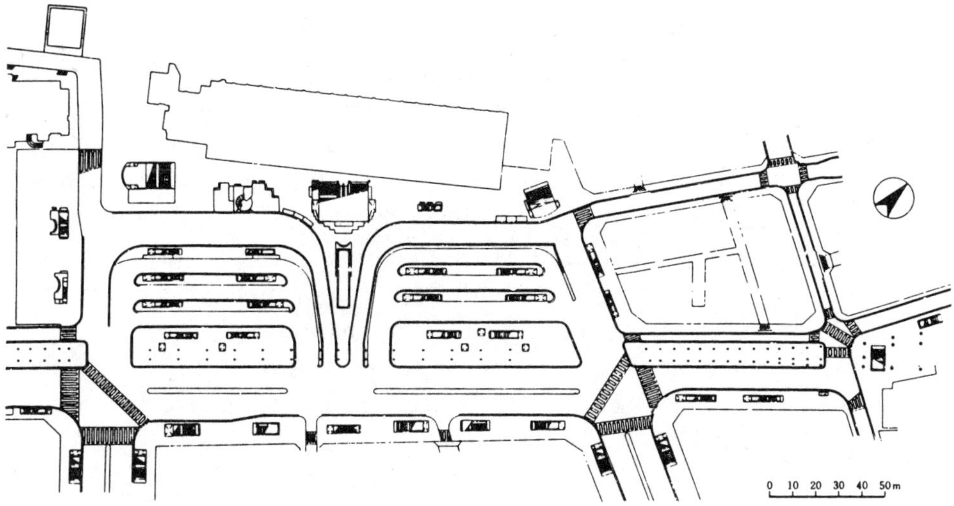

Site Plan

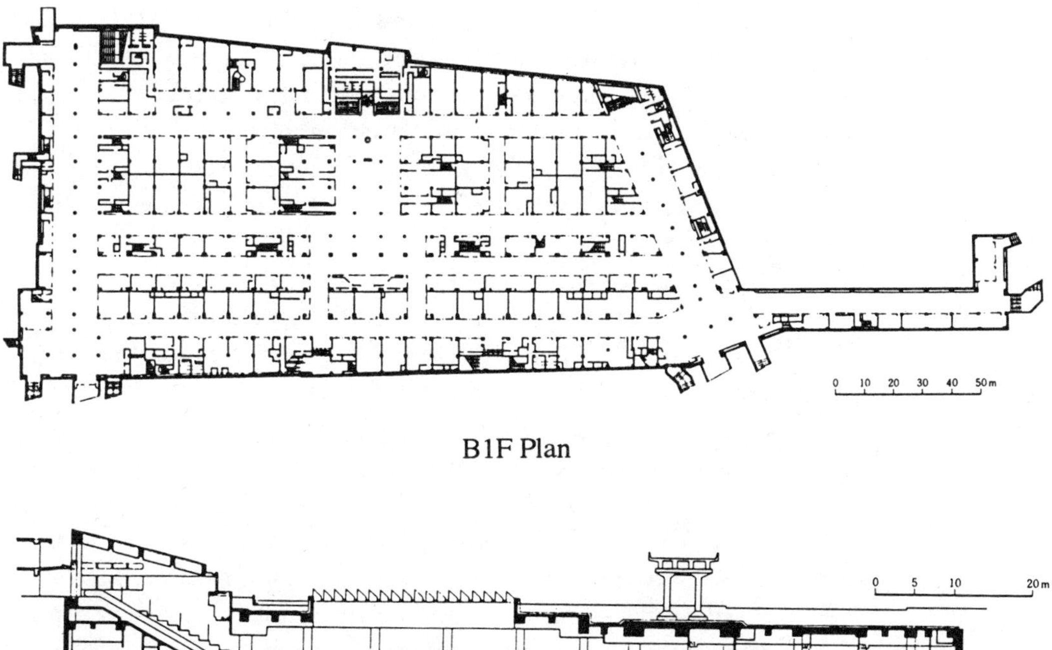

B1F Plan

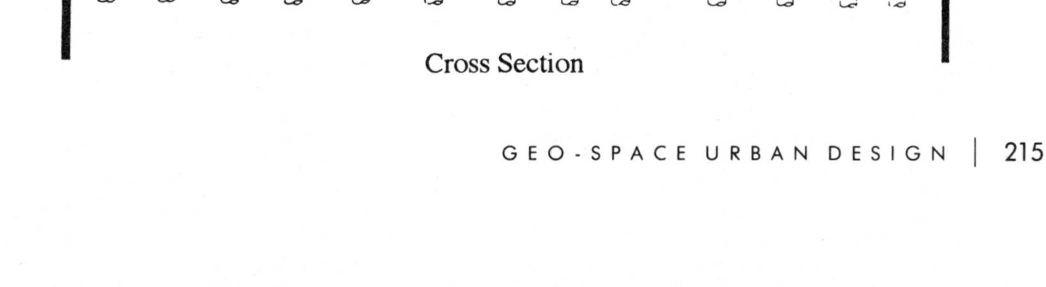

Cross Section

Design

Kawasaki Azelea Underground Shopping Center has a different space design from underground shopping centers constructed before the passage of strict building regulations (Fig. 3.25). A large square of 1,400 square meters directly connects to the neighboring railway station. The floor area for the underground shops is less than one-quarter of the whole area. The ceiling is very high, about 4.35 meters. An opening at the top of the ceiling allows the natural light from the square to come in, so visitors can forget that they are underground. The center is composed of seven passages, five squares, and 154 shops located on floor B1, and parking for 380 cars on floor B2. The widths of the passages range from 22 meters to 6 meters, with a public square located at the end of each passage.

The shops are separated into seven different zones to provide circulation and convenience for visitors. Kawasaki Azelea contains not only shops, but also many other facilities, such as a police station, information center for the city hall, a bank, a tourist information bureau, and a first-aid station, just as would be found in an aboveground city. As these facilities are located in a logical order, the overall plan is easy to understand.

Because of the size and the many shops in the center, it has the highest building standards in the world for its fire protection system. There are five private entrances for firefighting teams. The 23 restaurants that use fire are concentrated in one area. In the disaster prevention center, a remote sensor continually searches and controls the environment.

To accomplish the construction of this design while adhering to the new building standards, the construction cost came to $4.5 billion, which was a high cost to offset. However, Kawasaki Azelea is safe and comfortable, and it now attracts 63 million people every year.

Management

Generally, it is very difficult to financially manage underground shopping centers under the notification. Revenues from the Kawasaki Azelea shops and parking lot total $236 million every year, with the biggest support coming from the parking lot. However, the expenses are $316 million a year, creating an annual deficit of $80 million. The overall positive influence to the whole area, however, can compensate somewhat for the yearly deficit.

SELECTED READINGS

Japan Project Industry Conference, *Current State of Underground City,* Japan Project Industry Conference, Tokyo, 1986 (written in Japanese).

Kawasaki Underground Shopping Center Co. Ltd., *Underground Shopping Center Azelea,* Kawasaki, pamphlet (written in Japanese).

Toshio Ojima, "Management System of Underground Shopping Center," *Journal of Civil Engineering Society,* Tokyo, 1987, pp. 37–39 (written in Japanese).

Toshio Ojima et al., "Environmental Research in Underground Shopping Centers," *Summaries of Technical Papers of Annual Meeting, Environmental Engineering,* Architectural Institute of Japan, Tokyo, 1987, pp. 609–610 (written in Japanese).

Toshio Ojima et al., "Research for the Energetic Consumption of Underground Shopping Centers," *Summaries of Technical Papers of Annual Meeting, Environmental Engineering,* Architectural Institute of Japan, Tokyo, 1987, pp. 611–612 (written in Japanese).

Toshio Ojima et al., "Research on Living of Workers in Underground Shopping Centers," *Summaries of Technical Papers of Annual Meeting, Environmental Engineering,* Architectural Institute of Japan, Tokyo, 1987, pp. 613–614 (written in Japanese).

Toshio Ojima et al., "Environmental Estimation Factor in Underground Shopping Centers," *Summaries of Technical Papers of Annual Meeting, Environmental Engineering,* Architectural Institute of Japan, Tokyo, 1988, pp. 1069–1070 (written in Japanese).

Toshio Ojima et al., "Research on Air Pollution in Underground Shopping Centers," *Summaries of Technical Papers of Annual Meeting, Environmental Engineering,* Architectural Institute of Japan, Tokyo, 1988, pp. 1065–1066 (written in Japanese).

Toshio Ojima et al., "Energy Consumption and Thermal Environment in Underground Shopping Centers," *Summaries of Technical Papers of Annual Meeting, Environmental Engineering,* Architectural Institute of Japan, 1989 (written in Japanese).

Toshio Ojima et al., "Research on Operators in Underground Shopping Centers," *Summaries of Technical Papers of Annual Meeting, Environmental Engineering,* Architectural Institute of Japan, Tokyo, 1990, pp. 923–924 (written in Japanese).

Toshio Ojima and Shuichi Miura, "Environmental Study on Underground Shopping Center," *Urban Underground Utilization '91, 4th International Conference on Underground Space and Earth Sheltered Buildings,* Tokyo, 1991, pp. 471–477 (written in English).

Toshio Ojima and Shuichi Miura, "Study on Thermal Environment and Energy Consumption in Underground Shopping Centers," *Journal of Architecture, Planning and Environmental Engineering,* 430, Tokyo, 1991, pp. 13–22 (written in Japanese).

Osaka Underground Shopping Center Co., Ltd., *Introduction of Osaka Underground Shopping Center Co., Ltd.,* Osaka, pamphlet (written in Japanese and English).

Public Corporation of Sapporo City Development, *Pamphlet of Sapporo Underground Shopping Center,* Sapporo, pamphlet (written in Japanese).

Research Center of Land Development Technology, *Survey of Existing Technology of Underground Space Utilization,* Tokyo, 1989, p. 36 (written in Japanese).

Shinjuku Underground Parking Lot Corporation, *Introduction of Corporation,* Tokyo, pamphlet (written in Japanese).

Tokyo Fire Defense Board, *Fact-finding Report of Underground Shopping Center,* Tokyo, 1979, p. 5.

Underground Space Center of Japan, *Designing Underground City,* Daiichi Houki, Tokyo, 1991 (written in Japanese).

4 GEO-SPACE TRANSPORTATION FACILITIES IN JAPAN

T. Ojima

SUBWAY

Development

Cities have many modes of transportation, including automobiles, buses, and taxis. But subway transportation has become a unique and useful form of transportation in Japan. Previously, the streetcar was a widely used form of transportation and coexisted with the automobile. However, as the number of automobiles increased faster than roads were built to accommodate them, their coexistence with streetcars became difficult. Congestion worsened, and the average speed of streetcars decreased. These same phenomena affected buses and taxis. As mass transportation became possible with high-speed railways, this form of transportation became favored over other types of transit, and took two forms: elevated railway systems and primarily belowground subway systems.

The elevated railroad has eight times the transport capacity of a bus per meter of road width. The major disadvantage of elevated railways, however, is the difficulty of finding space in the center of a city to lay the track. Also, elevated railways create many environmental concerns. Their transport capacity, which uses mainly the space above roads, is limited. In comparison, the subway, which has become the most suitable form of transportation, has overcome some of the problems of elevated railways.

The first metropolitan subway was built in London in 1863. This subway consisted of a passenger car pulled by a steam locomotive, which polluted the surroundings with its smoke. In order to overcome the pollution, transportation facilities were constructed underground. A new technique enabled spiral stairs to be dug down to an underground station without destroying the buildings above ground. However, constructing new underground basic facilities was a painful effort, which brought aboveground city development to a standstill. In 1890, the electric locomotive revolutionized transportation.

In 1927, the first Japanese subway was built, 60 years after Europe and America had first built them. Today, approximately 60 Japanese cities use subways, many of them constructed after 1970. Worldwide subway development is shown in Fig. 4.1. In 1983, the average length of a subway route was about 400 km in New York and London, about 300 km in Paris, and about 200 km in Tokyo and Moscow. However, the number of passengers per year is sometimes the reverse of the length of the route, with 24 billion in Moscow and 22 billion in Tokyo. Especially after the war, the pace of Japanese subway expansion was remarkable. In Tokyo, for example, during the 30 years from 1955 to 1985, the total subway doubled (Fig. 4.2).

Expansion of Subway Station Concourses

In high-density areas, underground space is made for pedestrians with subway station concourses. The building of these concourses is different from building underground in cities. In order to make space for the subways, the upper tracks and platforms are used as a part of the concourse. The concourse does not contain shops, but serves only as an underground

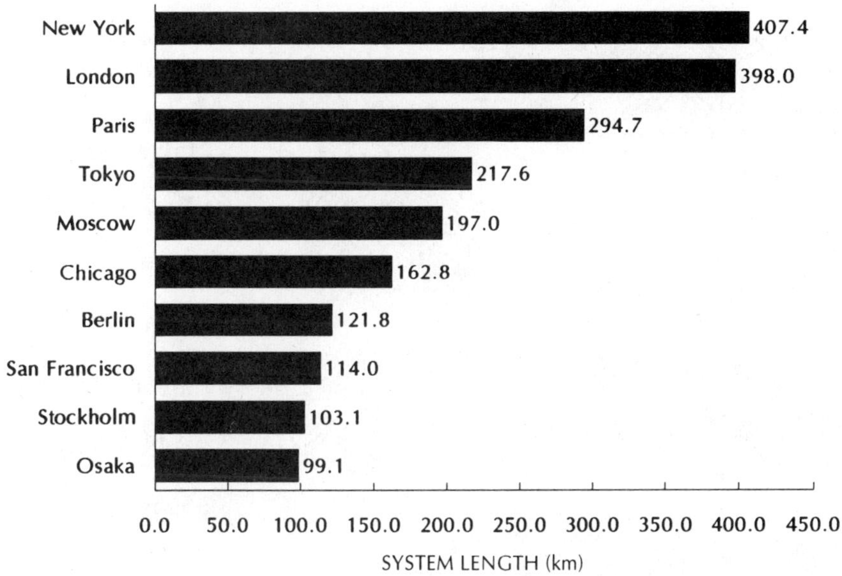

Figure 4.1. Subway development internationally (Teito Rapid Transit Authority, '88 Eidan Subway Handbook, *Tokyo, 1988, p. 69).*

passage. In the subway stations of Tokyo, the Otemachi, Ginza, and Shinjuku concourses are connected by joining the basements of the surrounding buildings to form large underground pedestrian networks. In these areas, since the aboveground sidewalks are heavily used, this kind of underground pedestrian network plays an important role in supplementing aboveground walkways. Moreover, the underground network protects pedestrians from bad weather. Because of this convenience, many shoppers and commuters use the underground passages.

The underground network of the Otemachi subway station is connected with that of the Ginza subway station by the Niju Bridge. The total floor area of both buildings is about 2.4 million square meters. This is similar in scale to Montreal's underground network, which consists of about 3 million square meters.

Currently, however, this underground space for pedestrians is not always comfortable. Since the upper space of subway station platforms is frequently used for passages, the passages are often designed to fit the platforms, and sometimes lack sufficient width to accommodate many pedestrians. In other places, it is necessary to build stairs, which

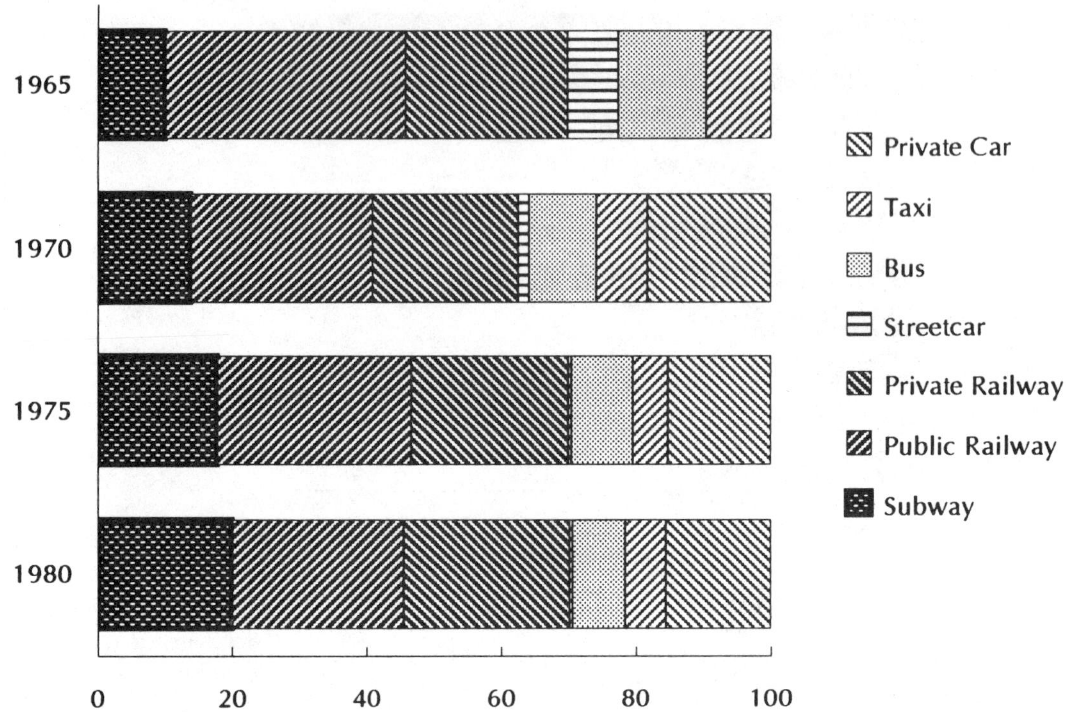

Figure 4.2. Comparison of use of different modes of transportation in Tokyo (Committee of Editing Handbook of Underground Structure, Handbook of Underground Structure, *Construction Industry Research Association, Tokyo, 1987, p. 43).*

require people to descend one flight and then climb another because there is no way to build the passages around existing pipelines. These problems resulted from not utilizing underground space systematically.

Forms of Construction

Generally, subways can be divided into two types by their construction form—a box form or a round form (Fig. 4.3). Generally, the box is the most practical shape for the excavation method of construction. In the case of the shield construction method, the round shape is preferable. Since the construction cost of subways is very high, a minimum-sized cross-sectional area is planned. Although almost all of the subways constructed in the early days were constructed by the excavation method, the shield construction method is now being used more frequently. In the case of the shield construction method, the largest round-shaped subway is roomier than the box form and is often used instead of the

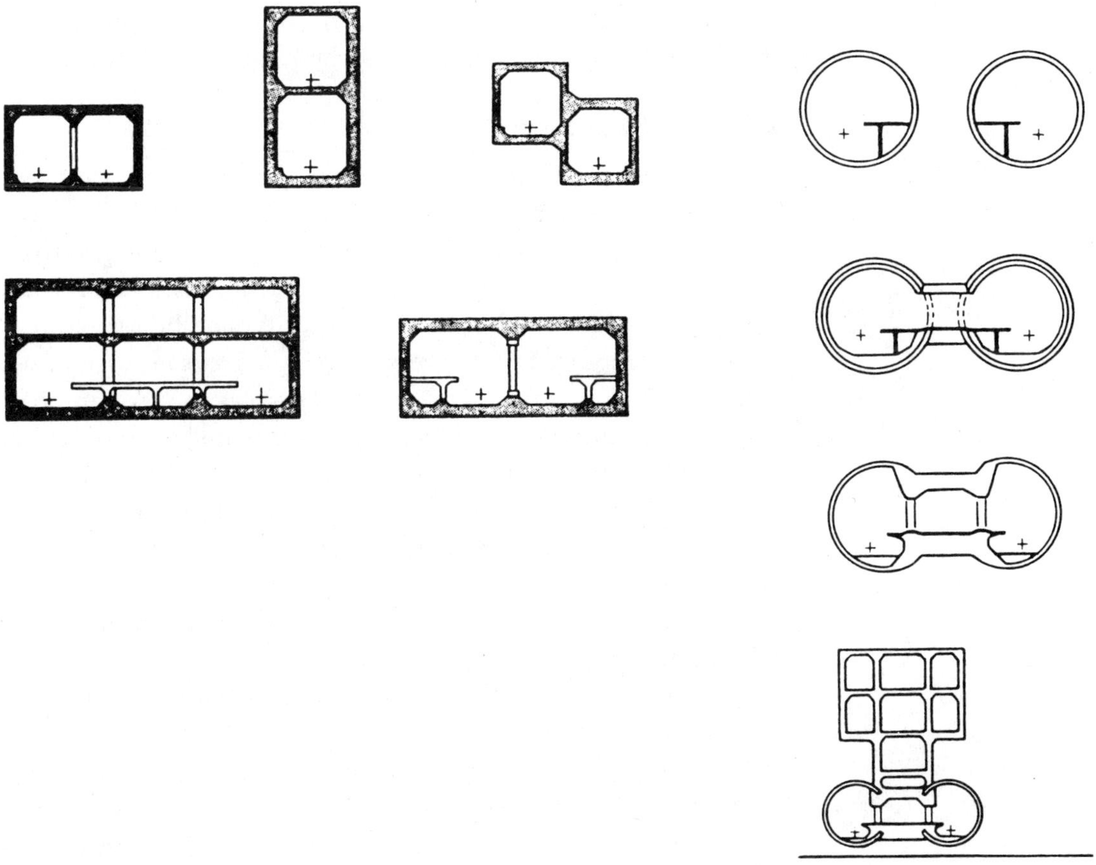

Figure 4.3. Sectional views of subway forms (Committee of Editing Handbook of Underground Structure, Handbook of Underground Structure, *Construction Industry Research Association, Tokyo, 1987, pp. 64–67).*

box shape to allow a larger space for the flow of passengers and for building facilities. It is difficult to extend or reconstruct structures already existing underground, so stations expecting a future increase in the number of users must be planned with extra space to allow for expansion.

Usually the subway tunnel is planned under a road. But when the tunnel veers away from the course of the road, it could fall under private land. Although it is easy for the subway train to make curves with a wide radius, the private land under which it would pass would cost much more to acquire and negotiations with the citizens would be difficult. Considering the advantages of building an underground structure along the road and the difficulty of negotiating with the citizens for private land, the tunnel should be planned to lie under the middle of the road, with the entrances and exits being accessed from both sides of the road.

Depth of Subway Routes

Subways should be constructed in as shallow a location as possible to cut construction costs. A shallow depth is also more convenient for users. The height from the ceiling to the floor of the tunnel should be at least 2.5 meters. Generally, however, because of the many objects lying underground, the height is required to be greater. Many factors necessitate a tunnel height greater than 2.5 meters, such as the crossing of underground passages, underground shopping centers, rivers, and crossings of the subways themselves. The need to build subways deeper increases every year. Figure 4.4 shows the Ginza Line, which was constructed in the early days, as being comparatively shallow. As time passed, subways have been constructed deeper in places, with some lines almost 40 meters underground.

The same holds true for building stations. Subway stations built 30 meters under ground are not rare. In this situation, when the tunnels are deeper than usual, the shield construction method would be used, since this method is not as influenced by the depth of the tunnel as is the excavation construction method. The excavation construction method is used to dig the tunnel from the surface of the ground, and its con-

Figure 4.4. Depth of Ginza and Hanzomon subway tunnels (Ryouichi Shimojima, "The Present and the Direction of Underground Utilization," Symposium of Underground Space Utilization, *Civil Engineering Society, Tokyo, 1989).*

struction costs are high in proportion to the depth of the tunnel.

Health

In Japan, subways were thought to be cool in summer and warm in winter because of the general characteristics of the space underground. Below 10 meters in depth, the temperature is 15° C and remains steady throughout the year. When the subways were first built, they were cool in the summer and warm in the winter. Recently, however, the temperature has been rising constantly, at a rate of 0.3° C per year (Fig. 4.5). The peak temperature has also increased; it was 27° C in 1951, 30° C in 1961, and 34° C in 1971. It was estimated that, by 1981, the peak temperature would be 37° C, or higher than body heat, and would become too warm for comfort. These thermal developments caused the Association of Countermeasures for High Temperature and Humidity to be organized by the Tokyo Metropolitan Subway in 1970. Since cooling systems were introduced after 1971, thermal conditions have changed for the better.

The association proposed that some of the exhaust heat in

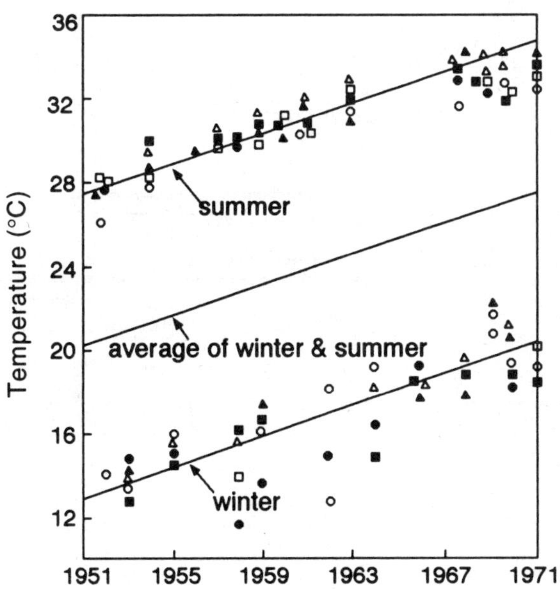

Figure 4.5. Annual temperature fluctuation in the subway (Tadaaki Zouga et al., "Analysis of Thermal Environment in the Subway," The Society of Heating, Air Conditioning and Sanitary Engineers of Japan, *Vol. 47, No. 6, Tokyo, 1973).*

the tunnel could be discharged into the wall spaces, and other exhaust heat could be discharged into the air through ventilation systems. When the amount of exhaust heat and discharged heat would be equalized, the temperature of the tunnel would become fixed. With electric consumption increasing, the increase in the number of passengers, and the fall of the level of groundwater, however, the problem of discharging heat has become more complicated.

It is shown in Fig. 4.6 that roughly 70 percent of the heat produced is generated by the trains; 16 percent by the lighting; and 14 percent by people. The existence of a large heat source, such as the train in a closed space, is a great load on a system. In this case, the characteristics of the underground

Figure 4.6. Heat sources in the subway (Teito Rapid Transit Authority, '88 Eidan Subway Handbook, *Tokyo, 1988, p. 42).*

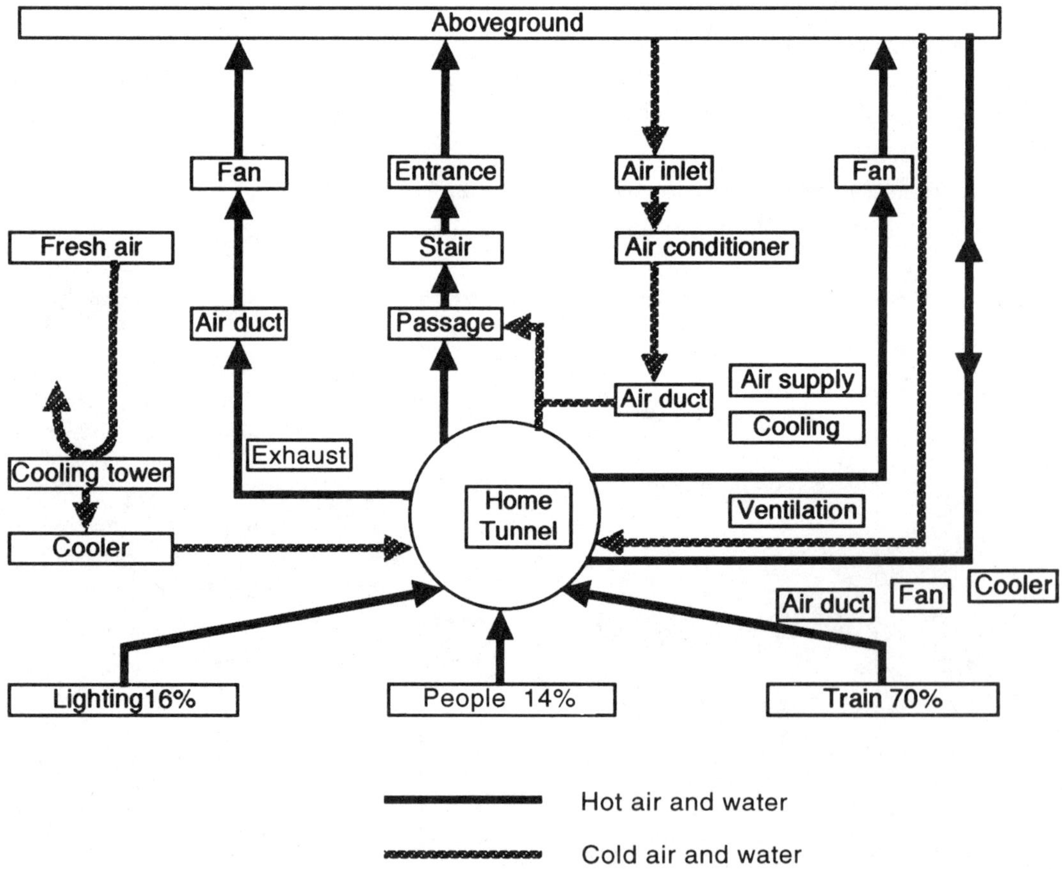

space serve as insulation in holding the heat and in accumulating the inner heat load. To prevent the temperature from rising, a ventilation and cooling system would be a basic countermeasure. Previously, a ventilation system would have been sufficient to cool the area, but recently this has not been the case—a cooling system is also needed. For a cooling system to be effective, there must actually be two cooling systems—one for the tunnel and a separate system for the train. By adopting a separate cooling system for the train, however, the electric consumption of the train would increase by 30 percent to 40 percent. At the same time, the heat discharge would also increase, and another system would be required to deal with this. To date, a cooling system has been adopted for the stations but not for the trains. By 1971, air conditioning began to be used; by now, air conditioning has been extended into 52 stations and 33 tunnels. Recently, as a train with a lighter-weight body and a more effective electric motor has been developed, energy consumption has been reduced 30 percent. Ten percent of the trains also have been equipped with cooling systems.

Environmental planning for subways also depends on the condition of the cities' climate. In some cases, only mechanical ventilation facilities are necessary. Compared to Tokyo, most other cities in the world using subways are located in a high latitude. Thus, the high temperature and high humidity are not as serious a problem in other cities as they are in Tokyo.

Thermal Condition of Concourses

As the center of a pedestrian network, the concourse, as well as the station platform, plays an important role in a highly populated district. Therefore, the concourse must be developed as a comfortable place. Until recently, space designs and environmental designs had not been planned for sufficient comfort. The temperature of the concourse is higher than the platform in the summer and lower in the winter, because the concourse is nearer to the aboveground surface than is the platform (Table 4.1). Also, the difference between outdoor and indoor temperatures is greater in winter than in summer. And with the wasted heat from the train, the temperature in

TABLE 4.1.
Comparisons of Subway Station Temperatures with Outdoor Temperature (°C)

			STATIONS	
			A	B
August		Mean	26.7	29.2
	Outdoor	Max.	33.6	31.8
		Min.	24.5	25.6
		Mean	24.0	27.1
	Platform	Max.	25.2	29.5
		Min.	20.2	24.7
		Mean	26.2	27.7
	Concourse	Max.	30.0	31.9
		Min.	21.6	24.7
November		Mean	16.4	20.5
	Outdoor	Max.	17.2	21.5
		Min.	15.7	19.1
		Mean	27.7	27.3
	Platform	Max.	29.5	29.2
		Min.	24.8	25.8
		Mean	23.8	26.9
	Concourse	Max.	26.7	28.7
		Min.	18.5	24.6
February		Mean	9.7	6.5
	Outdoor	Max.	13.1	11.5
		Min.	−0.7	1.9
		Mean	21.1	21.2
	Platform	Max.	23.0	23.9
		Min.	14.0	17.2
		Mean	17.3	19.0
	Concourse	Max.	21.0	23.3
		Min.	4.5	9.8

Toshio Ojima et al., "Thermal Survey of Subway Concourse," *Summaries of Technical Papers of Annual Meeting, Environmental Engineering,* Architectural Institute of Japan, Tokyo, 1990, pp. 925–926.

the station is generally high. Temperature distribution for the concourse is shown in Fig. 4.7.

Air quality also is not high. During the rush hour, the density of CO_2 generated by people amounts to more than 2,000 ppm (the environmental standard in Japan is 1,000 ppm). At present, the air conditioning system is not enough to mitigate this condition if operated according to the number of passengers. Moreover, the air conditioning system

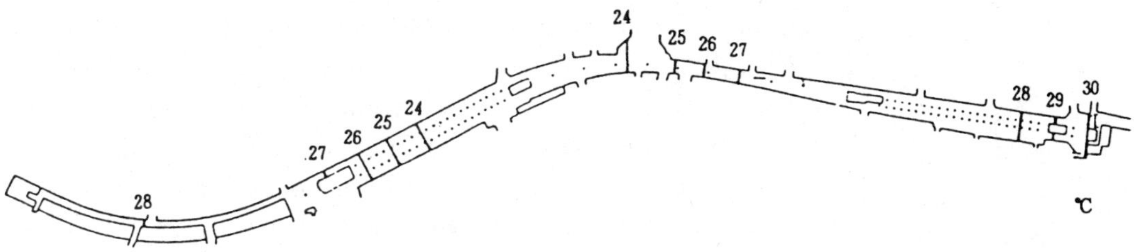

8/17/89 12:00 Outdoor temperature: 30.8°C

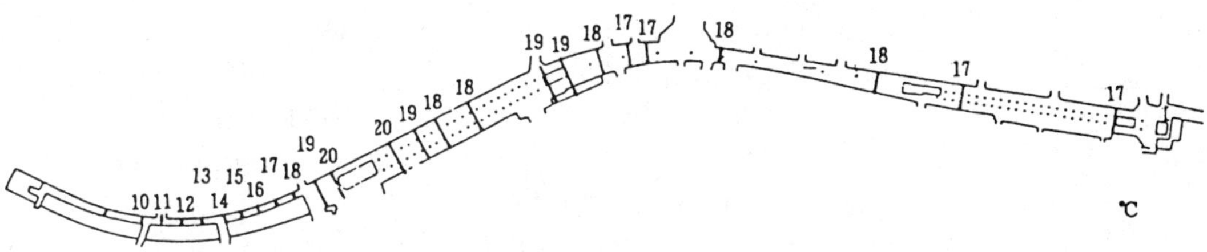

1/18/90 18:00 Outdoor temperature: 8.0°C

Figure 4.7. Temperature distribution of subway compared with outdoor temperature (Toshio Ojima et al., "Thermal Survey on Subway Concourse," Summaries of Technical Papers of Annual Meeting, Environmental Engineering, *Architectural Institute of Japan, Tokyo, 1990, pp. 925–26).*

does not have sufficient capacity to respond to the peak load. Even if its capacity could be enhanced, it would be very difficult to find additional space for the cooling equipment, and also very difficult to extend the space in the underground. If the capacity could be increased, the equipment could not be used efficiently. As a result, the subway must be designed with a long-term view for its future use, and with the use pattern considered as a fundamental priority.

Safety

Of all disasters that could possibly occur in subways, the most likely one is fire caused by the train. At present, the Ministry of Transport encourages the use of noncombustible materials for subway rolling stock whenever possible. Also, it requires the use of nonflammable materials to replace the flammable materials currently being used. The Ministry of Transport has also established standards to protect the main circuits, electric current, and heating equipment, and has installed broadcasting equipment and fire extinguishers. Because of these new standards, train fires can be prevented.

Another source of subway fires, and currently the greatest source, is the improper disposing of burning cigarettes inside the underground space. Moreover, in many cases, the openings for subway ventilation are installed near the sidewalks above ground, into which passersby can throw burning cigarettes. Thus, attention must be directed not only to the safe disposal of burning cigarettes in the subway stations, but also to the careful disposal of the cigarettes of smokers above ground.

Another safety problem is the possibility of flooding from a tsunami, typhoon, or heavy rain because subway stations are below ground level. Where this possibility exists, doorways with low boards must be installed in the upper space leading to the sidewalks to create a buffer and prevent water from flowing in. When flooding is severe, doors can close off the entire doorway. Ventilation openings are also an invasion route for water, but there, too, doors can be installed that shut automatically by flood detectors.

Efficiency

Construction Cost

The subway is a mass transportation facility that is indispensable for alleviating the cities' traffic. But its biggest problems are its high construction costs (Fig. 4.8) and the long time required to build a subway. Although the subway has merit over aboveground transportation, aboveground transportation requires no initial expenses for land acquisition. Subway construction costs are also increasing every year, as commodity and labor prices rise, as well as the costs for cutting the tunnels, as they are made deeper. A recently built line cost more than 2 million yen per meter. It generally takes up to 10 years to construct a subway line. Recently, this time frame has been increasing, as the underspace, which previously had not been used, is now becoming widely used for many purposes.

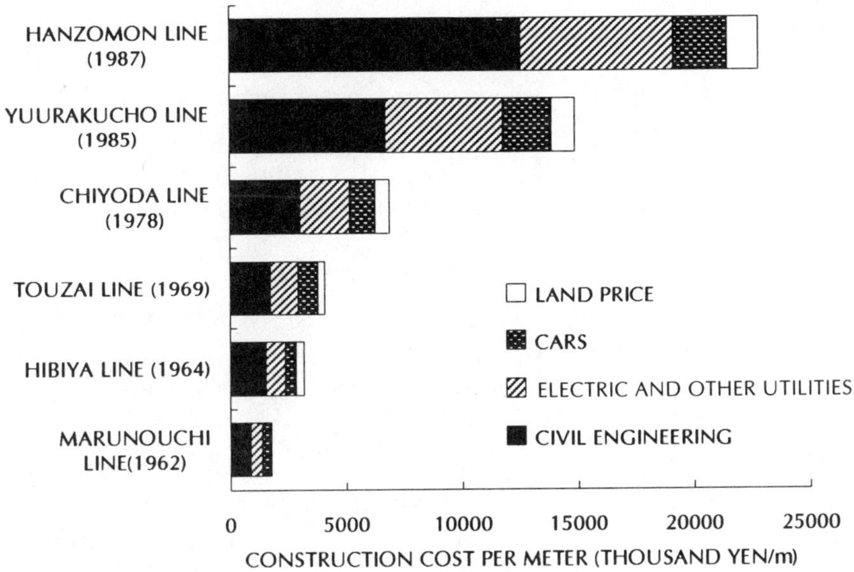

Figure 4.8. Construction costs per meter for six subway lines (Teito Rapid Transit Authority, '88 Eidan Subway Handbook, *Tokyo, 1988, p. 58).*

Energy Consumption

As mentioned previously, because the countermeasures for high temperature and humidity were of serious concern, a high-capacity cooling system became necessary. The cooling system is used to ventilate and cool the station and tunnel. For these cooling systems to be effective, the capacity of the equipment must be four times larger than that for the general buildings located above ground. In the new system for cooling the platform, a partition is built between the platform and the line to prevent the spread of waste heat escaping from the train. In this partition, there are automatic doors, which open only when the train stops at the station. This system also prevents people from falling onto the subway tracks.

Function of the Subway

Current Condition of Subway Passengers

Currently, half of all subway passengers travel during the morning and evening rush hours. Although the number of daily passengers is only 60 percent of the trains' seating capacities, during the morning rush hour there are twice as many passengers as seats. This seating shortage during rush hour has created the notorious scenes of subway riding in Tokyo. If the carrying capacity were augmented during this peak time, the use ratio in nonrush hour times would be very low. The double load during rush hours, along with the heavy concentration of the population in Tokyo, will influence subway development in the future.

Deep Underground Subway Project

The congested condition of Tokyo's population is a problem not only for the subway, but also for the railway. In large cities there must be an additional 29 lines built, encompassing 530 km of railway, by the year 2000 to resolve this condition. To help alleviate some of the pressure to increase aboveground railway construction, more subways are being built. Primary construction of these subways has taken place beneath existing public roadways. However, this will ultimately prove to be insufficient because of the limitations to such routes, thereby necessitating the use of private lands for subway and railway expansion. Unfortunately, compensation of private landowners is costly and unpredictable.

Using this background information, the need for building deep underground subways is being examined today. One project would give subway entrepreneurs a right to develop unused deep underground space and to promote the development of the subway in urban areas. If this project can be realized, the subway would not always need to be developed in the space under the road. Building deep underground would also allow the routes to be constructed straighter, which would speed up the trains and make the passengers more comfortable. Although the construction costs for a deep station would be high, this project would be effective and economical for an express subway that would have few stations and extend long distances through the center of the city, where land prices are very high.

Protection against disasters must also be planned more carefully for the deep underground subway than in the common subway. For example, if a fire occurred in a station located deep underground, it would take a long time to evacuate people from the station. There would have to be a disaster prevention zone in the middle of an escape path. This zone would be planned as a safe place that could accommodate passengers until evacuation became possible. Although the train itself is now designed not to burn, it would have to be examined in regard to its exhaust of gas and smoke. As the distance between the stations becomes longer, the tunnel would have to be divided, with a partition at a midpoint.

Example: Tokyo Subway Line Number 12

The network of railways in Tokyo has been designed in a radial manner to extend outward from the heart of Tokyo. The outer loop of the line is considered a secondary line. The Tokyo Metropolitan Government has constructed Subway Line Number 12 that consists of a loop line 29 km long and an outer line that covers the outside periphery of 18 km. The total construction cost of this project is 946 billion yen.

Subway Line Number 12 uses a linear motor, which has been run on a trial basis. With the success of its trial run, the linear motor became miniaturized and the section of tunnel was reduced by half in size (Fig. 4.9). The volume of the station was cut 60 percent to 70 percent from the size needed for traditional motors. This reduction in size also decreased construction costs. However, the heat produced by the linear motor is greater than that produced by traditional motors. With the adoption of this linear motor and the reduction of the size of the tunnel section, a higher air temperature and a stronger wind were produced by the train. Unfortunately, the increase in land prices made it more difficult to install ventilation openings in the tunnels. Now, the whole subway is ventilated only by installing openings in the stations.

In planning the stations, two issues were especially considered. One is achieving the efficient use of underground space by developing other underground facilities as a unit. Subway Line Number 12 will have 28 stations and, among these, five stations will be developed together with underground parking lots, and 17 stations with underground bicycle parking lots. Subway Line Number 12 runs through a variety of unique regions, such as the business area, the waterfront area, the downtown area (which retains the feeling of the Edo age),

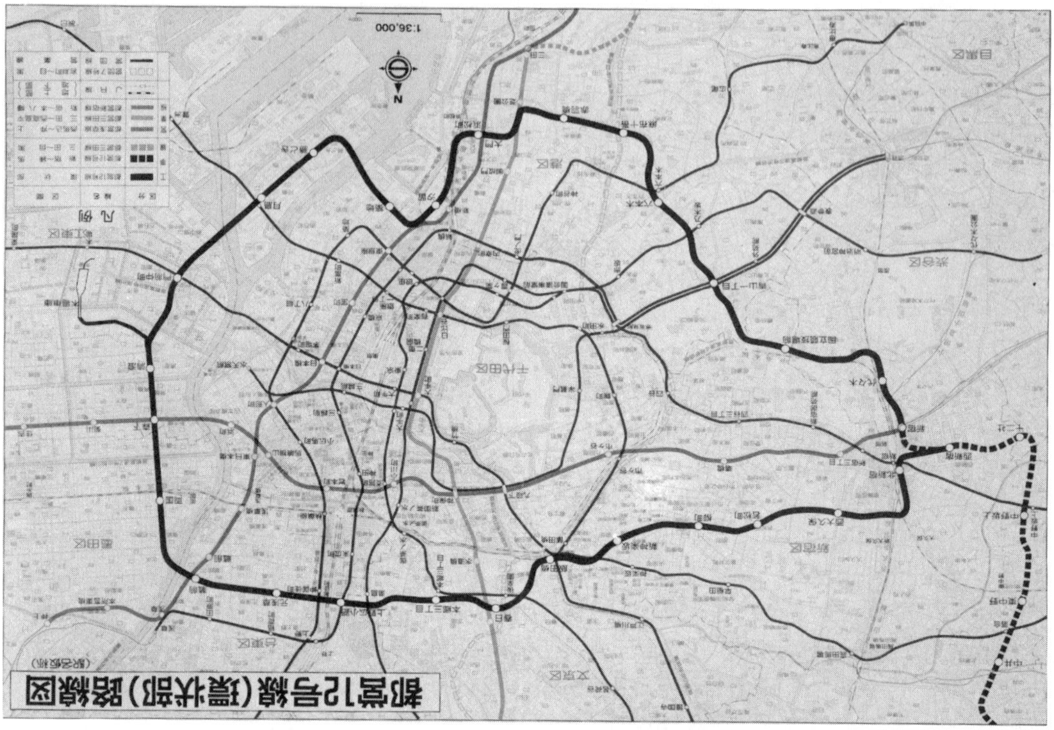

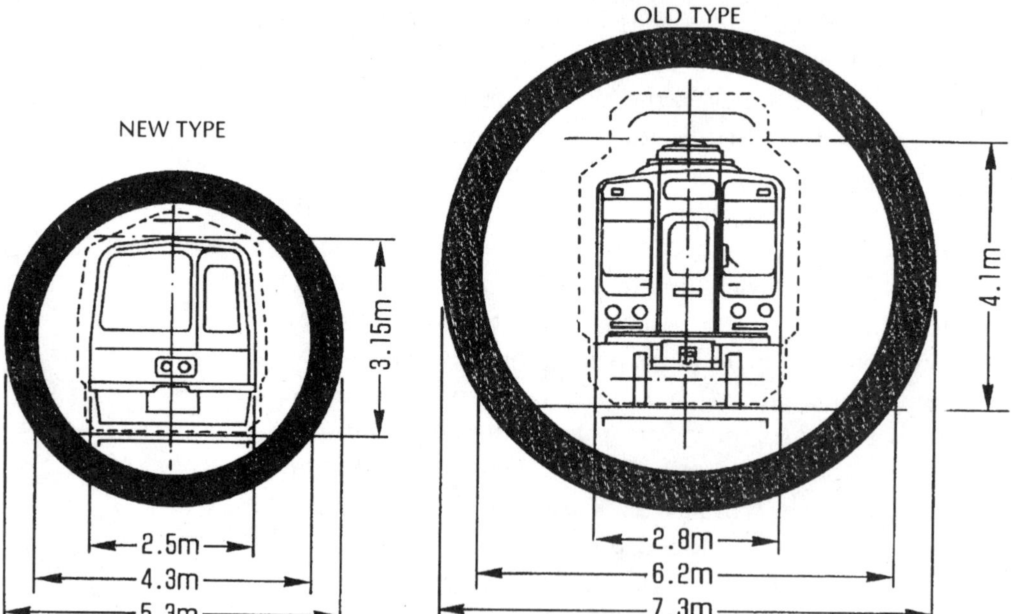

COMPARISON OF TUNNEL CROSS SECTION

and an old town constructed during the Meiji age, and preserves the specialty of these regions. Stations of the future should also be designed to preserve Japanese history and the regional characteristics of its industry and culture.

UNDERGROUND PARKING LOT

The need for developing parking lots is urgent in Japan today. It has been suggested that the construction of additional parking lots will increase the amount of automobile traffic, resulting in little, if any, improvement in current traffic conditions. But it is impossible to neglect the existing conditions because many districts with high urban activity do not have enough parking space with the present number of parking lots. Parking capacity is far below the minimum capacity required.

The problem connected with the development of urban parking lots is not only the increase in the number of automobiles, but also the increase in the number of people. The shortage of parking lots is causing an increase of on-street parking, which further increases traffic congestion and creates obstructions for pedestrians on non-sidewalk streets like access roads. Figure 4.10 presents the results of on-street parking research in Shinjuku Tokyo (whose district area is 22 hectares). There were 725 cars parked on the street during peak rush hour time, occupying about half the total length of the access roads. There were a number of double-parked cars on the street, as drivers tried to park closer to their destinations. Many pedestrians also use this district. Therefore, the danger was increased not only for pedestrians, but also for traffic, as accidents became worse.

In such districts, it is necessary to plan parking lots that will satisfy future demand. As estimated by the standards of the existing parking lots, Fig. 4.11 presents a comparison of the parking floor area, assuming eight floors of an aboveground parking lot, with the number of parking spaces on the street. This result showed that parking capacity is only about a quarter of the demand. The site area of the aboveground parking lots has a full-floor area ratio of only 12 percent of the total building area. It may not be effective

Figure 4.9. (facing page) Route of new Subway Line 12 and cross section of new- and old-type tunnels (Underground Space Center of Japan, Focusing on the Organic Combination of Aboveground Space and Underground Space, *Tokyo, 1991, p. 27).*

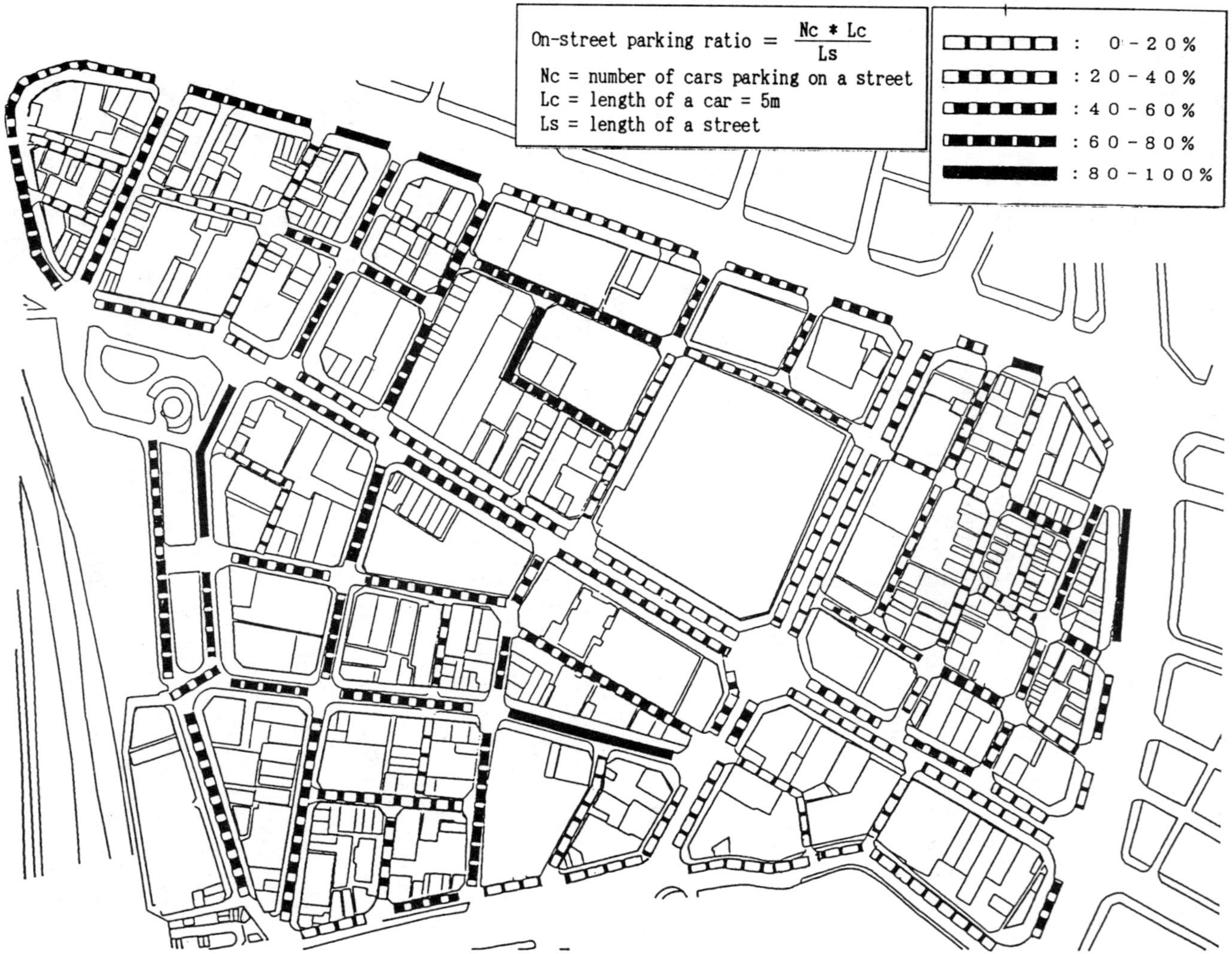

Figure 4.10. Streets with parking in Tokyo (Toshio Ojima, Katsuya Matsushita, and Shuichi Miura, "An Environmental Study of Underground Parking Lot Developments in Japan," Tunneling and Underground Technology, *Vol. 8, No. 1, 1993, pp. 65–74).*

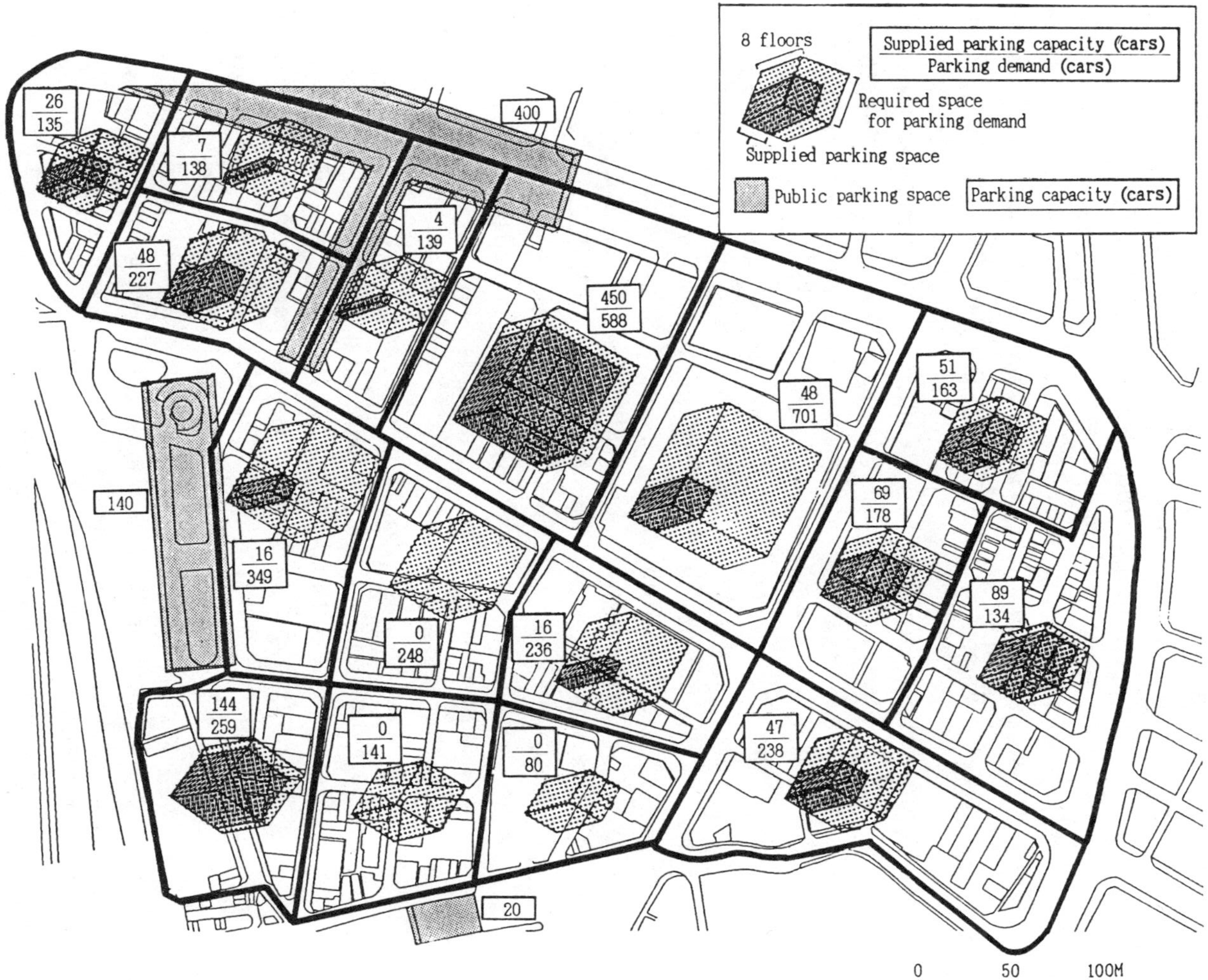

Figure 4.11. Parking demand in Tokyo (Toshio Ojima, Katsuya Matsushita, and Shuichi Miura, "An Environmental Study of Underground Parking Lot Developments in Japan," Tunneling and Underground Technology, *Vol. 8, No. 1, 1993, pp. 65–74).*

utilization of urban space to develop large aboveground parking lots, and the urban landscape would be sacrificed. Therefore, it is probably a better solution to develop underground parking lots.

The present number of parking lots in Japan is about 6,700, with a total parking capacity of about 950,000 cars. Of these, there are 417 lots with underground floors, with a parking capacity of 71,894 cars. A city parking lot is defined as a principal urban facility planned by the city. There are 311 existing city parking lots, with a parking capacity of about 68,000 cars (Fig. 4.12). There are 96 underground city lots, with a parking capacity of 28,080. City-planned parking lots are the special case that occupies underground public space, such as the space under roadways and parks. Many of the lots have a large parking capacity, since they were developed through deliberate urban planning and redevelopment. Also, many city underground lots are being planned in major cities, where the area for developing aboveground parking lots is extremely limited.

There are 36 city-planned parking lots in Tokyo, and their estimated parking capacity is 13,356. Of these, 28 are underground lots, with an estimated parking capacity of 9,698 cars. About 200 parking lots are planned according to the notification in Tokyo's 23 wards, with a capacity of about 28,000 cars.

The parking capacity of underground lots is 36.7 percent of their total capacity. Using both aboveground and underground lots, the parking capacity is 19.3 percent; that of the aboveground parking lot is 44.0 percent, with the underground floor capacity half of the total parking capacity. Figure 4.13 presents the parking capacity in each ward, with the percentage of three types of parking lots. In the center or subcenter of the metropolis, the parking capacity of underground space is more than its aboveground space.

Health

Environmental Pollution

Since there are few windows in the underground parking lot, ventilation cannot be obtained naturally. However, the emis-

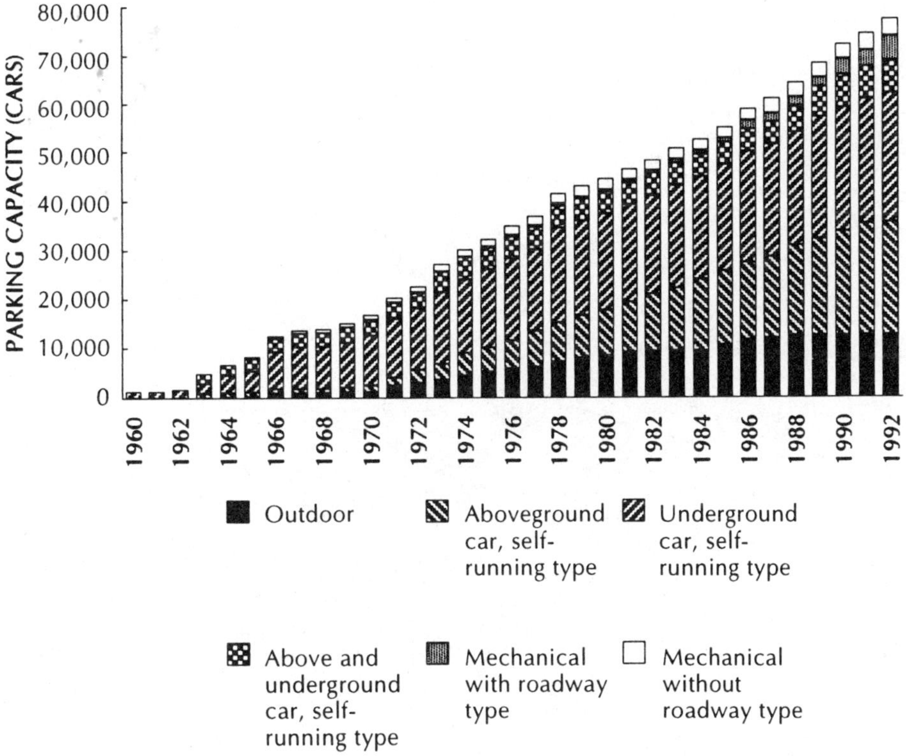

Figure 4.12. Increase in city-planned underground parking lots (Toshio Ojima and Shuichi Miura, "Study on Classification and Energy Consumption of Parking Lots," Journal of Architecture, Planning and Environmental Engineering *446, Tokyo, 1993, pp. 9–16).*

sions from motor vehicles is a great environmental pollution source in parking lots and must be controlled. Harmful elements contained in motor vehicle emissions are CO, CO_2, nitrogen oxides (NO_X), hydrocarbons (HCHO), airborne lead, and sulfur dioxide (SO_2), among others. It is well known that if the CO concentration were low enough to satisfy the allowable standard through dilution or ventilation, the density of other harmful elements would also generally fall below the thresholds standards set for safety. The allowable CO concentration for a working healthy condition is 100 ppm, but it is more desirable for health that the allowable concentration be reduced to 50 ppm. The regulation for the ventilation system is to cut the mean concentration per 8 hours to 20 ppm. On the other hand, the standard is only 10 ppm for a room that is not inhabited.

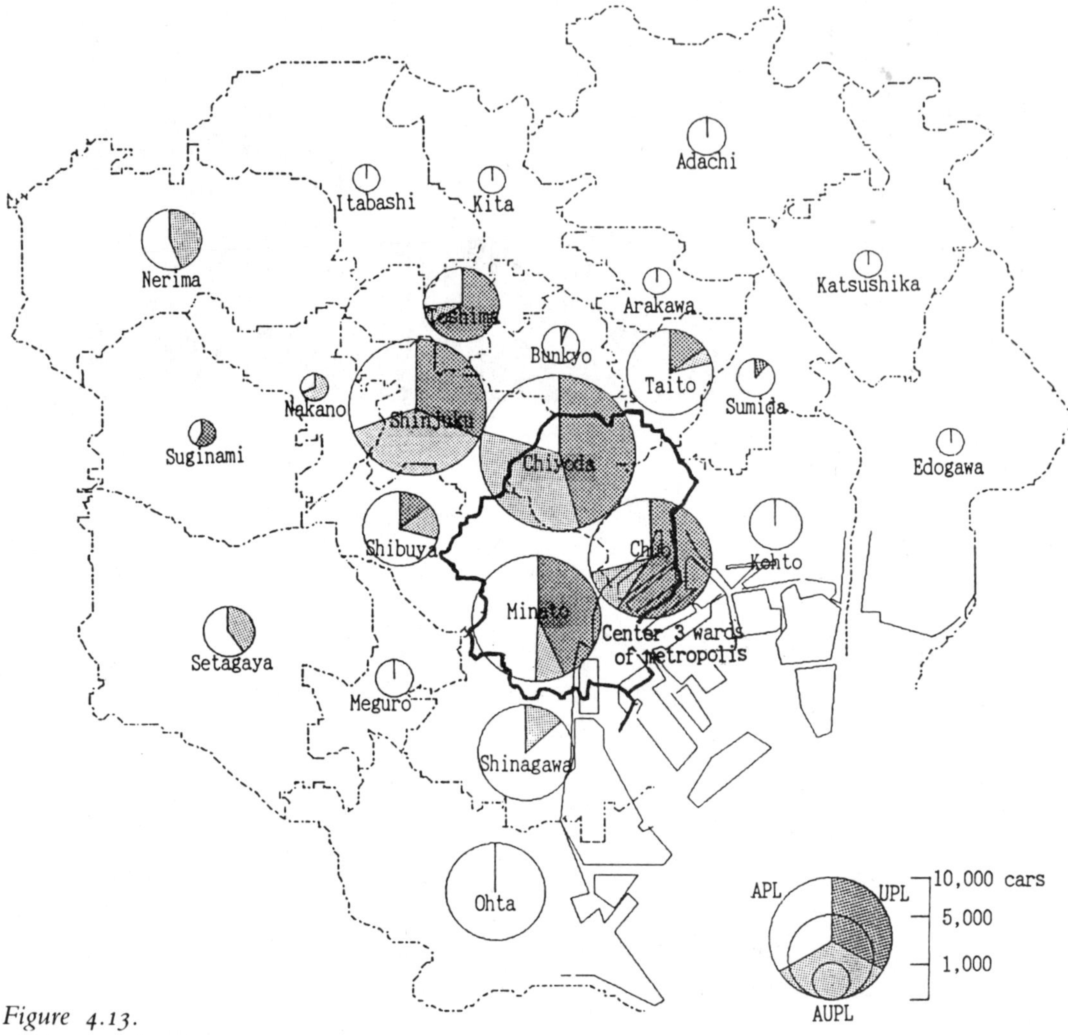

Figure 4.13. Underground parking lots in Tokyo (Toshio Ojima, Katsuya Matsushita, and Shuichi Miura, "An Environmental Study of Underground Parking Lot Development in Japan," Tunneling and Underground Technology, *Vol. 8, No. 1, 1993, pp. 65–74).*

Environmental Demand for Underground Parking Lots

The question is how or what kind of facilities could be built in urban underground space to meet the constraints of urban activity, climate, and conditions of the land. For the hot and humid climate of Tokyo, it seems unsuitable to develop a manned underground parking lot, except for some limited uses (Fig. 4.14). Moreover, effective or ineffective use of urban underground space will result in different levels of environmental pollution, depending on how heavy the traffic

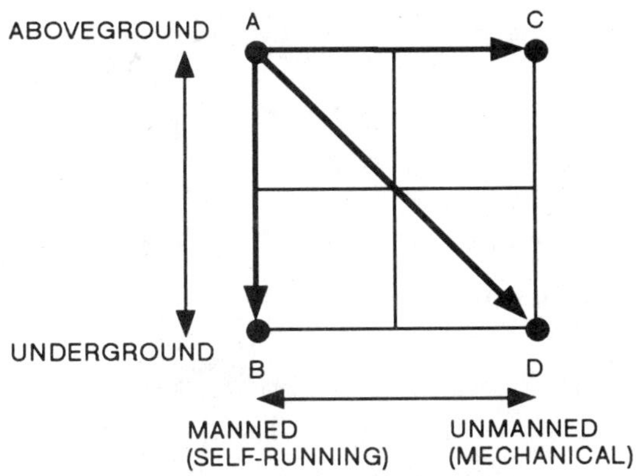

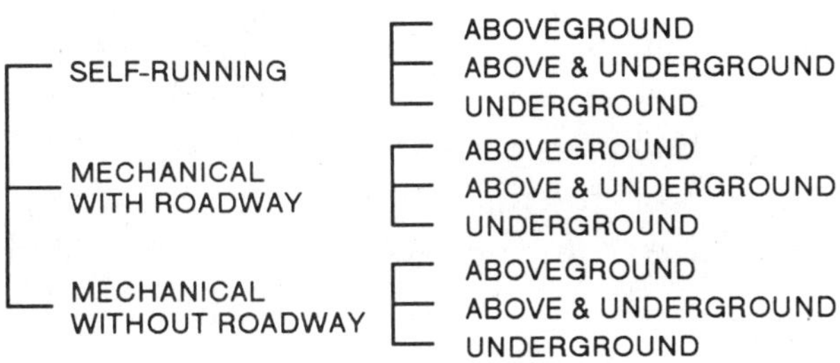

Figure 4.14. Classification of parking lots (Toshio Ojima and Shuichi Miura, "Study on Classification and Energy Consumption of Parking Lots," Journal of Architecture, Planning and Environmental Engineering *446, Tokyo, 1993, pp. 9–16).*

flow is. Because the load on the environment and safety in urban underground space is greater than it is aboveground, underground space facilities must maintain strict environmental and safety standards. But the efficiency of the underground facilities depends on management to a greater extent than that of aboveground facilities. Compared with the underground shopping center, for example, environmental standards are lower in the underground parking lot because human activity there is limited by the number of people and time spent underground.

Efficiency

Construction Cost

Most large buildings have basements to support the structural foundation that also serves as a parking lot. However, these parking spaces provide little income compared with general commercial parking facilities. Also, it is difficult to independently construct parking lots underground, since construction costs covered independently are very high. Most underground parking lots are constructed and financed through local governments.

A comparison of construction costs for different types of parking lot construction is shown in Fig. 4.15. Underground construction cost per car is 15 million yen, or four times more than construction costs for aboveground parking lots. This expense can be reduced by building the lot as a part of other facilities, such as an underground shopping center or a subway.

Energy Consumption

Operating the equipment needed to maintain a certain environmental quality requires a significant amount of energy. Underground buildings need not only a large initial financial investment for construction and equipment, but also a large operating cost as compared with aboveground buildings.

Figure 4.15. Types of parking lots and construction costs.

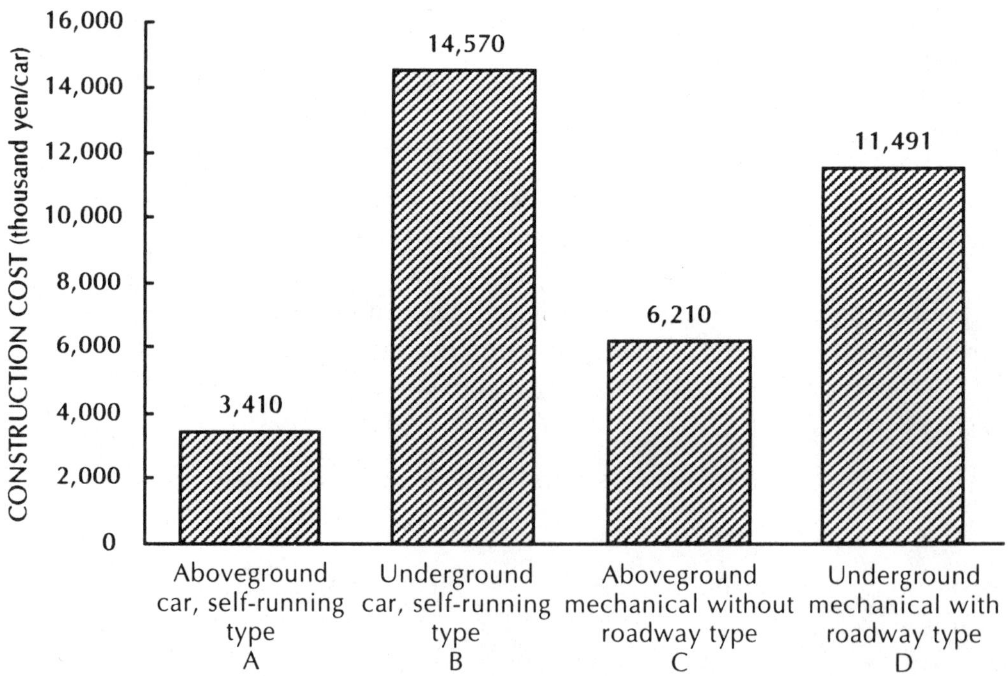

Moreover, underground structures must be planned after considering long-term maintenance, because underground structures are not easily modified and renewal is difficult. The required underground depth for parking lots is lower than that required for underground shopping centers. Also, extensive cooling and heating equipment is not necessary, and ventilation is not the principal source of energy consumption. In comparing underground parking lots with aboveground lots, however, the average annual consumption of electric power per square meter of floor area is still larger than that of aboveground parking lots (Fig. 4.16).

Comparisons were made between the average annual total energy consumption in underground facilities and energy consumption in aboveground structures (Fig. 4.17). The average annual total energy consumption for underground parking lots is about one-eighth that of underground shopping centers. And when comparing figures for aboveground buildings, the energy use of the underground parking lot is only about one-half that of offices and about one-third that of de-

Figure 4.16. Electric consumption of aboveground and underground parking lots (Toshio Ojima and Shuichi Miura, "Study on Classification and Energy Consumption of Parking Lots," Journal of Architecture, Planning and Environmental Engineering, *446, Tokyo, 1993, pp. 9–16).*

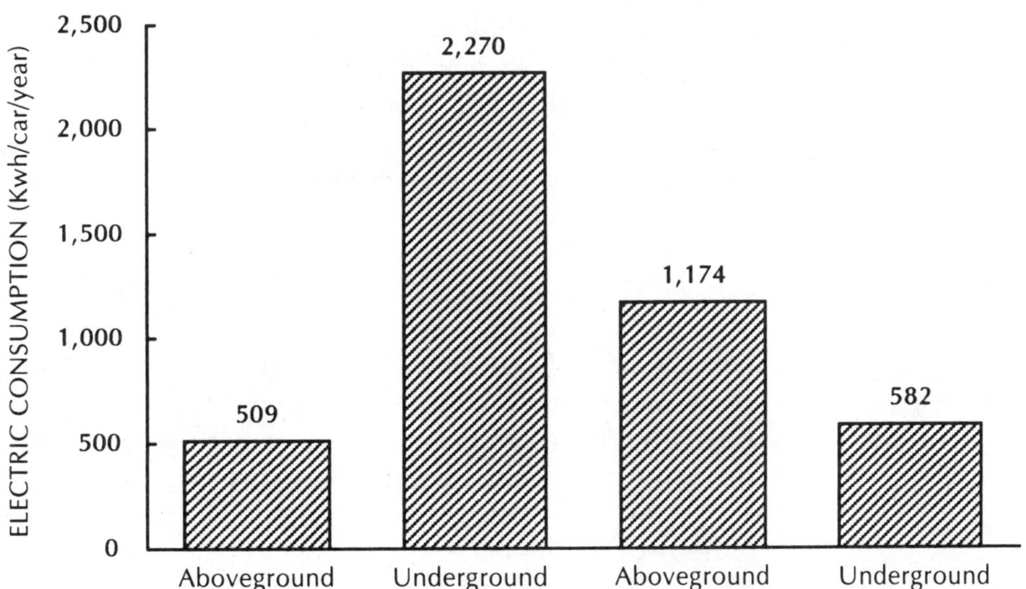

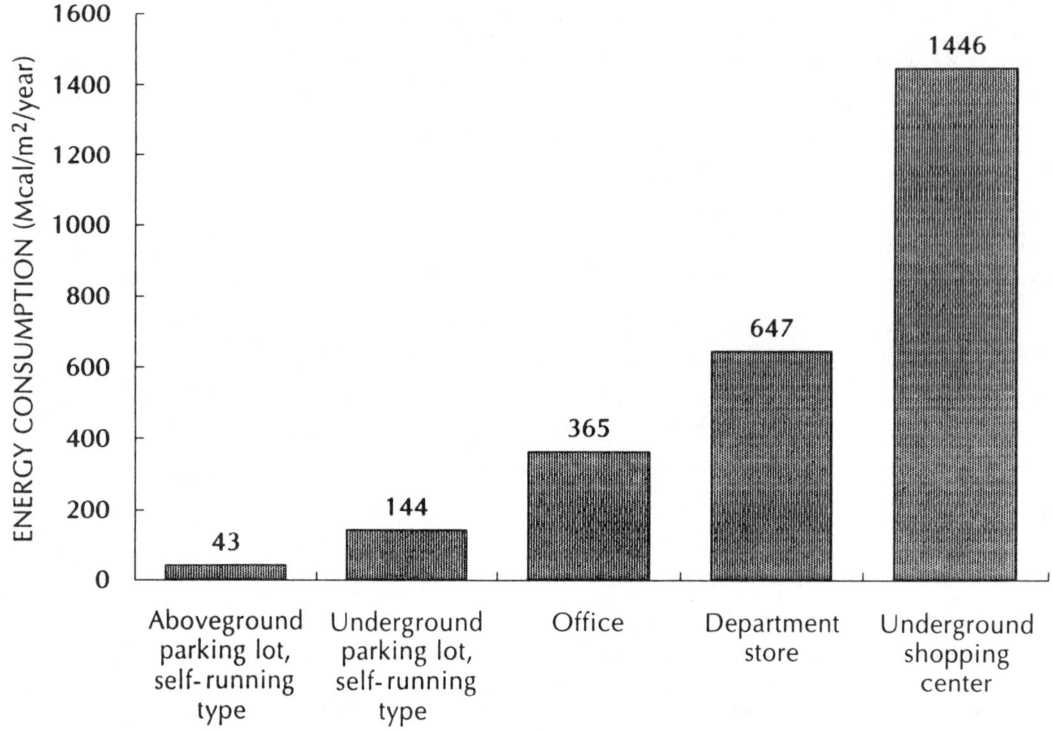

Figure 4.17. Comparison of energy consumption between underground and aboveground facilities (Toshio Ojima and Shuichi Miura, "Study on Classification and Energy Consumption of Parking Lots," Journal of Architecture, Planning and Environmental Engineering, *446, Tokyo, 1993, pp. 9–16).*

partment stores. While energy consumption of underground parking lots is 2.8 times greater than that of aboveground lots, this percentage is similar to that of the underground shopping centers, whose energy use is 2.8 times as great as that of aboveground department stores. It is obvious that a comfortable environment cannot be created without special equipment for ventilation and fire control. In underground parking lots, ventilation equipment should be capable of circulating more than 10 air changes per hour to maintain a satisfactory quality of air and temperature. Also, more equipment for fire prevention must be introduced into underground parking lots than in aboveground parking lots, because of the difficulty in controlling a fire in an underground area. This is particularly true in parking lots, since the rapid burning of automobiles and gasoline explosions and fires create large quantities of smoke. Mandatory firefighting equipment, therefore, would include foam heads, indoor fire hydrants, fire extinguishers, and automatic fire alarms.

Effect of Introducing Parking Equipment to the Parking Lot

There are two types of parking equipment used to maintain both aboveground and belowground parking lots: manned (self-running), and unmanned (mechanical). Mechanical parking lots require fewer roadways and ramps than do self-running ones because traffic is controlled with mechanized parking equipment. Consequently, space utilization is most efficient with the use of mechanical lots. However, a more accurate analysis of parking alternatives weighs both long- and short-term costs with regard to operational expenses, such as ventilation and maintenance, as well as land prices.

Finally, aesthetic and civic considerations must be factored into the cost analysis of parking lot construction. For example, if more gateways are used to the parking lots above ground, they may become obstacles for pedestrians and spoil the landscape. On the other hand, setting underground parking lots under a private site is difficult, since the planned placement of the building is done by the official organizations. Thus, joint public-private support is needed to design the parking lot with the building placement to use the public space effectively. Building officials and the public should contribute ideas to the plans for parking lots for the whole area (Figs. 4.18 and 4.19).

From the research of on-street parking in Shinjuku, 70 percent of truck use was during the peak time. At present, only large commercial facilities, such as department stores, have their own truck-loading facility in their building. The Shinjuku research showed that, to resolve the problem of on-street parking, it is important to develop special facilities for loading and unloading trucks. The time required to load and unload is usually short, but cargo can be cumbersome. This means that it is unrealistic to use a parking lot far from the loading dock. In order to park near their destinations, most truckers park on the street in front of the building.

Basically, the responsibility of designating loading spaces should be arranged by each building's management from the viewpoint of the building's needs. But in the existing city, it

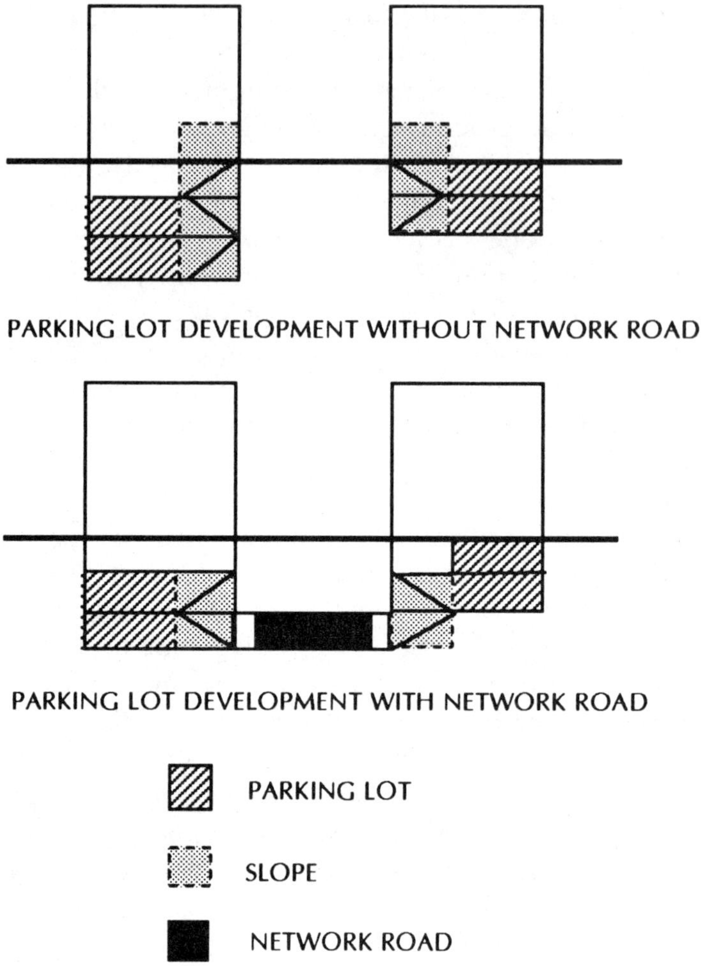

Figure 4.18. Concepts for underground parking lots (Toshio Ojima et al., "Study on Underground Utilization Plan of High-Density Area," Summaries of Technical Papers of Annual Meeting, Environmental Engineering, *Architectural Institute of Japan, Tokyo, 1992, pp. 1155–66).*

is difficult, if not impossible, to obtain surface space along the road near the building site. Even if such space were possible to obtain, the problem would still remain of mixed circulation of pedestrians with the confusion of loading trucks. In addition, as these trucks' loading facilities are different from the general parking lot, it would be difficult for small buildings to develop space for loading trucks privately. Because of these reasons, it might be more effective to designate space for truck loading and unloading underground.

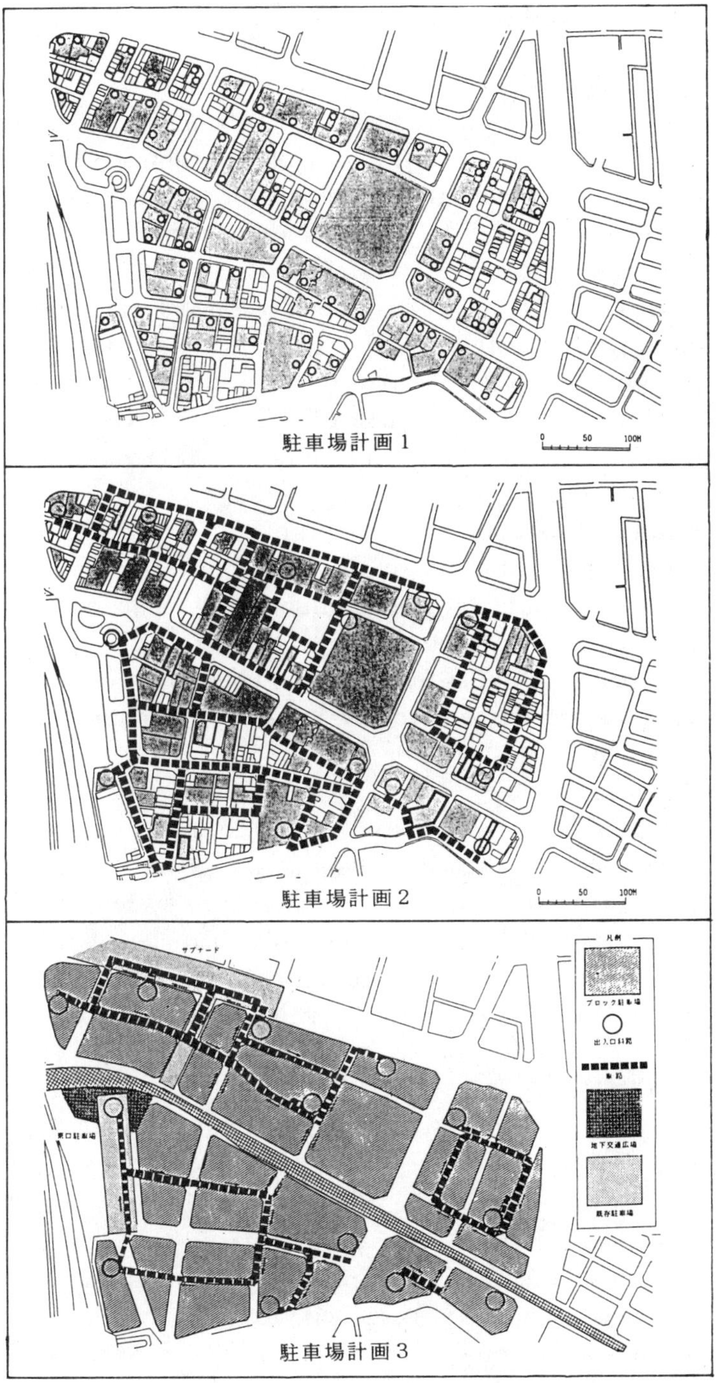

Figure 4.19. Planning of underground parking lots (Toshio Ojima et al., "Study on Underground Utilization Plan of High-Density Area," Summaries of Technical Papers of Annual Meeting, Environmental Engineering, *Architectural Institute of Japan, Tokyo, 1992, pp. 1155–66).*

UNDERGROUND ROADWAYS

Tunnels have been developed around cities to protect the landscape from disfigurement. These tunnels pass mainly through mountainous districts, while sea tunnels pass under coastlines. However, the necessity for developing underground roadways has recently been recognized.

For example, to facilitate road crossing in built-up areas, underground driveways are being planned from the viewpoint of traffic treatment and environmental conservation. It is also possible to plan a crossing of a railway by an underground driveway. In the case of crossing a river used as a route for ships, an underwater tunnel could be built in addition to a bridge.

A full underground roadway is not always necessary. Half-underground roadways are also effective for solving some environmental problems in neighborhoods. Since there is no roof to a half-underground driveway, there is the advantage of dispersing gas fumes by natural air circulation as compared to a full underground driveway, which would require electrical ventilation. Also, there are added advantages for safety and comfort.

The facilities required for a long-distance underground roadway are road space, a ventilation system, ventilation shafts, lighting systems, and an emergency system. Usually the road has two lanes, with an inner diameter of about 10 meters. (The longest underground driveway in the world is Gotthard Tunnel in Switzerland, which is 16,322 meters long.)

Examples

Boston

Recently, in Boston, many projects have been planned to give a rebirth to the city. Some of the proposals will use underground space efficiently. The Central Artery Project is particularly worth mentioning. Central Artery was an elevated speedway constructed in 1954, with its path crossing through the center of Boston and running from north to south (Fig. 4.20). But lately, this highway has become one of the most congested roads in the United States. In an attempt to resolve

Figure 4.20. Central Artery Project in Boston (Yuzo Onishi, "Urban Underground Utilization Plan of Boston," Journal of Underground Space Utilization No. 7, *Urban Underground Space Center of Japan, Tokyo, 1990, pp. 28–35).*

this condition, a plan was presented that would replace the old elevated speedway with an underground roadway, and another new highway would be built under the sea. After the elevated highway has been removed, the land that is left, about 80,000 square meters, will be redeveloped by private and public agencies for recreation, office space, residential areas, parking, commercial facilities, and open space. Construction was begun on this project in 1989 and will be completed in 1998, and will cost $3.1 billion. The idea behind the project is that the waterfront of Boston Harbor is rightfully the property of the citizens, and the passage linked to the bay must ensure that people can walk freely.

Tokyo

During the Edo age, there were many canals in Tokyo that were used for transportation. These canals were surrounded by a beautiful landscape. One of the bridges on this canal was Nihon Bashi Bridge, which was the starting point of all roads in Japan. Now, this historical bridge is entirely covered by a speedway (Fig. 4.21). During the period of high economic growth, development of the speedway was an urgent project in Japan. The space above the river and canal was one of the easiest and most economical spaces to obtain. In the same way, more than half of the canals and rivers in Tokyo have disappeared.

The Tokyo speedway is designed in a radial manner. But the loop road has not yet been constructed completely. Therefore, the network system of the speedway is not efficient, and many parts of it routinely cause traffic delays. The speedway loop line was planned to solve this condition. In the beginning, it was to be an elevated speedway. But the citizens objected to this, so only the eastern part of the elevated speedway has been constructed. The Shinjuku central loop line could not be constructed before now (Fig. 4.22). Recently, this line was designed to contain underground passages. It is located 7 to 8 km from the heart of Tokyo. Its total length is about 10 km, 8.7 km of which are planned as an underground structure. It will be the longest underground road passing through the city. Also, 3.3 km of this line are planned for construction together with the subway as one unit. In the future, this underground speedway will be extended to 20 km. On the surface, the present road of this loop is only 22 meters wide. It is planned to be extended to 40 meters in width, which will include a 9-meter center divider and a 5-meter pedestrian space on each side. Such a road has not yet been built in Tokyo.

Additional Underground Roadway Issues

Problems of Underground Roadways

An automobile with a full tank of gasoline becomes dangerous in an accident. If such an accident occurs in an underground roadway, it would be more serious than an accident occurring above ground. Not only would the other automobiles in the tunnel suffer damage, but the utility pipelines

Figure 4.21. Metropolitan speedway on the river (Toshio Ojima, The Reconstruction of Tokyo, *Chikuma Shobo, Tokyo, 1986, p. 11).*

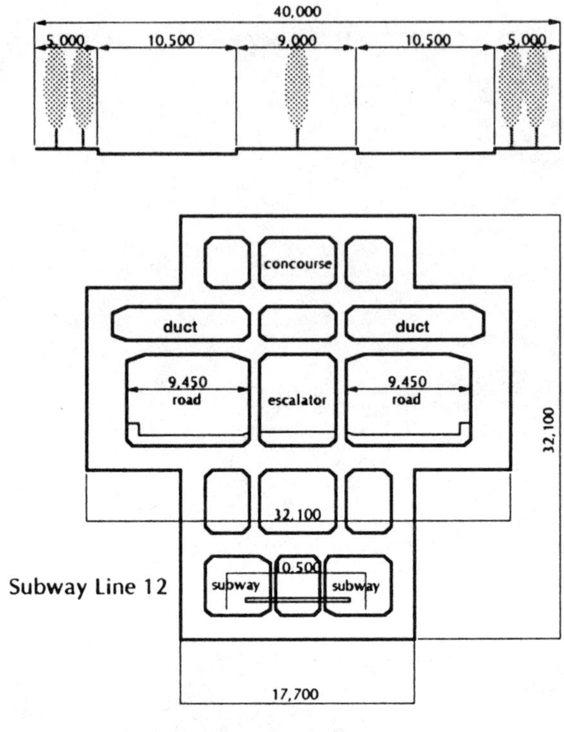

Figure 4.22. Shinjuku central loop line of speedway (Kou Horie, "Present and Future of Underground Utilization in Tokyo," Journal of Underground Space Utilization No. 8, *Urban Underground Space Center of Japan, Tokyo, 1990, pp. 4–17).*

and the buildings along the road would also be harmed. If the drainage system sustained damage, the underground road would be flooded. For controlling these disasters in long underground roadways, ventilation systems, information systems, escape facilities, firefighting systems, and other measures become critical in the planning.

Ventilating exhaust fumes from automobiles and replacing it with fresh air is a major problem in the underground super-highway of the Central Loop Line, now in the planning stage. It is being proposed to build an exhaust tower 2 meters wide, 10 meters deep, and about 50 meters high in order to blow the exhaust fumes aboveground. This tower will gather contaminants and discharge them to the outside air. If an exhaust tower would be too short to release the fumes above the district, however, the fumes would contaminate the immediate area. This would be particularly harmful on a road that passes through densely populated districts. In order to avoid such a situation, a high exhaust tower that would diffuse the gas fumes far above the road is needed. However, this kind of tower would probably be out of harmony with the surrounding streets. Many of these problems can be solved only by a vehicle that does not spread pollutants, such as an electric automobile.

Underground Distribution System

Traffic can be separated into the transportation of people and that of commodities. Although Japan is a country that relies on mass public transportation to a great extent, commodity distribution by truck is becoming greater every year. Especially in metropolitan areas, trucks are depended on for almost 90 percent of commodity distribution. However, trucks cause noise and air pollution in cities. They also contribute to the congestion of Japan's already overcrowded roadways, significantly slowing the distribution of commodities. Thus, a more efficient mode of transportation needs to be developed. Because the limits of aboveground roadways have already been reached, geo-space provides alternative venues for commodities transport exploration.

An underground distribution system can play an important role in solving this transportation problem. Although safety and health problems may develop when people use underground transportation for long distances, fewer problems exist for transporting commodities underground, since most commodities do not require a high environmental quality. If an underground distribution system is used to reduce the quantity of goods being transported by trucks, the number of trucks above ground will decrease, as will the gas fumes they produce. Also, if gas mileage would be improved, a further decrease in gas fumes would result. And, equally important, aboveground traffic congestion would be eased.

CONCLUSION

Transportation is one of the biggest problems of the urban area. It can increasingly be said that no space remains aboveground for transportation facilities. Until now, we have been using underground space primarily for subways. But a large amount of cooling energy is consumed to compensate for wasted heat from the train. In addition, underground space near the surface has already been used to its practical limit; however, if deep underground space were to be used for subways, many more safety and health problems will occur.

Similar difficulties arise with motor transportation. There are two types of transportation: transportation of passengers and of commodities. Underground space is now used mainly for the former, but in the future it should also be used for commodity transportation, which does not pose the same safety and health problems as passenger transportation.

SELECTED READINGS

The All Japan Society of Parking, *Annual Report on Parking Lot,* Tokyo, 1990 (written in Japanese).

Committee of Editing Handbook of Underground Structure, *Handbook of Underground Structure,* Construction Industry Research Association, Tokyo, 1987 (written in Japanese).

Tadashi Kinbara, Shuichi Okumura, Eishi Yaze, and Keihiro Saito, *Planning and Design of Parking Lot,* Kajima Publication, Tokyo, 1978 (written in Japanese).

Toshio Ojima, *The Reconstruction of Tokyo,* Chikuma Shobo, Tokyo, 1986 (written in Japanese).

Toshio Ojima et al., "Research for Environment of Underground Parking Lot in Tokyo," *Summaries of Technical Papers of Annual Meeting, Environmental Engineering,* Architectural Institute of Japan, Tokyo, 1989, pp. 1351–52 (written in Japanese).

Toshio Ojima et al., "Research on Workers' Conscious to Underground Parking Lot by Questionnaire," *Summaries of Technical Papers of Annual Meeting, Environmental Engineering,* Architectural Institute of Japan, Tokyo, 1989, pp. 1349–50 (written in Japanese).

Toshio Ojima et al., "Thermal Survey on Subway Concourse," *Summaries of Technical Papers of Annual Meeting, Environmental Engineering,* Architectural Institute of Japan, Tokyo, 1990, pp. 925–26 (written in Japanese).

Toshio Ojima et al., "Research on Effect of Introducing Parking Equipment to Underground Parking Lot," *Summaries of Technical Papers of Annual Meeting, Environmental Engineering,* Architectural Institute of Japan, Tokyo, 1991, pp. 1383–84 (written in Japanese).

Toshio Ojima et al., "Research on Facility Management of Underground

Parking Lot in Japan," *Summaries of Technical Papers of Annual Meeting, Environmental Engineering,* Architectural Institute of Japan, Tokyo, 1991, pp. 1381–82 (written in Japanese).

Toshio Ojima et al., "Study on Underground Utilization Plan of High Density Area," *Summaries of Technical Papers of Annual Meeting, Environmental Engineering,* Architectural Institute of Japan, Tokyo, 1992, pp. 1155–66 (written in Japanese).

Toshio Ojima, Katsuya Matsushita, and Shuichi Miura, "An Environmental Study of Underground Parking Lot Developments in Japan," *Tunneling and Underground Technology,* Vol. 8, No. 1, Tokyo, 1993, pp. 65–74 (written in English).

Toshio Ojima and Shuichi Miura, "Study on Classification and Energy Consumption of Parking Lots," *Journal of Architecture, Planning and Environmental Engineering,* 446, Tokyo, 1993, pp. 9–16 (written in Japanese).

Yuzo Onishi, "Urban Underground Utilization Plan of Boston," *Journal of Underground Space Utilization,* No. 7, Urban Underground Space Center of Japan, Tokyo, 1990, pp. 28–35 (written in Japanese).

Ryouichi Shimojima, "The Present and the Direction of the Underground Utilization," *Symposium of Underground Space Utilization,* Civil Engineering Society, Tokyo, 1989 (written in Japanese).

Teito Rapid Transit Authority, *'88 Eidan Subway Handbook,* Tokyo, 1988 (written in Japanese).

Underground Space Center of Japan, *Focusing on the Organic Combination of Aboveground Space and Underground Space,* Tokyo, 1991 (written in Japanese).

Underground Space Center of Japan, *Designing Underground City,* Daiichi Houki, Tokyo, 1991 (written in Japanese).

Tadaaki Zouga et al., "Analysis of Thermal Environment in Subway," *The Society of Heating, Air-Conditioning and Sanitary Engineers of Japan,* Vol. 47, No. 6, Tokyo, 1973 (written in Japanese).

5

JAPANESE GEO-SPACE INFRASTRUCTURAL SYSTEM

T. Ojima

THE PRESENT AND FUTURE OF INFRASTRUCTURE

In general, the infrastructure consists of underground facilities along with their accompanying support facilities. Complementary to the infrastructure is the suprastructure, the facilities located above ground that are supported by their belowground infrastructure. Physically, the pipeline and the subway are the underground infrastructure, while the buildings on the ground level form the suprastructure. Sometimes, the visible structure above ground together with its supporting belowground facilities is called the infrastructure.

Another concept is to look at the suprastructure as the house and office. In this case, the infrastructure becomes the facilities that support daily life, such as schools, parks, or museums.

Specifically, then, what is the infrastructure? For this discussion, infrastructure will encompass the main service systems, such as water supply and sewage, transportation, energy, and telecommunications. Its life maintenance function will support the house or office and daily life, especially for emergency situations; this will be referred to as the life rope or lifeline.

Infrastructure of Japan

The infrastructure of Japan and its major city, Tokyo, is not as modern as that found in other developed countries in the world. Transportation, along with urban supply and dispersal facilities such as electricity, gas, water supply, sewage disposal, telecommunications, garbage collection, and district heating and cooling, are estimated to be 30 years or more behind the development of the infrastructure of various cities in Europe and the United States. Also, Japan's creation of parks, ratio of buried to aboveground cable, road construction, and housing have lagged in development (Fig. 5.1). Some of this delay can be attributed to the lack of uniform internationalization of urban homes, parks, museums, hotels, international conference centers, hospitals, and schools. However, Japan's development of its water supply and telephone infrastructure compares favorably with that of Western countries, although it is becoming more difficult in Japan to acquire land to build the infrastructure. This is due partly to the high price of land and to the difficulty in negotiating this use for the land with the landowner. In addition, Japan is below Western standards in the amount of public space per person used to support urban activities (Table 5.1).

To understand why the infrastructure of Japan has lagged in development, the reader first must consider the circumstances out of which Japan's infrastructure has emerged. Second, the reader must analyze the economic base by which Japan's infrastructure is supported.

Figure 5.1. (facing page) Comparison of the development of infrastructures in Japan and worldwide (Toshio Ojima, Infrastructure of Japan, *Nikkan Industry Newspaper, Tokyo, 1983, p. 10).*

After the restoration during the Meiji era, the importance of the lower or underground infrastructure was underestimated, perhaps because of the urgency of building the upper suprastructure to catch up with other countries. Japan's policy of a rich country and strong soldiers forced Japan to become abundant in production and strong in a visible way. A noticeable suprastructure was constructed to pursue that policy. However, the infrastructure, or that which is under-

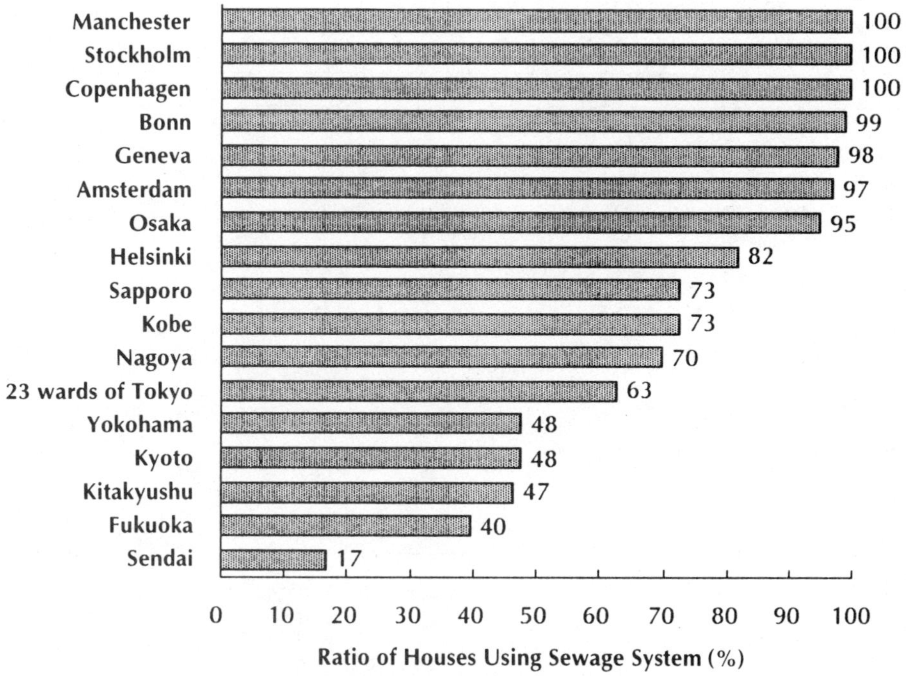
Manchester 100
Stockholm 100
Copenhagen 100
Bonn 99
Geneva 98
Amsterdam 97
Osaka 95
Helsinki 82
Sapporo 73
Kobe 73
Nagoya 70
23 wards of Tokyo 63
Yokohama 48
Kyoto 48
Kitakyushu 47
Fukuoka 40
Sendai 17
0 10 20 30 40 50 60 70 80 90 100
Ratio of Houses Using Sewage System (%)

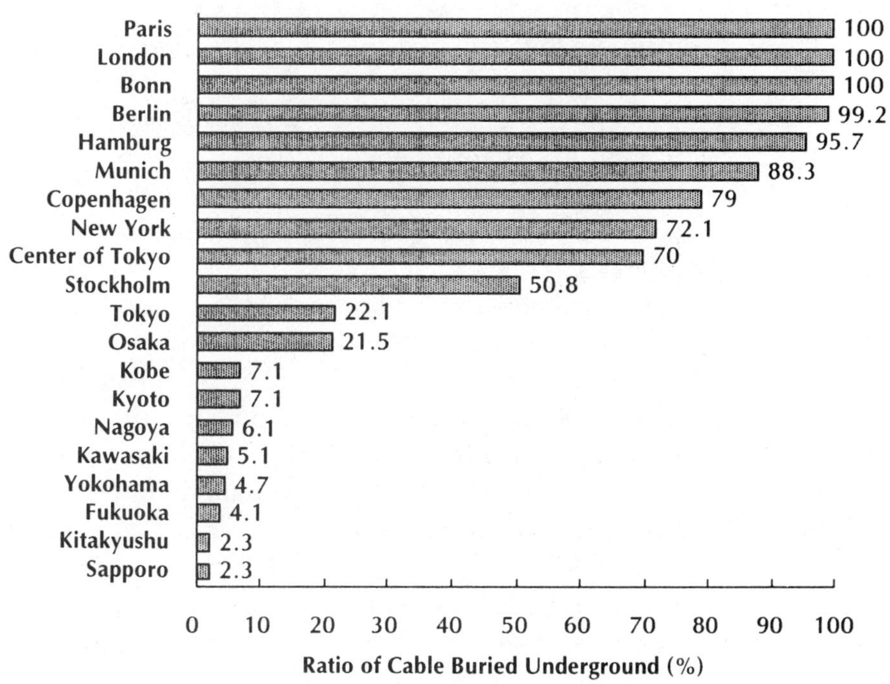
Paris 100
London 100
Bonn 100
Berlin 99.2
Hamburg 95.7
Munich 88.3
Copenhagen 79
New York 72.1
Center of Tokyo 70
Stockholm 50.8
Tokyo 22.1
Osaka 21.5
Kobe 7.1
Kyoto 7.1
Nagoya 6.1
Kawasaki 5.1
Yokohama 4.7
Fukuoka 4.1
Kitakyushu 2.3
Sapporo 2.3
0 10 20 30 40 50 60 70 80 90 100
Ratio of Cable Buried Underground (%)

TABLE 5.1.
Population and Public Space Use in Major World Cities

		TOKYO	OSAKA	NEW YORK	LONDON	PARIS	TOTAL
Area	(hectares)	59,973	21,308	78,658	157,900	10,540	328,379
Population	(10,000 people)	835	264	707	675	230	2,711
Public Area	Total (hectares)	11,226	4,475	33,893	35,412	5,084	90,090
	Road (hectares)	8,751	3,742	23,600	18,201	2,252	56,546
	Park (hectares)	2,475	733	10,293	17,211	2,832	33,544
Public Area Per Person (m^2/people)	Total	13.4	17	47.9	52.5	22.1	33.2
	Road	10.5	14.2	33.4	27	9.8	20.9
	Park	2.9	2.8	14.5	25.5	12.3	12.3
Population Density	(people/hectare)	139	124	90	43	218	83
Ratio of Public Area to Total	(%)	19	21	43	22	48	27

Japan Development Bank, *Direction of Deep Underground Utilization,* Tokyo, 1989, p. 86.

ground, was disregarded from the earliest time as being unimportant and was excluded from the overall process of planning and building the actual city. This attitude has continued even until today in creating the modern Japanese city. Therefore, the infrastructure was not heavily financed, even though it was for public use. Today, it is proposed that the network be built and maintained through a tax base supported by all the people through public businesses or public interests.

Certain ethics should be upheld in this aspect of city planning to maintain balance. Tokyo has become a huge, concentrated city, where an unprecedented number of people live and work together in factories built on the suprastructure. However, facilities to support these aboveground structures are extremely poor. The city, rebuilt out of the rubble of war, had been constructed as a labor-intensive city close to factories. It served as a barracks-type city and did not have a functionally balanced city structure.

As an analogy, a massive tree endures both the strong sunshine in drought and the heavy wind of a typhoon in the rainy season because its underground stalk and roots are firm and expand deeply into the soil to the underground water system. Its well-established root network is balanced with its leaf and trunk structure above the ground. This analogy can also apply to the ideal configuration of the city. A well-balanced, comprehensive city is not a city whose only constructions are well-served parks or roads, but one with a solid infrastructure to support it.

Development of Infrastructural Systems

Water Supply

The use of water and the water supply system is said to be a measure of the developmental stage of a civilization. Most urban city water supplies depend on a source out of the immediate city area. Construction of this source of water supply is not only expensive, but also destroys the natural water system. Therefore, an alternative, economical water supply source must be sought. One of the countermeasures Japan is using is development of a treated water supply through using reclaimed water for inhabitants living near the water source.

The concept of laying pipe to supply water was ahead of other infrastructural facilities in development. Most of this pipe was laid in shallow spaces under the roads. Although today, water purification plants are being built to resist damage by earthquakes, water pipes are, for the most part, very old and would suffer damage during an earthquake. This damage would cut off the water supply to a wide area. Therefore, it is important to back up the current water distribution system by strengthening it.

Sewage System

The sewage systems of Japan also lag behind sewage systems of the West, but Japan's quality of sewage treatment is high. As sewage flows naturally by gravity, the greater the distance that sewage must travel, the deeper the sewers must be located. When there were fewer underground facilities, sewers did not have to be located so deep. But recently, with many deeper facilities being built, such as subways and underground shopping centers, sewers have to be located at a lower level to accommodate these facilities. Also, treatment plants and pumping stations have to be located deeper, to a depth of about 60 meters.

There are two types of sewage treatment systems. One is a separate flowing system, to collect sewage and rain separately. The other is a combined system, to collect sewage and rain runoff together. Most large cities in Japan are adopting the latter system. But this combined system discharges the sewage without treatment in the event of heavy rain because the treatment capacity becomes overloaded. In cities, most surfaces are covered by concrete and other materials that cannot absorb rainwater. Especially in urban areas, where roads are paved, the runoff goes into the sewage system quickly. While there is no runoff into the river on a dry, sunny day, the runoff overtaxes the capacity of sewage treatment plants on rainy days. The rain runoff and the untreated sewage are not treated properly and flow into the river, causing pollution. In addition, the river level rises, causing floods (Fig. 5.2). To solve this problem, the construction of underground rivers is now being considered. In Tokyo, an underground river 12.5 meters in diameter and 4.5 km in length is now under construction (Fig. 5.3). Its estimated cost is about $1 billion.

Japanese land is dotted with large sewage disposal sites. In many cities, the parks, public squares for relaxation, tennis courts, citizens' halls, and gymnasiums have been constructed above the sewage disposal plants. Because of this proximity to the plants, citizens have the opportunity to develop an understanding of sewage systems and the need for them.

Treated Water Supply

A treated water system reuses circulated wastewater to reduce water and sewage. Today, one-quarter of the water used in Tokyo is nonpotable water for flushing toilets, car washes, sprinkling grounds, washing and bathing, and water cooling towers. These uses do not demand as high a quality of water as does drinking water. However, the same high standard of purification is applied to nonpotable water. The wastewater

Figure 5.2. Flood area (Tokyo Metropolitan Government, Master Plan of Second Generation Sewerage, *Tokyo, 1991, p. 53).*

is treated to be of high quality, but is still used as nonpotable water. In office buildings, it is mainly used for toilets and air conditioners, and in residences, it is mainly used for toilets.

There are three types of water supply systems—one that supplies a building, one that supplies a district such as a large redevelopment area, and one that supplies a broader area. Treated water is a very effective means of saving water. It reduces the fresh water demand by 25 percent. But treated water also requires the expense of double piping: one pipe for pumping common water and the other pipe for pumping

Figure 5.3. Underground river (Tokyo Metropolitan Government, Underground River, *Tokyo).*

treated water. The operating cost and space for this equipment become a great burden for the user, compared with a common water supply system. Because of this, only 0.4 percent of the total volume of water supplied is treated. To promote the use of treated water, a wide area network is required.

Electrical Supply

Electricity is a form of energy that is very easy to use, is safe, and is clean. It is predicted that the demand for electricity will increase as the use of air conditioners, computers, and other electric products becomes widespread. But problems exist for supplying this increased demand for electricity. One difficulty is in constructing new power stations in spite of the environmental movement against them. Second, the peak demand is becoming acute. From the expanded use of air conditioners, the peak demand in summer is higher and, therefore, the capacity of power generation is reaching its upper limit. Third, there is the problem of space for transmission facilities.

Recently, with the increasing demand for electricity, the capacity of power transmission has been extended fully. If electric transmission lines were to be constructed above ground, many more transmission towers and lines would be required. This would pose a problem for land acquisition, execution of work, and spoiling the landscape over people's heads. Therefore, underground transmission lines are being adopted for urban areas, but final power distribution is still above ground because of its easier accessibility to the end user.

To supply the electricity that is transmitted from the city to the user, a substation with decreased voltage is needed. Many substations are already constructed underground, saving the space above ground for buildings and more effective use of land. Thirty percent of the substations in Tokyo are constructed underground. One substation is even located under a traditional temple (Fig. 5.4). This particular substation has 7 floors underground that reach 36.4 meters in depth.

Figure 5.4. Underground electric transformer (Tokyo Electric Company, Takanawa Electric Transformer, *Tokyo).*

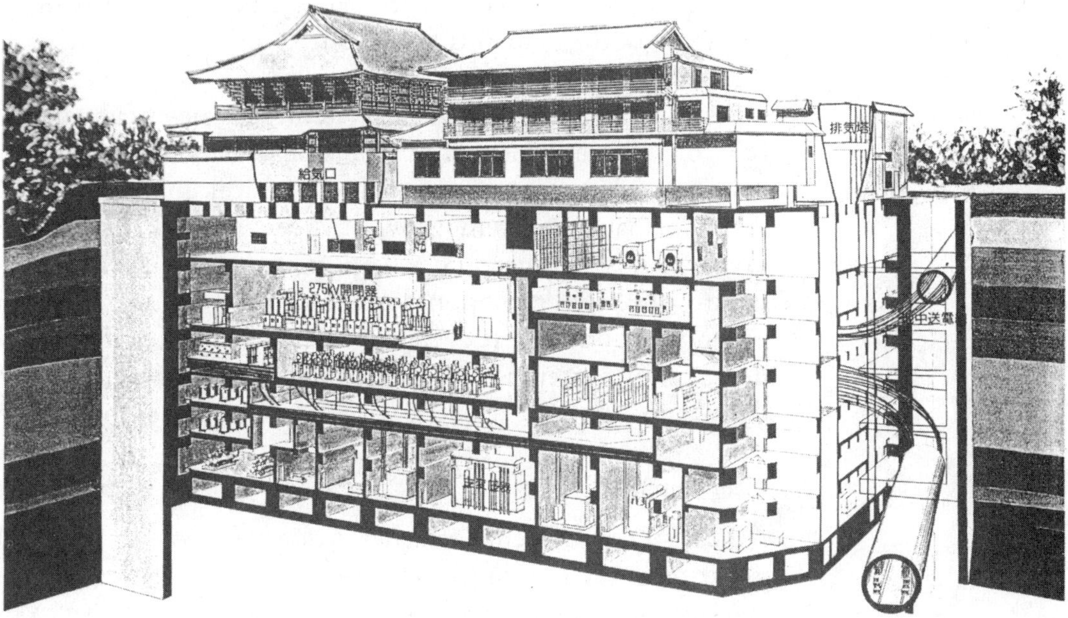

City Gas Supply

Gas is easier to use in the city and is cleaner than fuel oil. Previously, city gas had been produced mostly from coal, but recently, liquid natural gas (LNG) is being used. LNG has double the heat output of the usual city gas. Therefore, it could be possible to increase the supply capacity without changing the gas pipelines. Also, LNG emits few air polluting substances and is an efficient source of energy that can be used in place of oil.

The gas pipelines are buried under the road except in special cases, such as crossing a river. The maximum diameter of the pipeline is about 1 meter. Since this does not include maintenance space, however, repairs have to be carried out by digging up the road.

Information Transmission

It is said that we are in an information age, where the transmission of information is a vital function. The information system can be considered the brain and the nerves of the city. Although the basic media of broadcasting and telephone transmission have become national, it is becoming difficult to satisfy all of the demands for advanced information functions for the city.

As the current information system becomes national, the lack of countermeasures available in the event of sudden damage to the system becomes more critical. In the future, as the information system grows larger, this lack of disaster preparedness will become a weak point for the society, as a failure of the information system could bring paralysis to urban functions. To prevent this situation, protection of the information system is one of the most important concerns for the future city.

Heat Supply

A heat supply system furnishes steam, hot water, and cold water to many buildings. This service has the effect of preventing air pollution, saving energy, and saving space. In cold cities in the West, district heating systems were adopted because of their convenience, comfort, and safety. In Japan, a district heating system was introduced as a countermeasure for air pollution in urban areas. However, since Japan is very hot and humid, the system developed further into a district heating and cooling (DHC) system. The following are important components of a comprehensive DHC system:

- Effective use of energy can be achieved by using waste heat from power plants, garbage incineration plants, and factories.
- Stability of the DHC system would be sustained by the use of various energy sources and a highly developed operating control, which would not be hampered easily by energy shortages.
- Air pollution would be prevented by upgrading the operating method and by effective use of the unused wasted heat.
- The landscape would be improved by making an individual chimney and cooling tower in each building unnecessary.
- Calamities would be prevented by eliminating the need to handle and store dangerous articles in each building.
- Efficient use of the equipment would be achieved by sharing the equipment on a large scale.
- Capacity of the equipment could be expanded by sharing the equipment during peak energy demands and leveling off the differences in energy use by buildings. (The total capacity of equipment for peak energy demands of the general system could be reduced by sharing the equipment with different types of buildings, which have different peak times of energy use.)
- The size of the machine room for each building would be reduced by sharing the equipment. This would free space for other uses.

Waste Treatment

There are three concerns in the treatment of waste. The first concern is the increase in the amount of garbage and the problem of disposing of it properly. One solution is building a construction site for a new garbage disposal facility on reclaimed land. A second problem is the traffic jams caused by garbage collection vehicles and the illegal disposal of refuse, which attracts vermin, mice, or birds, creating hygienic problems. The third concern is the need for research and development of a different type of garbage collection system, such as a pipe distribution system, urban incinerator facility, or a recycling system.

Underground land can be used for pipelines for garbage transportation and disposal sites. Pipeline facilities installed under ground can receive garbage thrown from the edge or middle of the pipelines by using the "vacuum-cleaner" principle of sucking the refuse to the collection center. Pipeline facilities are used in many areas, such as in redevelopment areas, in new communities, and in business buildings. These facilities can save labor for collecting garbage, improve the environment and beautify the cities, and decrease traffic from collection vehicles.

Another solution to the shortage of land on which to build disposal plants could be locating disposal plants in underground space. A garbage plant 130 meters wide, 30 meters deep, and 60 meters high has already been constructed in an underground setting. Only its chimney appears above ground. The advantage of constructing such a plant is that space can be obtained even in the center of Tokyo, where it is difficult to secure land because aboveground space can be put to better use. If this plant had been constructed far from the urban area, where there are few buildings demanding heat, waste heat could not have been used; however, it is easy to use waste heat from the incineration of garbage in the depths below ground where there is a large heat demand, and the noise and vibration are less than above ground.

Utility Tunnel

A utility tunnel is often the form of the infrastructure in the city. These tunnels were built above and below the roads before the city became congested and complicated. Today, there are 2,000 to 3,000 cases of construction and repair of the national roads in Tokyo, or approximately 150 km, every year. This construction contributes not only to traffic jams, but also to the weakening of road structures and increased noise pollution for residents. To decrease the traffic jams, it was mandated in 1963 that utilities were to be buried low under the road. Thus, the utility tunnels became significant not only for preventing the need to dig up the roads for repair, but also for the efficient use of the valuable space under the road. A cross section of multipurpose utility tunnels is shown in Fig. 5.5.

The type and quantity of infrastructural systems in the city are apt to expand according to the progress of civilization. Now, in addition to the traditional infrastructure of water supply, sewage, electricity, and gas, new infrastructures—such as district heating and cooling systems and garbage transportation systems—are being introduced. Also, the river, roads, and pedestrians are being considered in constructing under ground with respect to countermeasures of safety and preservation of the landscape. But underground space near the surface is already being used, so it would be difficult

Figure 5.5. Section of a multiservice tunnel (Toshio Ojima, Infrastructure of Japan, *Nikkan Industry Newspaper, Tokyo, 1983, pp. 112–13).*

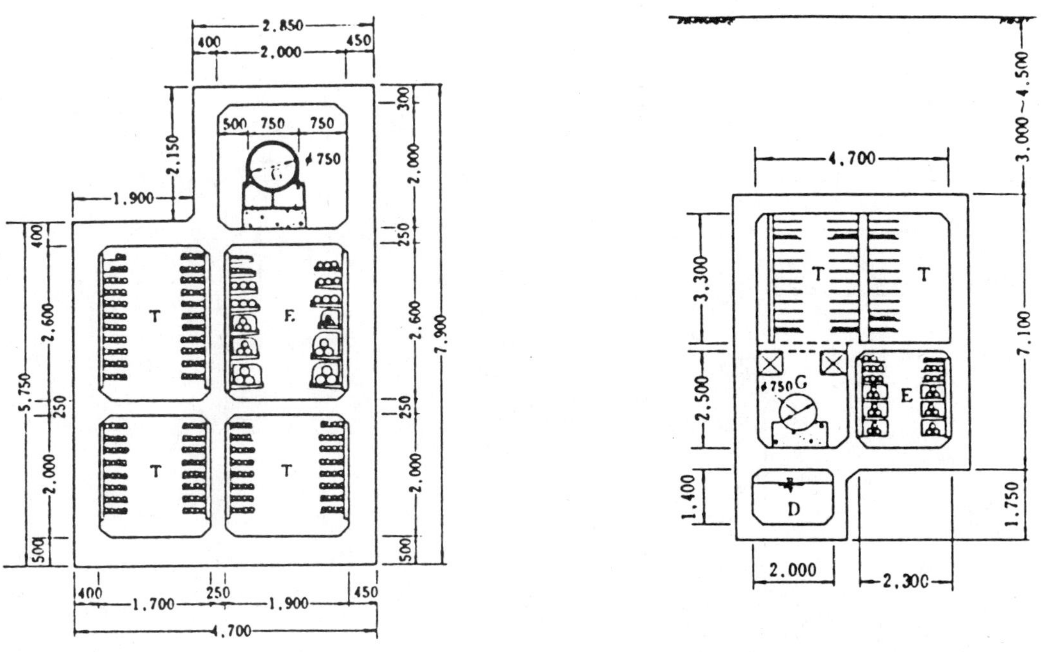

to increase the amount of infrastructure there. Utility tunnels, however, are indispensable for solving the lack of space under the road because they can be constructed at a lower depth.

Energy Tanks and Baserock Storage

An underground energy tank is dug vertically from above ground. The structure employs reinforced concrete walls and floors and a dome-shaped roof (Fig. 5.6). The tank is lined to contain the liquid. The underground tank is more expensive than an aboveground tank, but the distance between tanks under ground can be less than above ground. Thus, the site can be used more efficiently. As the level of liquid is lower than the ground level, it has a strong resistance to earthquakes. There is no danger of liquid flowing into the neighboring facilities to cause damage. Moreover, it does not damage the surrounding landscape. The biggest tank now in use in Japan has a holding capacity of 350,000 kiloliters, with an inside diameter of 97 meters and a height of 48 meters.

Baserock storage is constructed by digging horizontal

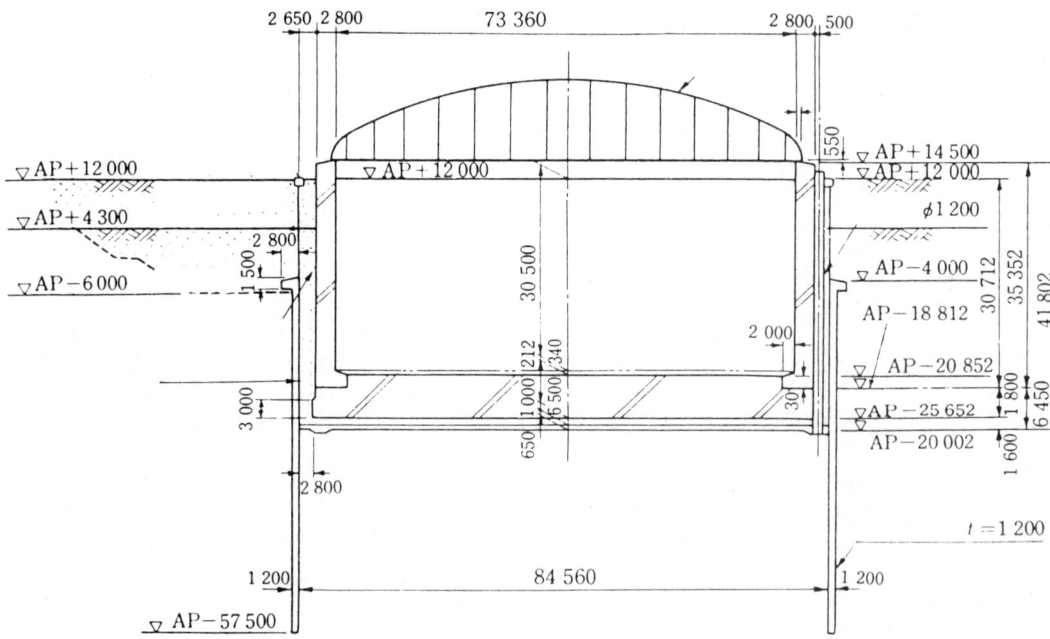

Figure 5.6. Underground tank of liquefied natural gas (Civil Engineering Society, Geo-Front New Frontier, *Gihodo, Tokyo, 1990, p. 136).*

holes into the baserock under the water level. Liquid is prevented from leaking by natural or artificial water pressure. A petroleum storage station constructed by this system is composed of 10 tanks. Each tank is 18 meters wide, 22 meters high, 555 meters long, and has a capacity of 1.75 million kiloliters.

Burying Electrical Wire Underground

The technique of burying electrical wire underground has been more slowly accepted in Japan than in Western countries. Germany and England already have buried 50 percent to 60 percent of their electrical wire; Belgium (and Malaysia as well) has buried 30 percent; France has buried 10 percent; but Japan has buried only 2 percent. In comparing cities, Paris, London, and Bonn have buried 100 percent of their electrical wire. Other major cities have buried more than 70 percent, while Tokyo and Osaka have buried only 20 percent. In England, it was the principle to bury the electrical wire from the beginning of the electric enterprise to preserve the landscape, increase safety, and decrease maintenance costs. Recently, in Japan, the pursuit of comfort and aesthetic values for the city has generated interest and quickened the pace of burying electrical wires.

The positive effects of burying electrical wire are the improvement in amenities, safety, and traffic conditions. Little damage occurs to electrical wire buried underground during an earthquake, as compared with significant potential damage to elevated electrical wire strung on poles. Damage to overhead wires causes an obstacle to all aspects of urban functions. Wires and transformers falling from the poles pose obstacles to an evacuation of people. In case of fire, electrical poles and wires would hinder the use of a ladder truck for rescue.

However, with the removal of poles for electrical wires, all forms of nonmotorized traffic—pedestrians, bicycles, baby carriages, and wheelchairs—could move more easily, and be safer and more comfortable. Also, in snowy areas, snow removal would be facilitated by having unobstructed roads. From the aspect of preserving aesthetics, city streets free of wires and poles would conserve the beauty of the landscape.

Newly planted trees could replace the poles and make the view from the street clear and open.

Underground Utilization

As mentioned previously, because of the highly developed services for urban systems and aided by the progress of technology for underground development, the use of underground space in the urban area is growing in importance (Fig. 5.7). In looking at the current state of underground use, many urban facilities are concentrating on construction under the mainline of roads, but are doing so without long-range planning. As a result, some underground passages and subway stations have been designed in irregular lines to avoid interfering with the existing pipelines. Some of them are under a main sewer or utility tunnel that does not have to be near the surface. These conditions burden pedestrians. In the future, new infrastructural systems, such as district heating and cooling systems, garbage transportation systems, and underground distribution systems, will be developed to resolve the environmental problem. These systems are based

Figure 5.7. Utilization of underground space.

Depth	Private Land		Road Area	Public Land
	Basement	Underground shopping center	Underground shopping center Subway station Underground road Underground passage Underground parking lot Main route of utility tunnel Branch route of utility tunnel Multiservice tunnel	Underground shopping center
-5m	Underground parking lot Pumping station Transformer station		Subway tunnel Underground road (bypass) Underground distribution system Main route of utility tunnel Underground river	Underground parking lot Pumping station Transformer station
-10m	District heating and cooling station			
-30m	Foundation of high-rise buildings		Main basic route of utility tunnel Subway tunnel	Main plant for utilities

mostly on the premise of underground use. Furthermore, as information technology and environmental technology progress, there is a possibility of new urban systems that could use underground space. Then, the underground usage in urban areas would become even more complicated than it is at present.

The cause of this current haphazard state of the infrastructure, as noted above, is the lack of long-term planning. There must be not only city planning for aboveground construction, but also a master plan for underground usage. This underground plan must deal not only with the public areas, but also with private areas, and, in addition, must be coordinated with the plans for aboveground land use. Here, also, the use of the deep underground and the superficial underground must be separated. The superficial underground should be used as a part of aboveground space to supplement the activity of people above ground. This space should be safe and healthful. Deep underground space should be used mainly for facilities where only specially trained people would enter the infrastructural system, for safety purposes.

Technology for Underground Utilization

To utilize deep underground space safely and comfortably, some technical issues must first be addressed. First is the subject of survey and design technology. Second is the execution of new technology. Third is the matter of management and operation technology.

The load-bearing foundation for the underground structure is a resource of pressurized sand, has moderate vibration, and is a potential source for disaster. Especially in the deep underground, both the skeleton and the foundation play roles in bearing the structure. It is very important to understand that the foundation must preserve the strength, stability, and durability of the structure and its influence on the surrounding space. Damage to the surrounding area and its natural components must be prevented before construction by surveying the composition of the layers of soil, the geological features, underground water, and amounts of oxygen and methane gas.

In the past, surveys for the foundation were carried out mainly by boring vertical holes. But the vertically drilled holes gave only one-dimensional information. Recently, two-dimensional information and three-dimensional information are being gathered to more fully grasp the inner condition of the foundation.

One of the biggest problems in surveying is that the current state of underground use is often not clear. There are no maps to show the details of foundations, their geological features, and various existing facilities. Especially in urban areas, where the underground use is complex, underground maps dealing with various information must be developed. In Japan, many institutions have already mapped their own underground information systems. But, since the maps are developed for their own purposes, they do not deal with other information. Another problem is that the format of data and the system of mapping are not universal among institutions.

In Japan, the execution of tunnel technology is reaching a high level. Tunnels, such as the Seikan Tunnel, are being laid as underground structures penetrating mountains and the bottom of the sea. Generally, the open-and-cut method is used for building superficially underground, but the shield method is also used for deep underground work when there is a limitation on land use above ground (Fig. 5.8). The New Austrian Tunneling Method (NATM) is used for deep and solid underground tunneling. Although there have already been achievements in Tokyo in construction below 50 meters depth, as for subways, there have been no attempts to construct large and complicated structures in foundation conditions.

In the shield method, a shield is propelled into the underground construction area to prevent a landslide. A tunnel is dug like a mole inside its space. In a variation of the shield method, the shield itself supports the land pressure. A propeller digs the tunnel. This process of construction is described as follows:

- A shaft is made in which the shield excavator and propeller start to move.

Figure 5.8. Multiface tunnel machine (Kumazai Gumi, Shield Tunneling Technology*).*

- While a hole is dug with a round cutter that revolves cylindrically, the shield becomes the excavator and the propeller advances horizontally.
- Each block, called a segment, is assembled in the space made by the advance of the shield excavator and propeller.

The tunnel is expanded by repeating this process, making it larger in diameter, deeper, and longer. At present, the biggest tunnel constructed in this manner has a diameter of 14 meters. Plans are under way to use the shield method in the tunnel construction of the Tokyo Bay crossroad. With this method, it is hoped to increase productivity, save labor, hasten the speed of construction, and develop the automated system.

The NATM method is a process of digging a hole to increase the strength of lands and mountains by driving bolts into rocks. Landslides are prevented by spraying concrete into the area. The NATM is regarded as a standard method for tunneling into mountains. Fundamentally, it is the method to be used when the ground is solid and would not collapse. Therefore, the NATM is not as adaptable to soft grounds as the shield method. Also, the NATM is not suitable for use where the underground water pressure is high. The speed of construction is slow. However, the NATM is economical in the long run because its construction cost is $^1/_3$ to $^1/_{3.5}$ that of the shield method. Also, the NATM is more flexible than the shield method in terms of the size of the construction site and its adaptability to its shape.

The vertical pile method has been used to construct deep tunnels. One of the problems with this method, however, is the construction technology for the vertical pile, which is used for the starting station to put the shield machine into the tunnel from aboveground and for carrying construction materials. Now, the continuous underground wall method and pneumatic caisson method have been adapted to construct deep tunnels.

For constructing a round shape using the continuous underground wall method, a vertical hole is dug and the surrounding underground wall is held by the compression of a ring to create pressure on the side walls. The walls are thin. This technique is economical to use in the deep underground. In one case, where a river flowed underground, a tunnel was constructed 59.9 meters in depth with a wall thickness of 1.2 meters and an inside diameter of 28.2 meters. In using this method, a countermeasure to keep underground water from leaking in is very important, since the deep underground water pressure becomes very high and the water can seep into the tunnel from the bottom of the wall. To protect against this seepage, the bottom of the continuous underground wall must be made watertight either by a natural impermeable layer or by an artificially created impermeable layer.

In the pneumatic caisson method, a strong concrete pipe is first constructed above ground. Then the pipe is sunk into the ground. Air pressure is used in the working space at the bottom of the caisson to prevent underground water from flowing into it. However, from the aspect of the workers' safety, the air pressure that can be used is limited. Recently, an unmanned caisson method, which is automatically controlled, has been developed. But even with this system, workers must go in for repairs and routine inspection, so only a limited use of automation exists.

The method of underpinning is used in urban areas where many other structures have already been dug into the foundation and where the new structures must be built without disturbing the existing ones. Although small utilities, such as pipelines, can be hung, large structures, such as bridges, subways, and buildings, can be effectively built by using the underpinning method.

Generally, the underpinning method is functional for building in a narrow space, and the process of the construction is intentionally planned. There are essentially two processes involved with this construction method. One process is to build a temporary foundation that will sustain the existing structure while the new construction is superimposed over the old structure. The second process is to completely reconstruct the foundation of the existing structure before building the new structure.

Deep Underground Utilization

Legal Issues

High expectations are held for the use of the deep underground. In Japanese civil law, land ownership theoretically encompasses the space from the sky to the deep underground. This right of ownership is required to be protected. Although there have been many court cases dealing with the noise from airplanes flying low over habitats, there has yet to be a case brought to trial for airplanes and artificial satellites flying high in the air, because this is considered too high to violate the right of private ownership. Similarly, it can be considered that space that is deep enough underground also does not violate the right of private ownership. It can further be argued that the deep underground space should be used to serve the public and not held for private use.

Usually, to legally use the underground space under privately owned land, a leasehold and easement condition must be set up. Therefore, it is first necessary to establish the rights of the owners and compensate them for their buildings above ground. However, this process would be so complicated that, in many cases, the project would be delayed, especially when dealing with linear facilities where there are multiple private owners.

The Japanese government has begun to examine the feasibility of the law for using the deep under ground. One alternative under consideration is to restrict the private ownership of the deep underground and make its use possible without compensation to the private property owners above ground. But the legal problem of defining ownership of the deep underground is that the precise depth of ownership has not been clearly defined. Currently, underground space deeper than 30 meters is not used often. Thus, it could be justified that space below 30 meters should be given priority for public use. But a problem would exist when considering existing buildings built on a foundation of soft soil, where the piles for the building must be sunk to a depth of 70 meters.

Foundation Map of Tokyo

The foundation maps of Tokyo show that Tokyo's wards extend from the plateau in the west to the low ground in the east (Fig. 5.9). Although their superficial foundation is thin, the foundation just below the superficial surface is very solid, as in most of the Western cities. The superficial foundation is

Figure 5.9. Foundation of Tokyo wards (Toshio Ojima, Imaginable Tokyo, *Process Architecture, Tokyo, 1991, p. 62).*

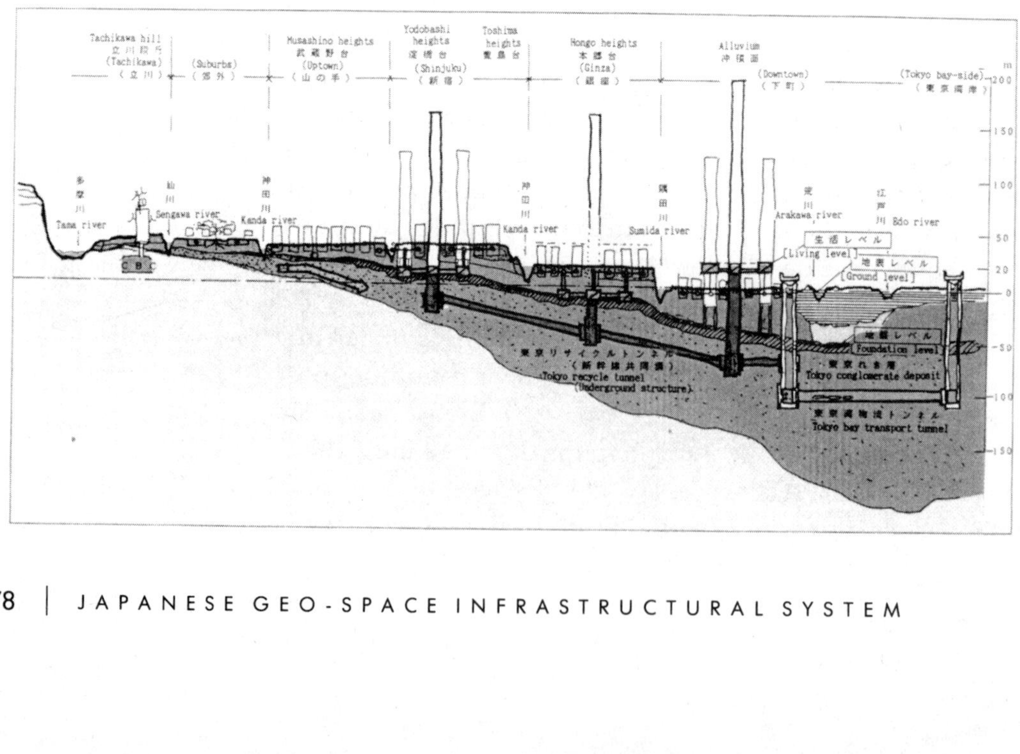

composed of soft alluvium. With this surface condition, the support structure of a large building must be sunk into the solid foundation under the alluvium surface to prevent sinking and leaning of the foundation piles. This type of solid foundation beneath the alluvium layer is about 30 meters to 60 meters deep from west to east, and is the foundation base for Tokyo. Primarily, it is space that has not been used yet and can be considered the deep underground.

Underground Projects of the Government and Private Enterprise

Underground space has been used since ancient times for defense from outside enemies and protection from severe climate. Recently, however, considerable attention is being directed to underground space in Japan for its utility value because of the land shortage, the soaring prices of land, and the concentration of activities in the city. The considerable attention being given to underground space has produced the word *geo-front*, which was derived from the term waterfront. The horizontal space existing above ground adjoining a body of water is called the waterfront, and new vertical space is termed geo-front. Space deeper than what has been used so far is called the deep underground. This terminology has become the center of public attention.

During 1987–88, many research associations, representing private companies and government offices, were inaugurated to deal with underground space. Negotiations between some ministerial offices also increased. Among these offices, the Ministry of Transportation showed interest from the start by examining the Deep Underground Subway Project. Soon, the Ministry of Construction became interested in the Underground Bullet Road Project, which was to connect the two centers of Tokyo—Shinjuku and Marunouchi—without a stop for 8 km at more than 50 meters in depth. In addition to the two ministries that were examining linear use of the deep underground, the Ministry of International Trade and Industry began to look at possibilities of using a huge space of more than 50 meters in diameter. Further, the Ministry of Postal Services examined the Deep Underground Communication Project from the aspect of safety, and the Ministry of

Welfare studied the Deep Underground Water Supply Pipe Project. But legislation for deep underground use is still stalled, as the agencies are vying for power. On the other hand, private enterprises, as general contractors, are making a special team effort to study and take advantage of deep underground development.

Thus, underground utilization appears to be indispensable in remedying the paralyzed state of urban functions above ground, the overcrowded conditions, and the destruction of a comfortable urban space for people. However, the underground is a special space and cannot produce a healthy urban space if it is used solely to correct the failures resulting from haphazard urban planning. Entire cities must be intentionally reconstructed in a positive way by making the most of the underground space characteristics.

Underground Space Utilization Charter

There have been great construction projects in Japan, such as the Honshu-Shikoku Bridge, Seikan Tunnel, Tokyo Bay crossroad, and, finally, the airport on the sea, which is now under construction in the Kansai area. Until only recently, most construction took place outside of high-visibility areas, but recently, these developments have been appearing in the visual urban spaces as well.

Not long ago, the headlines of the *Japan Stock Newspaper* featured a story on the doubling of national land by depending on deep underground development as a solution for the extreme lack of land and the jump in land prices. Once the waterfront development held much attention, but now, geo-front development is becoming attractive. *Geo* means earth in Greek. Now, the term is used to refer to the underground or underground space.

The geo-front boom started in Kabuto-cho, which is the center of the stock market. Its success has been responsible for spurring underground space development. A land counterplan in June 1988 followed the basic recommendation of the Cabinet Council. The purpose of the discussion of rules and plans for coordination and utilization was to intentionally promote the public use of the underground space, and to make preparations to present a law guiding the public use of the deep underground to a meeting of the National Diet. This would facilitate the extension of the railroad into the center of the city and arrange the infrastructure under roads and waterways.

Contrary to the dream of the construction industry and the stock industry to develop the underground space, the actual use of this space is not so easy. It would not pay to engage in construction without favorable conditions. Although underground space has the possibility of superior uses, as for cemeteries, shelters, and storage, it has many disadvantages for the daily users' convenience and health. Therefore, general ethical principles for development of underground space need to be established.

Examples of principles that would need to be considered for using the deep underground are:

1. Develop the underground space only where it is unavoidable to facilitate redevelopment of the aboveground, as in the center of a large city.
2. Aim to revive nature for land over the underground development.
3. Use the deep underground only for public utilities and connect them safely in the heart and surrounding areas of Tokyo as the lifeline for emergencies.
4. Use the deep underground as a stable foundation for high-rise buildings.

The use of the deep underground will be an important proposal for Japan. Public welfare must be considered first before using the deep underground, followed by the consideration of private projects. A pilot project must be done first, from the viewpoint of technology and sociology, before starting an experimental project. This would ensure an important base for supporting various demands of the big city, especially in emergencies.

TOKYO METROPOLIS UNDERGROUND SPACE NETWORK PROJECT

Problem of Urban Environment

The urban area of Tokyo has been continually expanding (Fig. 5.10). Its commuting area now extends to 50 km. No urban sprawl as great as Tokyo's exists in a city in any other country. Even the urban area of Manhattan, where there are many skyscrapers, still has many natural areas among the buildings. In London, there is a beautiful green belt surrounding the city.

Japanese public spaces and roads, however, are narrower and smaller compared to roads of the West, which were built originally for coaches. Park areas per person in Tokyo wards, in London, in Paris, and in New York are 2.1 square meters, 30.4 square meters, 12.2 square meters, and 19.2 square meters, respectively, while the ratio of roads to area are 13.2 percent, 16.6 percent, 20.0 percent, and 23.2 percent, respectively. In Tokyo, during the last 20 years, while the number of buildings and cars has almost doubled, park areas and roads have increased only slightly. Utilities under the roads have become complicated, and space planning for cars and pedestrians has been delayed.

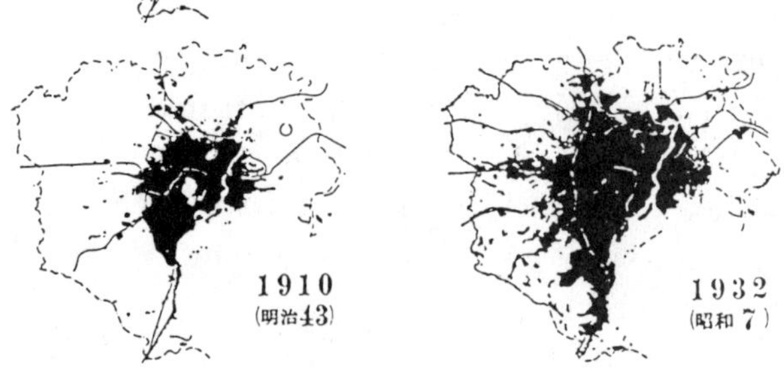

Figure 5.10. Expansion of Tokyo (Tokyo Metropolitan Government, White Paper of the City '91, *Tokyo, 1991, p. 27).*

Japan has the most expensive land in the world. Why have structures not been built more than one story high when land prices are so extraordinarily high? It is because the surface of the earth is covered with buildings, and there is little land left with natural surface. The lack of open surface restricts public space for roads and parks, and there is a shortage of water-front land as well. Thus, buildings are regulated for lighting, setback regulations, and construction. Construction of high-rise buildings is almost impossible, since such buildings would restrict sunshine and air circulation.

As mentioned previously, with the lack of capital, surface area, and manmade infrastructures to support new buildings, architecturally, space cannot be built up. By increasing these resources and capital, private space could be used for the first time.

Heat Island

Figure 5.11 shows the annual fluctuation of urban temperature from the mid-1880s to 1975. During this period, the temperature in Tokyo has risen by an average of 2° C, while humidity has decreased 15 percent. The phenomena of the heat island and dust dome have been continually increasing

Figure 5.11. Annual fluctuation of temperature in Tokyo (Toshio Ojima, Imaginable Tokyo, *Process Architecture, Tokyo, 1991, p. 94).*

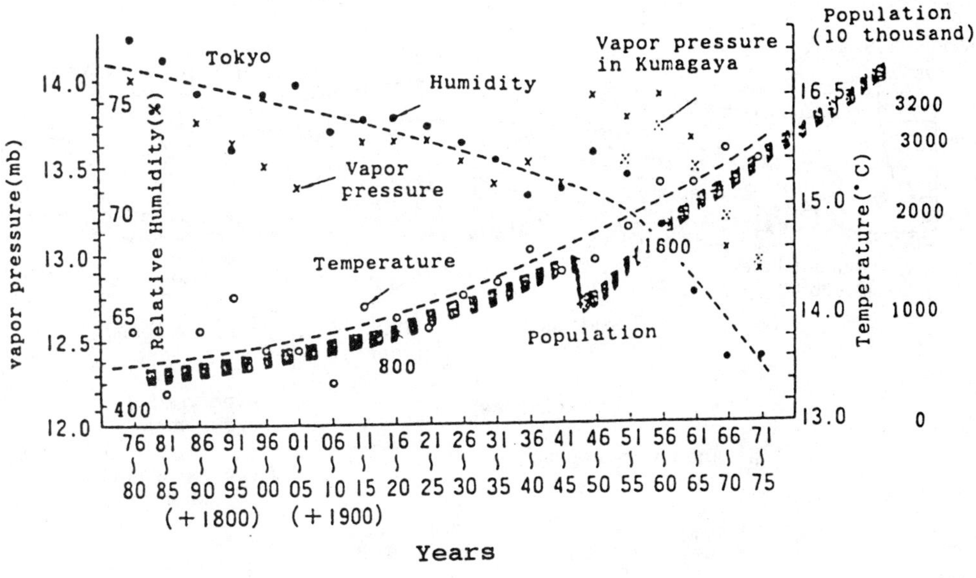

(Fig. 5.12). The air heated in the center of the urban area rises, is cooled, and then falls back into the suburban area, while cold suburban area air flows into the center of the urban area. This wind circulation pattern is the same pattern as that created by a bonfire. Wind circulation in itself is not a harmful phenomenon. But in Tokyo, the industrial area near the bay and the residential area on the opposite side allow the polluted air from the urban center to fall on the houses and green areas by pulling the smoke of the factories into the urban area, while the cold air falls into the bay. Thus, the heat island is not just a phenomenon of rising temperature, but causes other problems as well.

Energy consumption in Tokyo is 12 percent of the solar heat received into the area. It can be said that the climate of Tokyo is influenced by its energy consumption. From the east-west section of the Tokyo ward area, the amount of heat discharged in summer is three times greater than in winter, and it is concentrated in the center of the Tokyo ward area. Measures must be taken to deal with these situations.

With the development of Tokyo, green space has decreased as land surfaces have been paved with concrete. Concrete does not absorb rainwater, which flows to the river without being filtered through the soil. Also, concrete has a

Figure 5.12. Heat island (Toshio Ojima Lab., Tokyo Underground Project, *Kenchikubunka, Shokokusha, Tokyo, 1982, p. 63).*

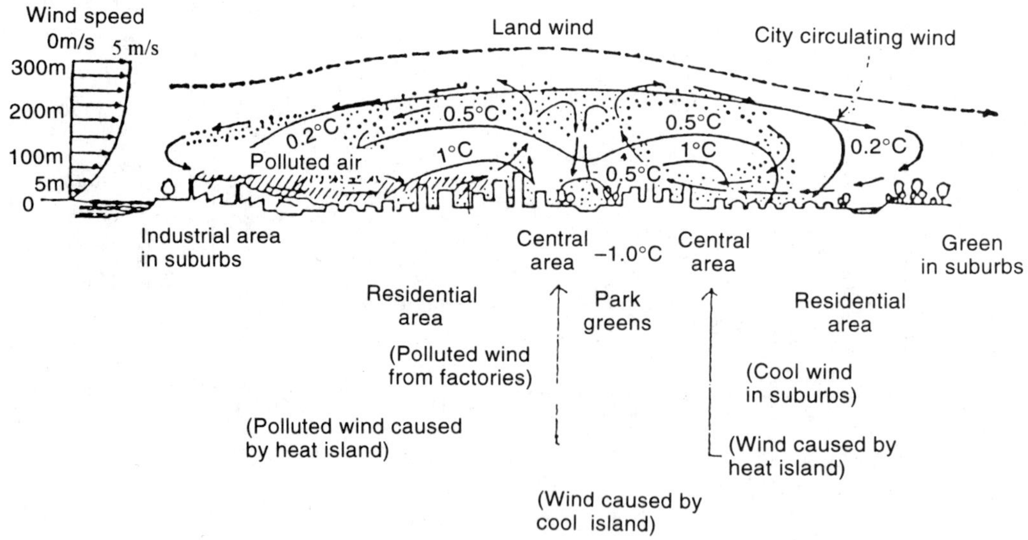

large heat storage capacity, and does not cool down even during the night. Thus, the temperature is never lower than the dew point. The heat capacity of the green area, on the other hand, is small. Even a potted plant contributes to nature's circulation system by making condensation: Dewdrops fall to the potted soil and evaporate again. But the city is moving away from this natural circulation, as the conditions allowing dewdrop condensation are diminishing. The whole city is getting drier, creating a desert phenomenon. In Tokyo, this phenomenon can be reduced only by introducing green space again that would separate the concrete areas of the city.

Potential Disaster in Tokyo

Tokyo is one of the most naturally dangerous cities in the world—serious natural catastrophes such as earthquakes, typhoons, landslides, and volcanoes can occur there. The Japanese themselves are not fully aware of the extent of this danger. But for people who have lived in countries such as the United States or China, Tokyo is like a city set on a bowl of Jello.

Tokyo's social environment is also in danger. The 134 disaster relief shelters in Tokyo are a good example of this. A shelter should be at least 300 square meters in size to avoid radiant heat in case of fire. As for the safety of individual persons gathered there, about 2 square meters are needed for each person. Tokyo's shelters are large enough to provide refuge for only 18 percent of the population. Also, for elderly people or children, a refuge area is needed every 2 km, but only 60 percent of the Tokyo area meets this requirement. The relief shelters are actually too few, too far apart, and too small in size to serve their purpose.

Edo used to be known for its frequent fire breakouts. During a 300-year period, starting in 1601, when Ieyasu Tokugawa came to Edo castle, there have been about 100 large fires, which destroyed approximately 10 percent of the town in each fire. This was equivalent to burning down the town of Edo every 30 years. Also, during this period, there were four powerful earthquakes as strong as the one in the Kanto area in 1923. When people protected their houses from fire by tiling roofs, earthquakes crushed them; when they roofed their houses with wood, it was easy for fires to spread. In spite of this destruction, Edo survived as the center of Japan.

After the Meiji restoration, the Japanese started making buildings fireproof. New buildings were constructed of reinforced concrete instead of wood. Therefore, during the severe Kanto earthquake, 140,000 people died or were missing, and 380,000 houses were destroyed by fire, affecting 60 percent of the people in Tokyo. In the large-scale air raids that struck Tokyo 120 times in 1945, 330,000 people died or were missing and 760,000 houses, or 50 percent, were burned by fire bombs.

After the war, city plans were made for reconstruction of the war-damaged area. One of the basic ideas was to make a compact center of Tokyo, with small towns surrounding it. By passing public works laws, a complete infrastructure was expected to be built, but under the economic control of the United States Dodge Line, the public works were cut back. The government employed many foreign architects, who were unfamiliar with Japan's terrain. As a result, Japan made fatal mistakes in design again, which led, for example, to the present situation of ineffective and insufficient shelters. Seventy percent of the buildings were still made of wood, and the huge amount of flammable material concentrated in the city of Tokyo is not very different from the dangerous conditions that existed in Edo.

Damage from the next big earthquake, which is forecast to occur in the 21st century, could cause three times the damage of the war. By interruptions of parts of the lifeline, such as electricity, gas, information transmission, traffic, and so on, damage would occur not only to the family and enterprise, but also to the nation as a whole.

Problem of Infrastructural Network System

The defined area of Tokyo has been a 60-km range, but is currently expanding to a 100-km range. If the Shinkansen commuting area is included, the range becomes 200 km. With this expansion, all of the infrastructure, such as the electric power network, the communications network, the water supply, drainage, and gas pipelines, extend without order or a preconceived plan, tending to follow the sprawl of the city. Some of these services are built under ground, while others extend over the largest plain in Japan—the Kanto Plain. Over the long term, however, it is difficult to maintain such a complicated mesh of infrastructure, even though it was efficient when it was originally constructed.

The infrastructure in the central city that was built just after the war is already more than 50 years old and in much need of renewal. New housing complexes developed in the suburbs lie in a continuous sprawl. The massive infrastructure of the third wave of industry to be constructed on the reclaimed land in Tokyo Bay is larger in scale than the infrastructure of the second wave of industry built on the other side of the bay. However, trying to accommodate the infrastructural demand of the third wave in an unplanned way would require a large amount of capital investment and a safety management system to maintain the facilities. Moreover, backup facilities to support the infrastructure would become difficult to depend on because of the same infrastructure that secures safety for the center of a city is also necessary in the suburbs and at the waterfront area.

This infrastructural situation would be similar to the condition that existed under the "rich country and strong soldier" policy. At that time, Japan expanded to greater East Asia by sending the military to the front and acquiring territory to bring prosperity to Japan, as the Roman Empire once had done. Ultimately, however, the Roman Empire destroyed itself because of its inability to maintain its vast territory.

Many resources are consumed to maintain, support, and control the network of an infrastructure that extends horizontally. For instance, it requires approximately 10 billion yen a year to maintain a water supply system with tens of kilometers of extended pipe with the necessary pressure for the electric power to pump water from Tokyo metropolitan areas. This creates a great financial strain on the government's budget. In addition, finances are needed to handle the drainage of rainwater in the metropolitan area. The rainwater flows into the area's network sewerage system, where it is pumped to a processing plant and then, in turn, pumped into Tokyo Bay. The heavy use of the central water network makes it costly to maintain in good condition. Another large-scale operation was used for retiring the Japanese National Railways. They were dismantled throughout the country from Hokkaido in the northeast to Kyushu in the southwest, excluding Okinawa. This operation also incurred a huge deficit.

Moreover, in an emergency, it would not be possible to back up the metropolitan network of an infrastructure made of numerous congested routes. Unfortunately, the probability is very small that the infrastructure network would function continuously. Once one part of the network shuts down, other parts become overloaded and also slow down. If the malfunctioning section could not be repaired, the whole system would stop functioning, since it is extremely difficult to reconstruct a problematic area. Electric power could not be restored for a long time; neither could water, gas, or drainage services, and the area would thus be without necessary services. Communication lines in high-rise building areas or business centers in the central city would cease functioning, and the area could not be used. How can Japan continue to depend on this large area network in such situations? The present expanded network cannot provide the answer.

In the United States, a similar expanded multinetwork was created by the Greyhound bus line and Amtrak trains, which connected every area in the country, but the lines were ruined because of their vast scale. This network was superimposed on the existing network of the roads, the railways, energy lines, and pipelines throughout the United States. The roads were well established to connect cities, but the country was forced to spend a lot of capital to maintain the bus and railroad lines. That infrastructure now is clustered and centralized on a regional basis.

In the present Clinton administration, the last large-area network being built is a system of optical fiber cables for community antenna television (CATV) to provide television reception nationally to rural areas that would otherwise not receive television. This large-scale infrastructure exists individually in each city so that the city may become independent and broadcast in a limited capacity to meet the demand of the region. The idea that this kind of infrastructure is friendly to the earth's environment is predominant at present in the United States. However, Japan would do better to change its policy and have a total infrastructural unit for the limited areas of Tokyo, instead of enforcing a huge-scale network system over many areas. In Japan, only a minimum system is necessary for each area needing to be covered.

It is more feasible to construct the infrastructure independently, based on the population density, the soil, the location, and the planned usage in each region. High-density and low-density network areas should be separated according to the characteristics of each.

Tokyo's Garbage Problem

The garbage problem in Tokyo is unresolved at present. The city's transportation and collection methods detract from the pleasantness of Tokyo in an obvious and embarrassing way. Finding suitable sites for garbage disposal is also a problem. The amount of garbage produced in Tokyo's 23 wards grew by 2.18 times from 1966 to 1986. In the 12 years after the oil crisis of 1973, the amount of garbage produced increased 1.18 times, and is continuing to increase. Today, Tokyo produces 6 million tons of garbage every year. This means that Japan is struggling to dispose of more than 180 thousand tons of garbage daily. Many factors explain this increase in waste material, but one of the main reasons is the change in lifestyle to mass production and mass consumption.

Ideally, the best way to solve various garbage problems would be to decrease the amount of trash generated. But it is impossible to return to the lifestyle of ancient times, where people's lives were suited to a natural cycle. With modern industrial production, waste has been increased from many sources and will continue in this fashion as long as Japan maintains its production activity. Once Japan recognizes how garbage is produced, it can understand that it is just as important to look for the way to reuse garbage as a way not to produce it.

In order to effectively reuse garbage, Japan has to grasp the way to collect it, its amount, and its contents according to the areas where it is produced, and then introduce an effective garbage disposal system. Recycling is a very important issue, especially in areas where an extensive amount of garbage is produced.

The "garbage war" among Tokyo's wards began in 1972, when Suginami Ward tried to dispose of its garbage in Koutou Ward. The dispute was settled by a common agreement that each ward would be responsible for building incineration plants within the ward. However, it is not feasible to build garbage disposal plants on high-priced land, which costs 15 million to 30 million yen per square meter, which would lead to a cost of at least 50 billion yen to construct one garbage plant that would serve one ward. Thus, due to the prohibitive cost, 11 wards still have not been able to construct their own garbage incineration plants. The construction expense for all 11 garbage plants would be about 600 billion yen. So it has been suggested that, instead of building separate plants for each ward, a large area network, as previously described, should be introduced at a construction cost of about 1 trillion yen.

Garbage can be used effectively both as a resource and as a fuel if it is suitably collected and properly disposed of. For example, paper can be recycled and reused. Kitchen refuse can be reused as a fertilizer by composting it. Incinerated garbage can generate heat of more than 1,900 kcal/kg, so it can be reused as fuel. Recently, the waste heat from a garbage disposal plant became the heat source of a district heating and cooling facility and neighboring public facilities.

The complexity of garbage transportation, collection, separation, processing, and disposing is a significant problem for the future city's maintenance. Disposed waste should eventually be returned to the ground. Currently, the resources are produced above ground; then, the consumed waste is returned to the underground or to the sea. But in the future, the consumed waste will ultimately be returned to the place of its origin. The plan of reducing garbage must take into account its distribution, inflow, outflow, disposal, and reduction at such large-scale levels.

Natural circulation of refuse originally was sufficient in Tokyo. In the Edo era, a million people lived with the area's natural features. In fact, the Japanese followed a philosophy which said that everything would circulate and return to the soil—even man. The Japanese had an instinct to live with the natural cycle and carried this system into all aspects of their lives. Even vegetables from the farm were returned to the earth as fertilizer. Today, many materials, such as oil and iron, are obtained from foreign countries and are not recycled. They are piled up into mountains of garbage because there is not space to dump these materials.

The same thing is happening to heat. In central Tokyo, great amounts of energy similar to energy from the sun are consumed by the wind or river or soil, but the sources for dispersing the heat are decreasing. Water also used to soak down into the soil naturally, but now, to safely use the water, it has to be treated.

Therefore, the use of materials must be considered in terms of input and output. With technology, only the methods for production were seriously studied. Today, there is the further realization that products are useful only when there are methods for their disposal, as well. As with the cyclical philosophy described above, industry, too, has a connection with ecology. From this idea, the concept of a recycling system is returning. Since surface absorption alone cannot solve the recycling problem, a multilevel treatment using wind, water, and any other resource or system has to be used to totally achieve recycling.

Project for Releasing Aboveground Space in the Metropolis

Utilization of Unused Energy

The consumption of energy for public use in the home, office, or shops, continues to increase, especially in the region surrounding Tokyo. As mentioned previously, after the 1973 oil crisis, Japan's consumption of industrial energy was curtailed by using the energy-saving technology of the advanced world. The rate of energy consumption by the public and the total energy consumption of Japan now are less than those of Europe and America. However, the forecast is that Japan's public consumption of energy will increase as the quality of life is upgraded with the modern conveniences that require additional energy. This load will vary by region and by season, and require huge supplies of energy. Therefore, it is necessary to develop an innovative advanced technology to save energy in the future.

Japan must depend on oil and fossil fuel imports from overseas. It currently imports 99.7 percent of its petroleum, 87 percent of its coal, and 95 percent of its natural gas. The availability of energy from these foreign sources has become more limited during the past several decades. Securing a steady, dependable supply of energy will become increasingly more critical in the future.

To begin with, there are many inefficient aspects in Japan's present energy supply system. For example, gas and electricity need to be uniformly supplied and a high-quality electric energy must be converted into thermal energy. If Japan burns gas and kerosene or uses electricity for thermal energy to satisfy public demand, there will be much waste, because the indoor temperature level is about 80 degrees F, at most. Energy consumption in a high-density city like Tokyo is enormous, and results in heat loss and waste. Therefore, it is necessary to exercise a more delicate, more premeditated use of the energy supply by eliminating energy waste. By adequately understanding the temperature level and by converting unutilized energy into a usable energy form, it becomes more efficient to have an energy supply plan that would match the heat consumption characteristic on the demand side.

It is also necessary to consider the influence of energy consumption on the environment. The inland side of the city is covered with concrete and asphalt, which causes a heat storage effect contributing to the heat island phenomenon. As a result, more energy is required for cooling. Also, a variety of air pollutants, such as CO_2 and NOx, are generated through the combustion of fossil fuel. These pollutants are one of the causes of global warming and acid rain.

Decreasing the energy demand, developing alternative sources of energy, and improving the efficiency of the energy supply system are considered strategies for decreasing the total energy requirement. However, there is a limit to decreasing the amount of energy consumed. Improvement in the efficiency of energy use and the development of alternative sources of energy are of a most urgent necessity. Time is needed to develop alternative sources of energy or improve energy efficiency. The so-called unutilized energy in the surrounding environment is considered a source of energy that has not previously been used. River water, sea water, heat generated from water and garbage processing, heat of thermal power generation and drainage, and rejected heat from buildings, subways, and factories, are important to use effectively because they all increase the temperature level of the city. This, in turn, creates a higher energy demand for cooling and increases even further the public consumption of energy. Improving the efficiency of energy use will be of merit on the economic side, the resource side, and the environmental side (Fig. 5.13).

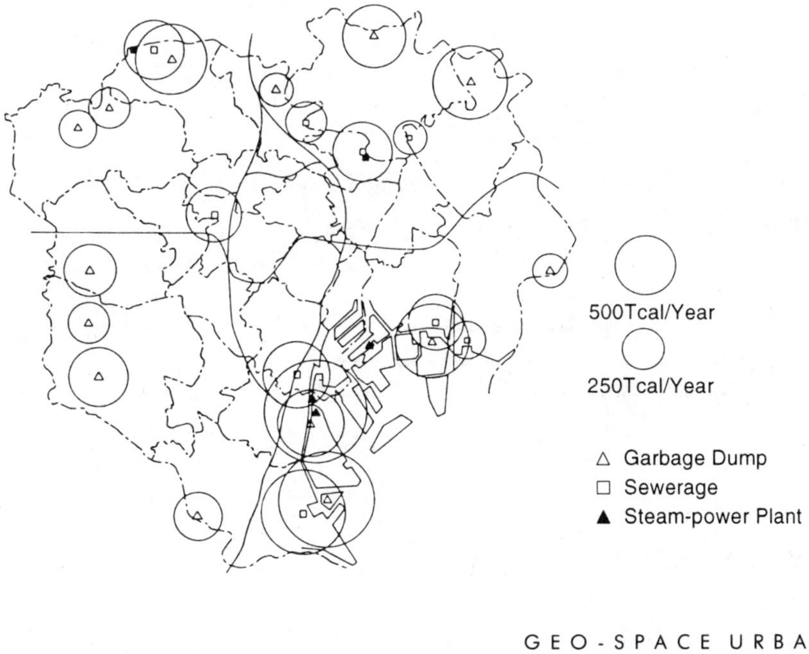

Figure 5.13. Unused energy storage in Tokyo ward area (Toshio Ojima, Advanced Scene of Tokyo, *Waseda University, Tokyo, 1995).*

Moreover, utilizing the unused energy produced from the temperature difference of a large-scale rejected heat source, as, for example, garbage disposal plants and thermoelectric power plants, and drainage and river water, needs a large investment in a system to collect the heat and to produce energy. Therefore, such a system would more effectively be used at a regional level than it is currently being used at the level of building units.

The district heating and cooling (DHC) system, as a method of effectively using this unutilized energy with high energy saving and increased utilization, can be achieved with a new infrastructure for the city (Fig. 5.14). The DHC system is a plan to install a plant in a certain region and to control the concentration of air conditioning. An aggressive approach is necessary to make the DHC network an effective user of energy in the future, not at one specified district level for pollution control, but also at the global environmental level.

The DHC in Tokyo has been aggressively promoted as a solution to the air pollution problem. This is still its main purpose, as promoted by the Tokyo metropolitan government. Actually, the area of the DHC system is being expanded according to the district where the amount of SO_2

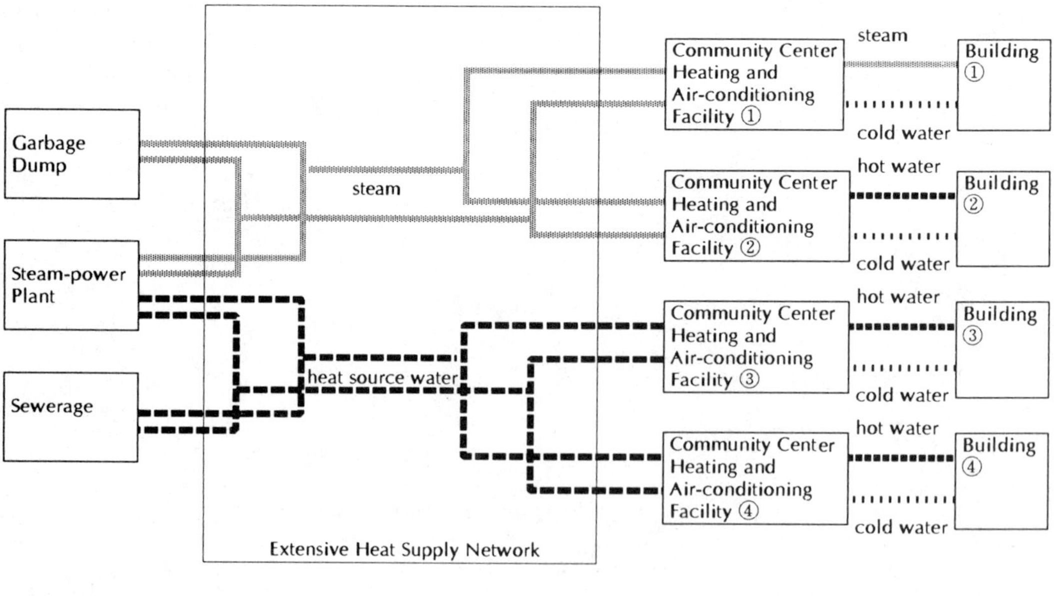

Figure 5.14. Extensive heat supply network system (Toshio Ojima, Advanced Scene of Tokyo, *Waseda University, Tokyo, 1995).*

measured in the air is under the regulation of the pollution control ordinance.

Western countries have started to introduce the DHC system by using waste heat. In Japan, recently, less importance is being placed on the prevention of air pollution as fuel use has increased. The emphasis of the DHC system is now being redirected toward saving energy and substituting alternative energy sources for oil. A new system for using waste heat was developed whereby half of the energy supplied was wasted without being recirculated. But recently, the technology for using waste heat is being directed toward conserving energy. The objective of supply for the DHC system is the buildings and houses, where not only high-temperature heat but also low-temperature heat can be used. The major sources of unused energy in urban areas are as follows:

- Fire stations
- Factories
- Incineration plants
- Sewage treatment plants
- Transformer stations and transmission lines

DHC systems were introduced in 27 places by the end of 1990. Considerable expansion occurred in 23 wards in Tokyo. However, the number of DHC systems using unutilized energy is low. Two of these systems use rejected heat from garbage disposal plants in the large-scale apartment complexes; one uses the energy from the difference in temperature between the air and the river water; and several areas collect the rejected heat from air-conditioning systems of transformer substations.

There are several additional sources of high-temperature rejected heat. Garbage disposal plants that are expected to be such heat sources are situated around the 23 wards. Electric power plants that also discharge effective high-temperature heat are located along Tokyo Bay. Sewage treatment plants that are expected to be a source of effective low-temperature rejected heat also lie in all 23 wards. Until now, these sources of unused energy were not tapped because their location and the areas of demand were not in proximity. As the part of the energy infrastructure maintenance of the third industrial phase, it is important to expand the network of the DHC to areas of demand. In the third industrial phase, the information industry plays an important role; this industry requires a great deal of office automation equipment, such as computers, copy machines, faxes, and many others. This equipment needs electricity, and exhaust heat and air conditioning would also become more important.

The DHC network should establish mutual support, for each DHC unit to be a backup unit to another in an emergency. Since the DHC facilities have considerable capability for backup, this will save energy by using highly effective equipment. It is possible to renew and update the equipment without interrupting the energy supply. This may allow the range of the energy supply to be expanded throughout the entire network. In addition, the leveling of the load change can be accomplished by tying into two or more DHC plants. There is also the possibility of increasing the estimated capacity of equipment within one plant.

Creating a network of DHC systems would be effective not only for current DHC facilities, but also for facilities where the supply is just being started. The city planning for this infrastructure to be networked is important. Enabling the DHC systems to effectively use wasted heat will ultimately help preserve the earth's environment.

Extensive Network System

Currently in Tokyo, there is an urgent need to use water and energy effectively, to establish a water-recycling system and a widely networked heat supply, and to study their impact on the ecology. A deep underground tunnel network, called the Cooperative Utility Space Network, is being proposed to alleviate these problems. This underground system would be composed of a vertical tunnel that would connect the surface of the earth with the deep underground base, and a horizontal tunnel that would join each of the underground bases. The recycling facilities for energy and water and the pipeline facilities also would be located deep under ground (Fig. 5.15).

Garbage, unused water from the river, rain runoff, unused energy and rejected heat, trunk line networks for dissemination of information, pipelines for oil and gas, and trunk lines for electric power can be placed in this deep underground space network (Fig. 5.16).

This new urban infrastructure would be most effective and meaningful when introduced into the central part of the city, where congestion is heavy. However, this project is in its proposal stage and is not fully planned. Also, no one has yet spoken strongly in support of its benefits. For example, the privilege of using space under roads has not yet been approved by law. Therefore, construction of a wide infrastructural network under roads has not yet been realized.

Yet, in order to extend the infrastructural network, Japan must seriously consider using space deep under ground. In terms of feasibility of using this space, there are many problems to be solved. Some of the obvious problems are financing a major project, providing ventilation, minimizing or avoiding the high risk of disaster, and security in a space ample enough in size for this project. However, companies that own their networks already partly use deep underground space. In order to extend the infrastructural network in the future, all of Japan must use the underground deep space. This result is becoming inevitable. As network demands on the infrastructure increase, new networks will be opened and existing ones will be renewed.

The country's transportation systems also are in desperate need of renewal and restructuring, and, with increased congestion, the deep underground again will be the most suitable place for development. To ease the traffic impact during the rush hour, a railway system deep underground around the central city would be economically feasible and highly practical. Currently, this transportation system is in the actual planning stage, even though there are some drawbacks. However, in terms of using the underground space below roads, there are many more problems to be solved. These include securing a large space, financing a major project, providing ventilation, and minimizing or avoiding the high risk of catastrophe. Still, the use of underground transporta-

- ● Network Subplant
- ■ Operating Area of Community Central Heating and Air-conditioning
- ○ Waterfront Redevelopment Area
- △ Garbage Dump
- □ Sewerage
- ▲ Steam-power Plant
- □ Propulsion Area of Community Central Heating and Air-conditioning

○ Length of Network Pipeline

	circumference ■	radius ■	intervals ■	TOTAL ■
A-LOOP	10,715	1,750	1,750	24,715
B-LOOP	21,430	3,500	1,750	35,430
C-LOOP	42,861	7,000	3,500	70,861
D-LOOP	64,291	10,500	3,500	92,291
E-LOOP	64,291	14,000	3,500	88,791
Grand total	203,588		14,000	312,088

• Total Length of Each Loop
= circumference + (Length of Access Pipes to the Inside Circle)

Figure 5.15. Basic project of extensive network (Toshio Ojima, Imaginable Tokyo, *Process Architecture, Tokyo, 1991, p. 99).*

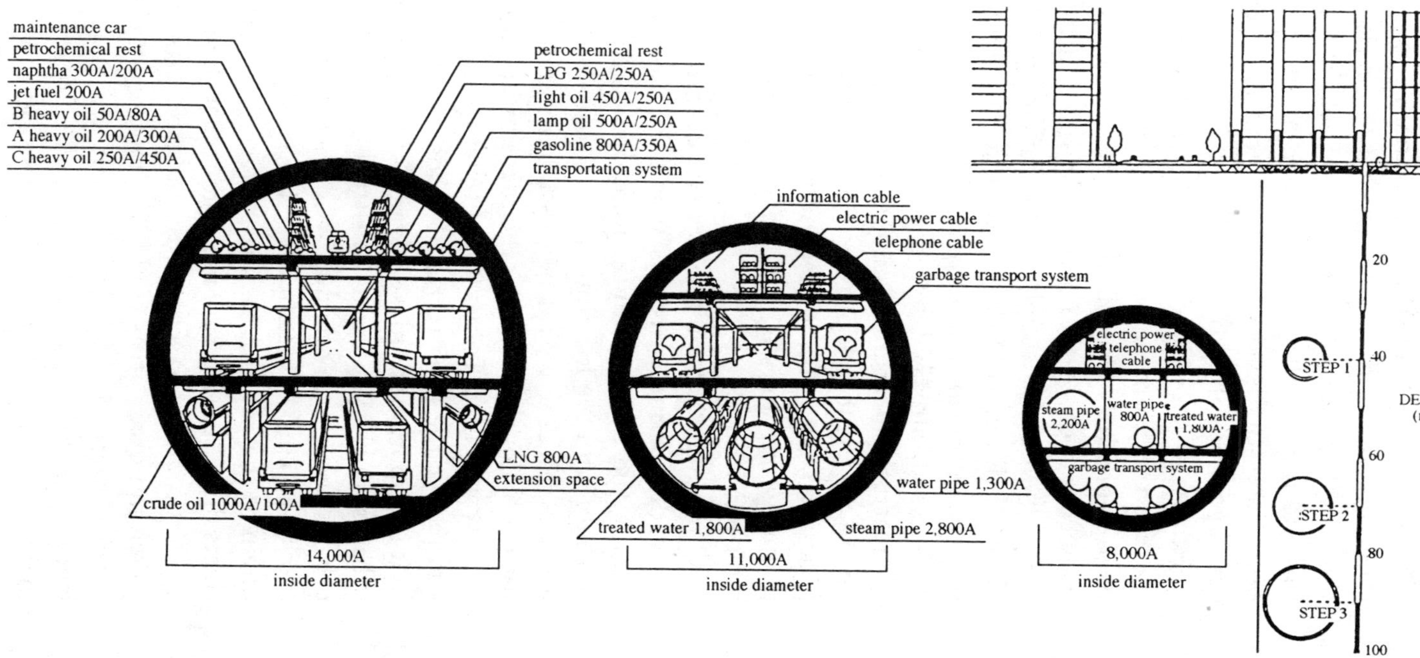

Figure 5.16. Section of extensive network (Toshio Ojima, Imaginable Tokyo, *Process Architecture, Tokyo, 1991, p. 99).*

tion systems for unattended commodity distribution is feasible, and would help to reduce traffic above ground. But if these underground facilities are left to develop without centralized planning, they will not be coordinated. Considering this and the fact that linear systems are bound to cross each other, uses of the deep underground in several layers also could be a possible solution.

Finally, it should be noted that the later construction is started, the more difficult it will become to accomplish a well-planned underground system. Therefore, to induce a planned use of deep underground space, it is proposed that the main lines of a supply-managing facility that has inherent financial difficulty, but is best suited for underground tunnels, be put inside a joint tunnel. Meanwhile, with the space under roads already being used, there is a growing need for the rearrangement of existing facilities that are wasting space due to a present lack of plan. For example, the underground pedestrian network, underground roads, electric and telephone wires, and so forth, have not been preplanned and built efficiently.

On completion of the deep underground infrastructure network, sharing of routes would increase, and more sublines would be placed in the joint tunnel. And, over a period of time, with the location of some main lines in deep underground space, it would be possible to improve the intricate use of shallow space under the roads by adding new facilities. This would give Tokyo some relief to its traffic congestion and reduce the need to repair roads.

It is important to make advances in improving the existing urban sprawl by keeping construction volume higher in the urban center (up-zoning), and lower in middle regions (down-zoning). Specifically, Tokyo's land slopes gently down to the bay and has many rivers. The points where the rivers divide the land are useful for redeveloping each district. Underground infrastructures could be planned here. Also, water and green areas could be strategically placed to divide the urban area in order to prevent fire disasters or heat islands. Establishing a cool island and a fire refuge on the land that is freed in this way would be possible by constructing an underground space network. The construction in surrounding farmlands would also stimulate urban redevelopment. The lower-density areas would serve as a backup when extra space is needed between the farmlands and the center of the city.

TOKYO BAY UNDERGROUND SPACE NETWORK PROJECT

Current Problems

Land Use

Heavy chemical production after World War II forced the redevelopment of Tokyo Bay by locating big industrial complexes, ironworks, power stations, and the like on the coastal area's reclaimed lands (Fig. 5.17). However, since this postwar transition period, these industrial lands have become idle. Figure 5.18 illustrates the land use today.

Now, a final decision on the Bay Area Redevelopment Project and on how to best use these lands is pending. For its present planning and development stage, more than 1 billion yen will be invested during the 21st century. A rearrangement of supply and disposal facilities, especially thermal power plants, sewage treatment plants, and garbage furnaces, will be necessary. Some of these facilities are old. In this case, it would seem more profitable to use the cheaper reclaimed lands to rebuild the facilities and use the more expensive land for new enterprises. Currently, however, some of these supply and disposal facilities are located at the connecting point with the inland area. In this case, from a practical viewpoint, it may be wiser to reutilize the land where the facilities are already built rather than relocating them to cheaper land.

Transportation

The Japanese trade pattern has changed in the last few years from export to import. Imports of farm products have increased especially rapidly. As the Japanese yen has grown stronger and imported products have become cheaper, NIES

HISTORY OF RECLAIMED LAND AROUND TOKYO BAY AREA

Figure 5.17. Reclaimed land of Tokyo Bay area (Toshio Ojima, Imaginable Tokyo, *Process Architecture, Tokyo, 1991, p. 76).*

countries have increased their exports, and the free market system has progressed.

Marine transportation covers 45 percent of all domestic transportation and 80 percent of the transport of basic industrial materials, such as petroleum, cement, and iron (Fig. 5.19). Currently, more than 500 million tons of products a year are transported by bay waters. The bay is crowded and has almost reached its capacity for handling vessels; as a result, many accidents have been occurring. Forty percent of

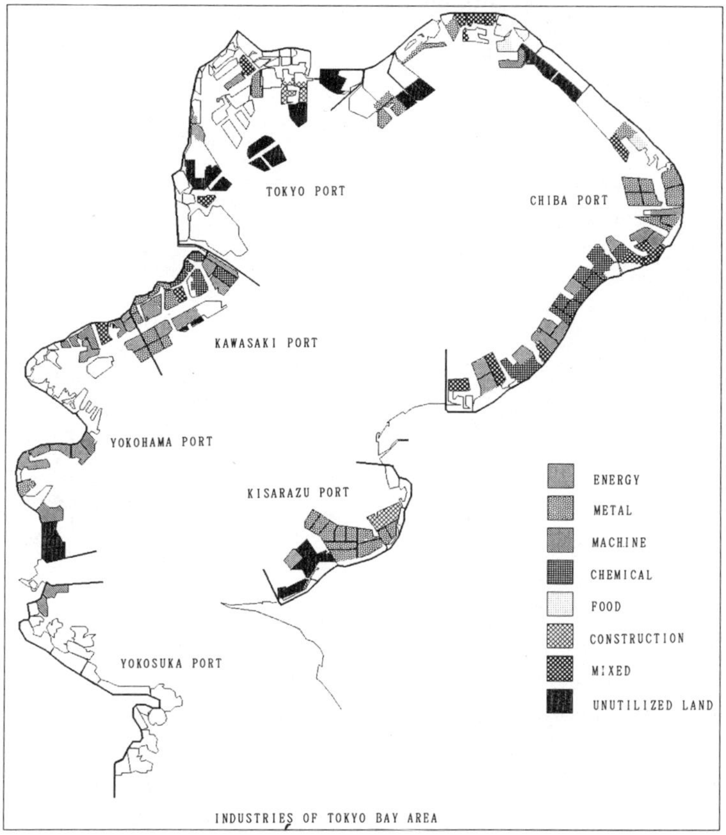

Figure 5.18. Present land use of Tokyo Bay (Toshio Ojima, Imaginable Tokyo, *Process Architecture, Tokyo, 1991, p. 76).*

the ships carry oil products, so there is an extremely high possibility of serious accidents happening in the bay.

Danger of Disaster

The waterfront areas of Tokyo Bay are extremely dangerous. Damage to buildings and facilities is much worse in these areas compared to the damage to buildings and facilities in other areas due to deterioration caused by their proximity to the water. Other potential disasters include:

- Deterioration or stopping the supply function in the central city area will affect the developing area as well.
- Water service, which is already in limited supply, will be strained even further.

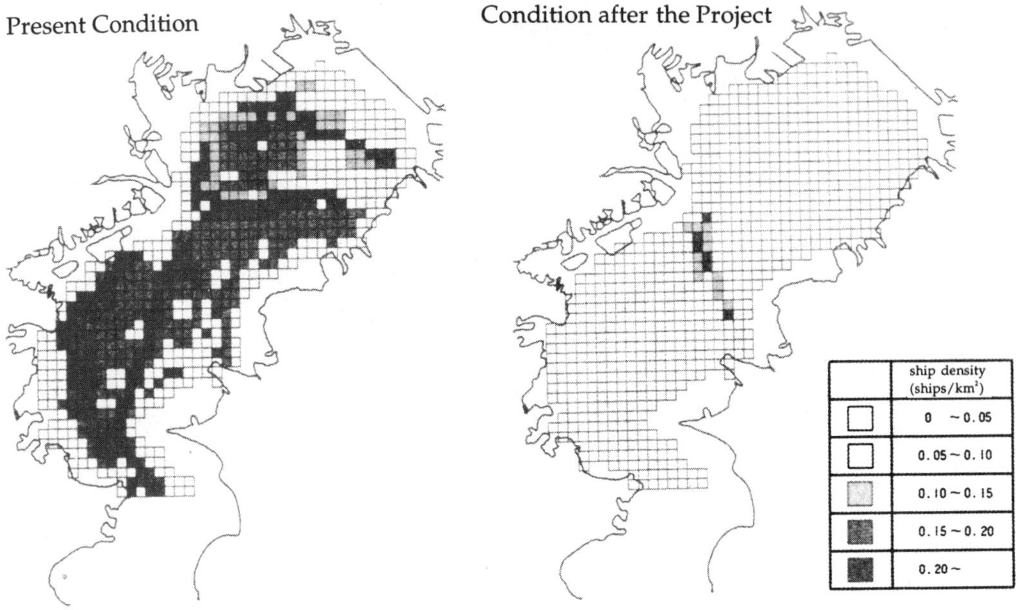

Figure 5.19. Ship density in Tokyo Bay (Toshio Ojima, "Creating New Urban Life through Free Underground Tunnel," IATSS Review, *Vol. 15, No. 1, 1989, p. 21).*

- The developing area is an island surrounded by water, with bridges and tunnels as the only means of protection. When these routes become blocked, it will be difficult to protect things and people. Disaster relief would be limited.
- Once the liquefaction happens in the reclaimed land, large amounts of water and sand will spout into the air and fall to earth, and eventually the ground will cave in. Generally, this surface collapse will not occur uniformly. The foundation support will be weakened. Heavy structures, which are supported directly by the earth's surface, will also cave in. On the other hand, lighter items in the ground, such as reservoir tanks, sewage tubes, and manholes, will surface.

Measures must be taken to solve these problems so that buildings will not be destroyed in the event of a disaster. This phenomenon can be prevented by using firm foundations or anchors to support and fix the structures of the buildings soundly.

Project for Releasing Space in Bay Area

Tokyo Bay Compound Infrastructural Network

Tokyo, as the center of Japan, supports one-fourth of Japan's population. The question becomes, then, what infrastructure should Japan build now in order to realize a future that would be most pleasant, safe, healthy, and efficient for the city?

First, Tokyo needs to accommodate the waterfront area with a specially designed infrastructure. The bay area infrastructure has to be unique and independent, since this area lies over a fault line. If an earthquake should occur, the risk is too great that underground pipe networks would be damaged severely and the city supply function would become limited or would stop altogether; that is, a liquefaction phenomenon could occur. Therefore, an independent infrastructure or a lifeline could be installed in the deep stable layer in the reclaimed waterfront area (Fig. 5.20). This lifeline would be

Figure 5.20. Section of multiservice tunnel (Toshio Ojima, Imaginable Tokyo, *Process Architecture, Tokyo, 1991, p. 80).*

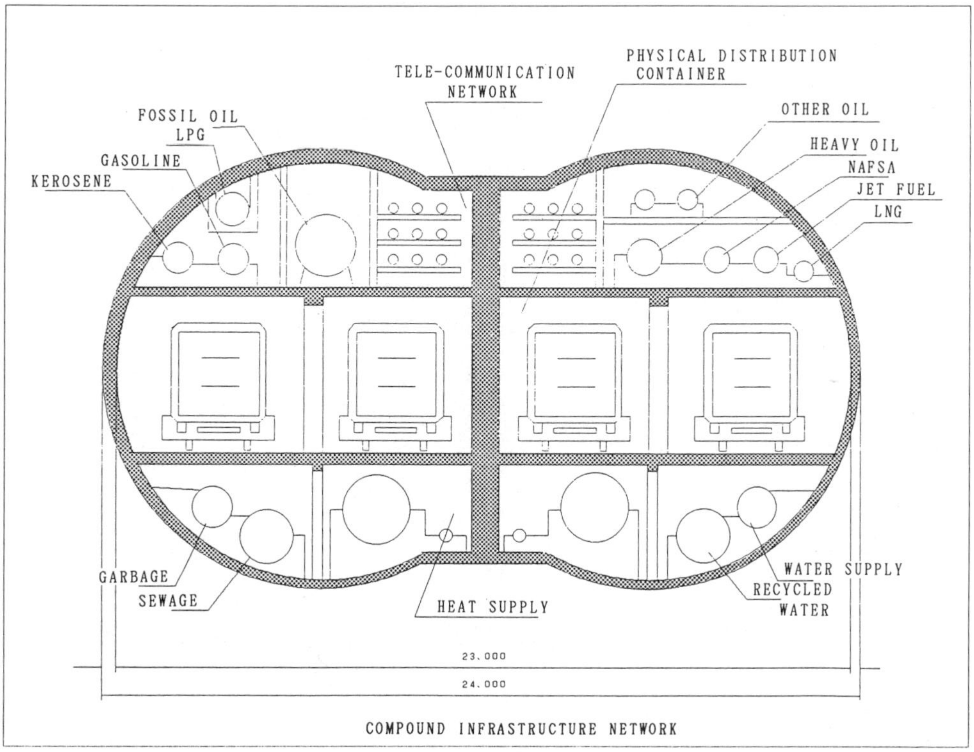

built in addition to the usual network of water, electric power, gas, and information transportation. This deep underground network of utility corridors in the Tokyo Bay area is proposed as the Life Anchor Project.

The proposed Life Anchor Project calls for an independent basic facility for electric power, water, and an information system (Fig. 5.21). Several anchors would be set on a stable foundation in the deeper underground in the bay area so that the infrastructure would not be damaged even during an earthquake.

The Tokyo Bay area would be further supported in case of disaster by designing backup facility bases on the deep and stable subterranean ground under the diverse area of the bay's shore. This facility would be able to maintain the city's function in the bay area for at least one week after an earthquake. Backup equipment, such as oil tanks, water tanks, heat storing tanks, private dynamos, boilers, refrigerating machines, and computers would be installed in these facilities.

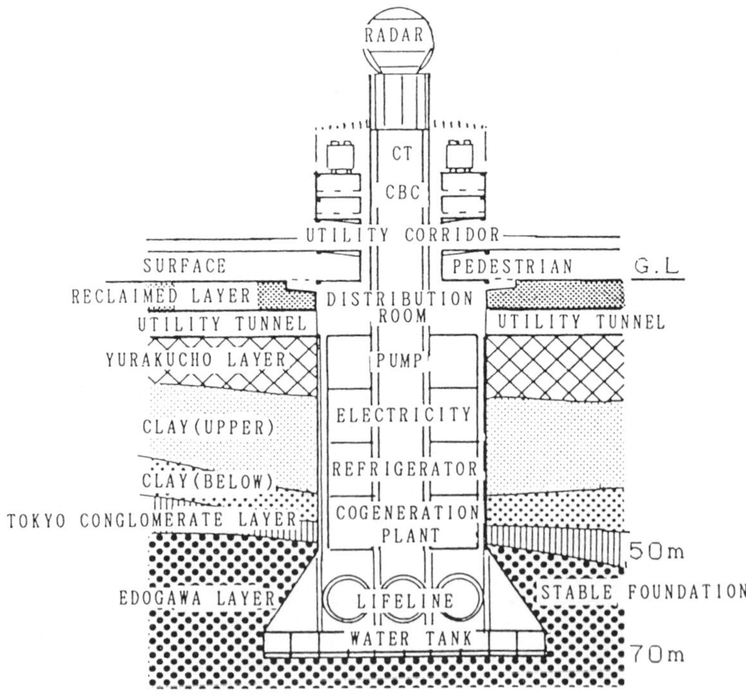

Figure 5.21. Section of life anchor (Toshio Ojima, Imaginable Tokyo, *Process Architecture, Tokyo, 1991, p. 79).*

The facilities' bases would be connected by deep underground tunnels for transportation. Oil-related articles would be shipped by containers through the pipeline from the entrance of the bay to the plants inside the bay. Other freight would be distributed in the same way. The oil transportation network and the physical distribution network would have to be located in these deep underground tunnels (Fig. 5.22).

About 40 percent of the ships that come into the bay are carrying oil. It would be most advantageous to eliminate these hazardous ships from the area and create an amenity with parks, beaches, yacht harbors, and so on. This could be accomplished by building an outport entrance outside of the bay to avoid congestion from ships that go into and out of the area. The outport would become the entrance to a newly installed oil transportation network. It would have to be deep enough for a large-scale tanker to moor and connect with the underground network inside the bay. The base for distributing oil should be constructed near related plants that use oil,

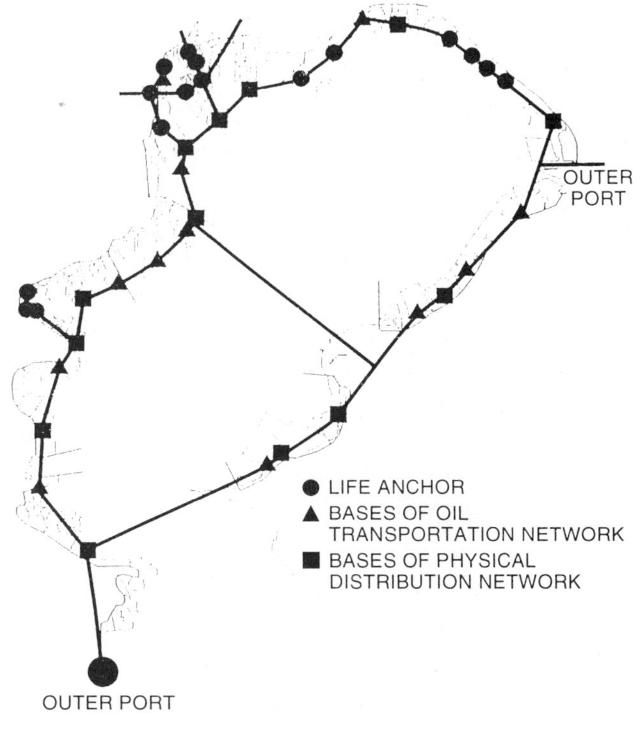

Figure 5.22. Route of Tokyo Bay compound infrastructure (Toshio Ojima, Imaginable Tokyo, *Process Architecture, Tokyo, 1991, p. 80).*

such as chemical factories, gas plants, thermoelectric power plants, oil refineries, and the like.

In addition, changes are being proposed in the distribution network to reduce the number of freight accidents and free some of the bay area for parks and recreation. It is planned to use 10-ton containers in place of freighters.

The bases of the distribution network should be located on the sites of existing or planned large-scale harbor facility areas. In the proposed network, 96.9 percent of the oil and 39.5 percent of the freight that is now transported by ships would be transported by this network. This would reduce the number of the ships by 82 percent from 280,000 ships to 50,400. In turn, this would also reduce the number of accidents, which are now a serious problem. For the second industrial-type facilities, 120,000 hectares of the Tokyo Bay area can be zoned for public recreation by relocating the harbors to seaside industrial complexes.

Road traffic would also be reduced and, through efficient transportation by the network system, energy use would be cut. As this system will be using the underground, there will be no problem with separating citizens from the waterfront, as is often the case in road development. The Tokyo Bay shore would become a beach park, where highly developed industrial and commercial facilities would be scattered. Hotels and residential buildings surrounded by trees and landscaping would be constructed in place of the factories and industrial complexes, which have stood as a symbol of modernization in the 20th century. Reconstruction of city facilities, which has come to a standstill, would be promoted to create beautiful bayshore scenery comparable to southern France or Rio de Janeiro.

CONCLUSION

Infrastructural systems have a long history of using underground space; these systems establish the environmental condition of the city. Tokyo's environmental problems are now becoming global problems, and technological city planning should be promoted. The basic plan for reconstructing the urban environment is now more important than the environmental assessment measures that evaluate the environment afterwards. The ideal future use of this region will become clear by making up a long-term, phased environmental plan for the infrastructure before the suprastructure is constructed.

SELECTED READINGS

Civil Engineering Society, *Geo-Front New Frontier,* Gihodo, Tokyo, 1990, p. 136 (written in Japanese).

Engineering Promotion Society, *Report of Underground Space Utilization,* 1991 (written in Japanese).

Japan Development Bank, *Direction of Deep Underground Utilization,* Tokyo, 1989, pp. 33, 86 (written in Japanese).

Toshio Ojima, "Creating New Urban Life through Free Underground Tunnel," *IATSS Review,* Vol. 15, No. 1, 1989, p. 21 (written in Japanese).

Toshio Ojima, *Heating the Big City,* NHK Books, Tokyo, 1975.

Toshio Ojima, "Imaginable Tokyo," *Process Architecture,* Tokyo, 1991, pp. 62–99 (written in Japanese and English).

Toshio Ojima, *Infrastructure of Japan,* Nikkan Industry Newspaper, Tokyo, 1983, pp. 10, 112–13 (written in Japanese).

Toshio Ojima, "New Underground City Project," *Journal of Underground Space Utilization,* No. 2, Urban Underground Space Center of Japan, Tokyo, 1988, pp. 8–19 (written in Japanese).

Toshio Ojima, *Planning of Urban Utilities,* Kashima Publication, Tokyo, 1973 (written in Japanese).

Toshio Ojima, *The Reconstruction of Tokyo,* Chikuma Shobo, Tokyo, 1986 (written in Japanese).

Toshio Ojima, *To Construct a Picturesque City,* NHK Books, Tokyo, 1984 (written in Japanese).

Toshio Ojima, *Twenty-First Century Design of Tokyo,* NHK Books, Tokyo, 1986 (written in Japanese).

Toshio Ojima Lab., *Tokyo Underground Project,* Kenchikubunka, Shokokusha, Tokyo, 1982, p. 63 (written in Japanese).

Tokyo Electric Company, *Takanawa Electric Transformer,* Tokyo, undated pamphlet (written in Japanese).

Tokyo Metropolitan Government, *Master Plan of Second Generation Sewage,* Tokyo, 1991, p. 53 (written in Japanese).

Tokyo Metropolitan Government, *Report of Underground Garbage Plant System,* Tokyo, 1993, p. 52 (written in Japanese).

Tokyo Metropolitan Government, *Underground River,* Tokyo, undated pamphlet (written in Japanese).

Tokyo Metropolitan Government, *White Paper of the City '91,* Tokyo, 1991, p. 27 (written in Japanese).

Urban Underground Space Center of Japan, *Challenge to Geo-Front,* Seibunsha, Tokyo, 1989, pp. 134–37 (written in Japanese).

6
INTERNATIONAL USES OF GEO-SPACE: INDIGENOUS AND CONTEMPORARY EXAMPLES

G. S. Golany

INDIGENOUS EXPERIENCE

The earliest natural shelter used by humans for habitation was the cave. Throughout the many millennia in which this form of shelter was used, humans came to recognize the significance of the thermal and process patterns of the soil. They used these thermal conditions to their advantage for living, food storage, working, and the protection of their well-being. The cave also provided an ambient protective environment against the harsh diurnal and seasonal climates.

Today, we can find three major indigenous communities that still use belowground space and that have accumulated thousands of years of experience living in this type of habitat. The largest concentration is located in the loess soil zone of northern China, where approximately 35 to 40 million people live in belowground dwellings (*yao dong*) in rural and urban communities. It is estimated that the Chinese have used belowground space for more than 4,000 years. The second largest concentration is scattered over the Matmata plateau in the southern section of Tunisia and the northern part of the Sahara Desert, where some 20 fortified rural communities exist. In Tunisia, the size of the communities ranges between a few thousand persons, as in Matmata village, to a few hundred persons, as in the very southernmost villages. The third region is Cappadocia, in central Turkey, where more than 40 towns and rural communities thrive. The practice of using underground space for living has been flourishing in this region for more than 6,000 years.

In addition to these three ancient-contemporary belowground communities, there have been similar practices throughout history in other parts of northern Africa, Spain, central and southern France, southern Italy (Trulli region and Sicily), Poland (Wieliczka, underground salt quarries and cavities near Krakow), southwestern United States, Canada (Eskimo dwellings), Iran, India (Elura and Ajanta in western India, and Bahaja), the Middle East (Avdat and Petra of the Nabataeans), Afghanistan (Bamiyan valley), Ethiopia (Adi Kado), Israel (Jerusalem, Beersheba suburbs, Megiddo), and Persia, and in many other scattered regions of the ancient world. The history of many of these sites goes back as far as the third millennium B.C. Only Cappadocia has developed a fully interconnected city network below ground.

The common denominators among these three major belowground community concentrations are their location in regions of arid and semi-arid climate, their lengthy accumulated historical evolution, their distinct architectural design, and their adjustment to the harsh environment. Their arid climates are characterized by temperature fluctuations from extreme highs by day to extreme lows at night, and by torrential, yet limited, rainfall. It seems that the need for protection against the climate, in regions where fuel resources are minimal, factored significantly in the selection of underground living. Availability of water, usually not needed for construction of belowground dwellings, is another factor in arid regions. In the three areas mentioned, the inhabitants adapted the "cut-and-use" method of construction, in part, because it is one of the few, if not the only, method that does not necessitate the use of water in the building process. Also considered was defense, especially in the cases of Tunisia and Cappadocia, where fortification of belowground communities was possible.

All life cycles revolve around the soil. With the development of technology, humans have been able to create and use the spaces below ground for a variety of land-use types, most of which were dwellings. The most common usage of the belowground space throughout ancient history was primarily, yet not exclusively, limited to dwellings, food storage (mainly grains), mines, defense, and burial. Cave dwellings have been used over many eras, and, in some places, even until modern times. Evidence of caves inhabited by humans during the Paleolithic period and afterward were found in southern France, Israel, northern China, South Africa, and many other sites throughout the world. People were, and still are, successful in using synthetic materials to build dwellings that imitate the cave environment. Throughout history, humans have been attached to their native soil, which produced their food, supplied their natural materials to exist, and furnished a final resting place.

China

Belowground space usage in China is the most ancient and extensive in the world in terms of scale and magnitude.[1] The housing distribution in this region is almost exclusively associated with a special, adaptable soil called *loess*. Among the five loess soil regions of the world, the Chinese one is the largest and covers 631,000 square kilometers, or 6.6 percent of China's total land area. It also supports around one-fifth of the country's population. The total region accommodates approximately 200 million people, of which about 40 million live below ground in what are commonly called cave dwellings (*yao dong*). The highest concentration is located in the eastern part of Gansu Province and reaches more than 83 percent of its population. Stretched over a zone of more than 2,000 km in length and around 800 km in width along the northern section of China, this loess soil zone is crossed by the Huang He (Yellow River). Five provinces fall within this region: Henan, Shanxi, Shaanxi, Gansu, and Ningxia. The significance of the soil lies in its firmness, stability when dry, suitability for agriculture, absence of stones, and uniformity. Yet it is sensitive to erosion and requires special treatment.

Two major types of cave dwellings exist in China. The first type is the pit cave dwelling (Fig. 6.1); the other is the cliff cave dwelling (Fig. 6.2). The pit dwelling is commonly used on flat land.[2] It is composed of a large pit, exposed to the sky, with its opening averaging 8 by 10 meters and around 8 meters in depth. It is usually square or rectangular in form. The patio of the pit is surrounded by two to three rooms on each side. Each room is a separate vaulted space dug into the side of the patio, with an average depth of 6 to 7 meters below the surface. It has a width of 2.5 meters and a comparable height. Pit dwellings consume a great deal of land but receive limited light and sunshine. Some farmers use the land above the cave dwellings as a threshing floor. The pit

Figure 6.1. Bird's-eye view of Liu Xueshi family pit dwelling in Zhong Tou village, near Luoyang city, Henan Province (Golany, Design and Thermal Performance: Below-Ground Dwellings in China, *U. of Delaware Press, 1990, p. 121).*

Figure 6.2. Cliff cave dwelling complex at Gaomaowan Cun village, 12 kilometers east of Yan'an (Golany, Chinese Earth Sheltered Dwellings: Indigenous Lessons for Modern Urban Design, *U. of Hawaii Press, 1992, p. 55).*

is approached by underground stairways, beginning on the site near the pit opening and leading down to the patio.

Cliff dwellings are cut from sharp cliffs, with the rooms facing a wide terrace of leveled soil. The size of cliff dwelling rooms is similar to that of the rooms in the pit dwelling; yet, they receive much more sunshine and light throughout the day when they are positioned facing south. Some of these cliff dwellings are located on the narrow strip bordering the zone between the plateau and the cliff. The plateau is used for grain production, while fruit trees are planted on the cliff, using soil that would not have been used otherwise. Cliff cave dwellings consume less space than pit cave dwellings. Rooms are usually accessible only from the patio or the terrace and are not connected with each other. Some cliff cave dwellings combine belowground space with aboveground space.

Cave dwellings are used in both rural and urban areas. The Chinese currently use belowground space for food storage, shopping centers, educational and cultural facilities, restaurants, workers' clubs, hotels, and hospitals (Figs. 6.3, 6.4, and 6.5). Hospitals, especially, derive significant benefits from belowground space. It has been found that patients with external wounds recovered from their operations 10 percent to 30 percent faster in a belowground space than in an aboveground space.[3]

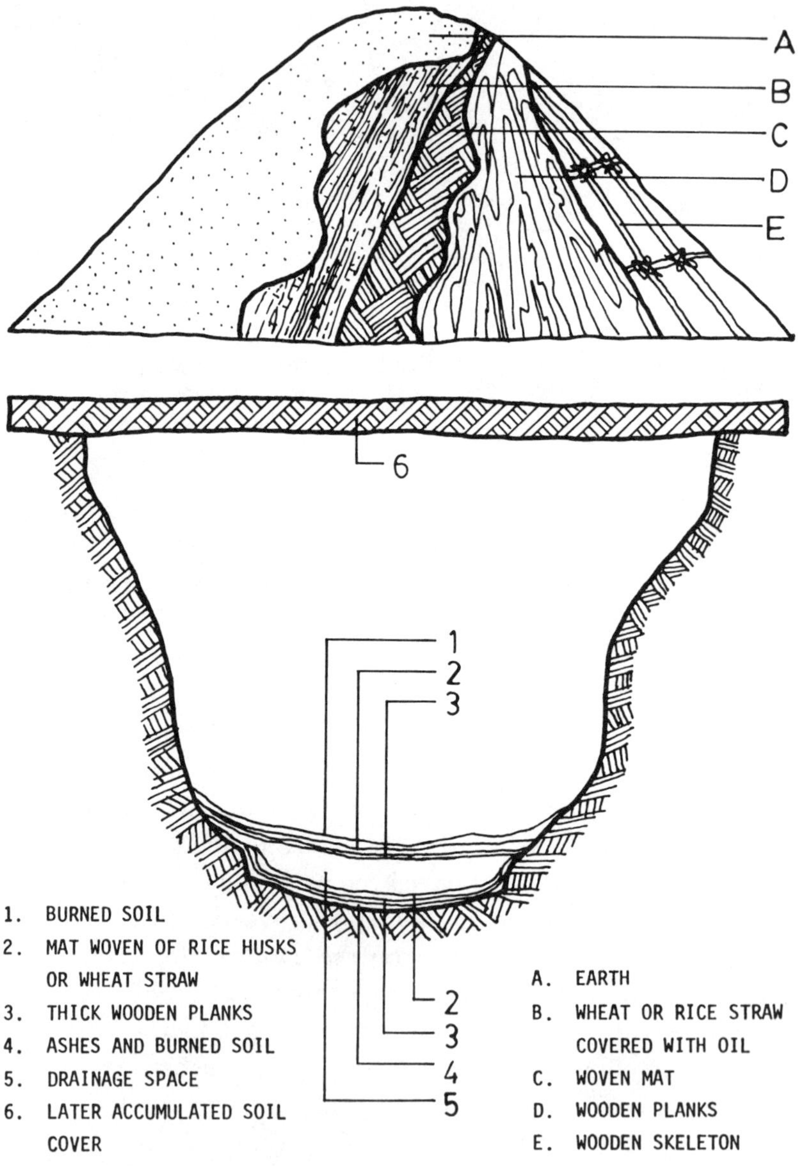

Figure 6.3. Cross-sectional view of an ancient grain pit in Hanjia, Luoyang city (Golany, Urban Underground Space Design in China: Vernacular and Modern Practice, *U. of Delaware Press, 1989, p. 36).*

Figure 6.4. Conference room in Qianmen, Beijing. The room is part of an underground urban network originally designed for civil defense (Golany, Urban Underground Space Design in China: Vernacular and Modern Practice, *U. of Delaware Press, 1989, p. 86).*

China is certainly a remarkable laboratory for experimentation and investigation of belowground housing. The lessons that can be learned cover soil thermal performance, health conditions, and food storage. The historical Chinese experience, in all its magnitude, is not confined to dwellings only. The Tang and Song Dynasties (approximately 7th century to the 13 century) used belowground space extensively for grain storage. Grain storage had been recently introduced at that time and was found to well preserve most stocks. One remarkable site was in the town of Hanjia, which was a granary town with 412 storage pits.

Cappadocia

Belowground space utilization in Cappadocia, Turkey, is unique among all the other indigenous subterranean spaces. Its distinction is found in its construction in tufa soil, accompanied by a sophisticated system of tunnels, and its develop-

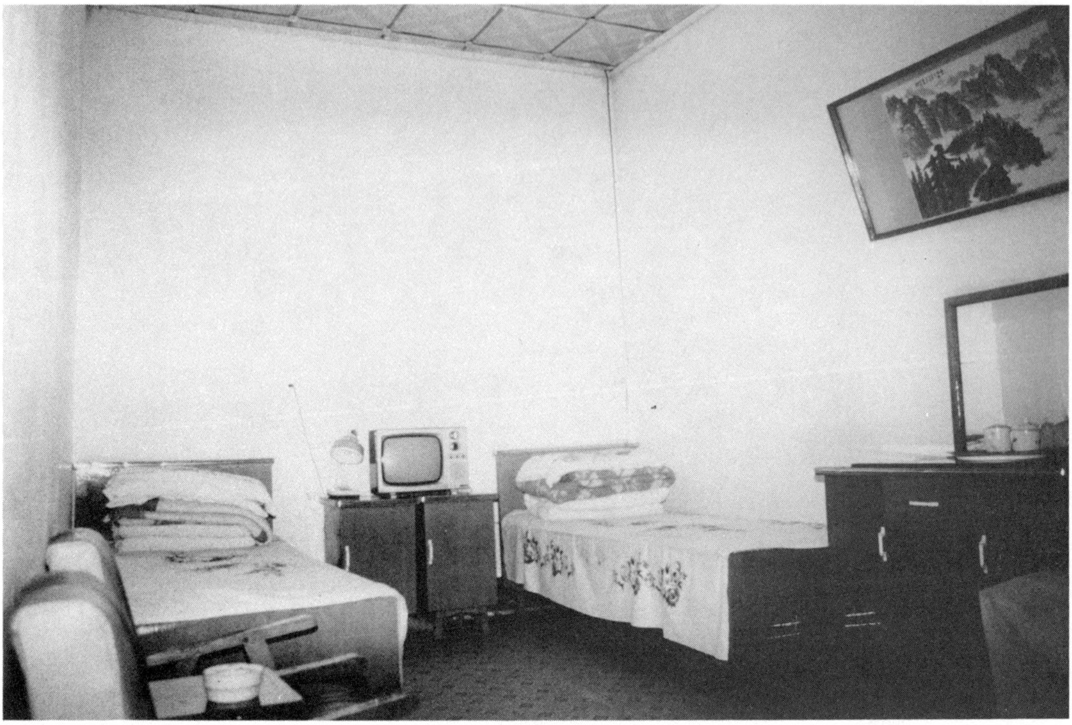

Figure 6.5. Belowground hotel room in Chongqing city of Sichuan Province in southern China (Golany, Urban Underground Space Design in China: Vernacular and Modern Practice, *U. of Delaware Press, 1989, p. 62).*

ment of cities and villages. Cappadocia is located in the central part of the Turkish plateau, about 400 km southeast of Ankara, the capital of Turkey.

The name Cappadocia derives from Persia, the ancient district in eastern Asia Minor, which stretched from the Taurus Mountains in the south to the Black Sea in the north. The region is a rugged tableland with poor soil. The plateau was formed of deposits of tufa and volcanic ash, which were eroded by water and wind over time to produce a large number of conical forms composed of relatively soft stone (Fig. 6.6). Its climate is warm in the summer and cool in the winter. It is dry and does not support agriculture. Most of its precipitation comes in the form of snow in the winter.

Historically, people began to use the belowground space in this area around the fourth millennium B.C. Since then, it has been used by the Hittites, Romans, Greeks, Byzantines, and, finally, the Turks. The uses of belowground space

Figure 6.6. Cliff dwellings in Grome Valley, Cappadocia, Turkey. Landscape of a tufa rock where wind deflation creates a forest of cones. This landscape has been used by people for more than six millennia, creating geo-space for diversified land uses (courtesy of the Turkish Embassy in Washington, DC).

ranged from dwellings, cemeteries, catacombs, and defensive structures to churches, monasteries, villages, and cities. This belowground network is still used for storage and living and as the national center for citrus refrigeration (Fig. 6.7). The Byzantines made very extensive use of the belowground space to build monasteries and churches decorated with artwork.

Extensively developed cities and villages were constructed during the Hittite period, especially from the middle of the second millennium to the middle of the first millennium B.C. Other underground settlements were built between 3000 B.C. and 100 B.C. Today there are 155 underground villages, towns, and cities in Cappadocia, and many more underground caves. The number of underground shelters in the area is very high, and it is almost impossible to survey all of them.[4] Nevsehir and Kayseri are the two major centers of underground cities, but there are others as well.

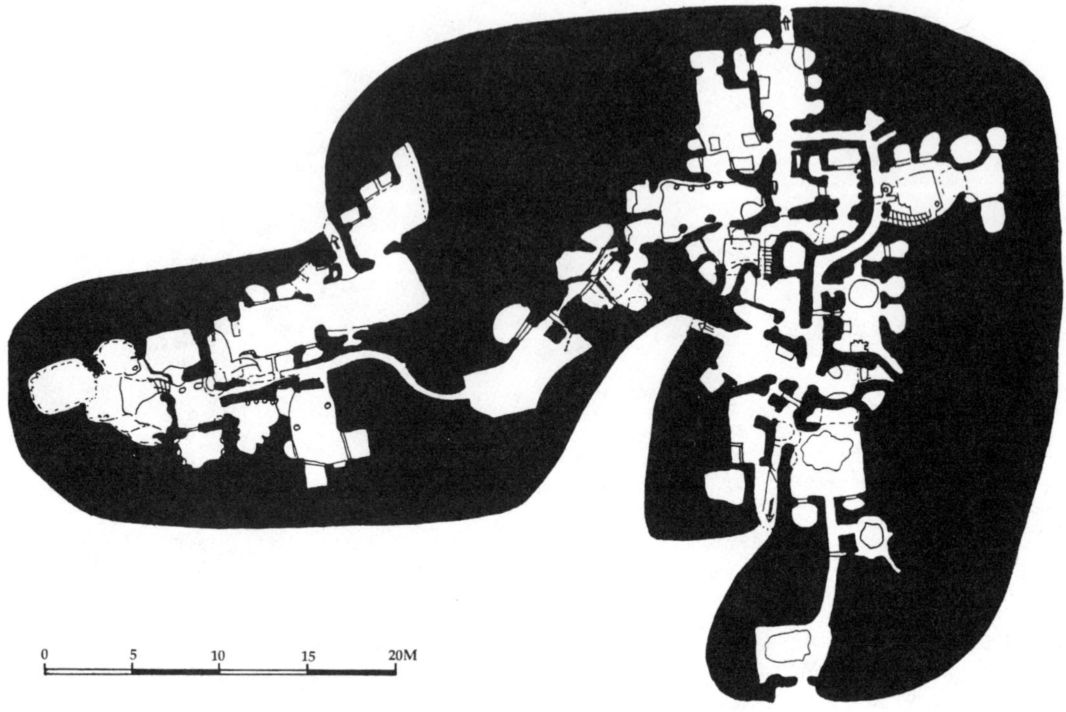

Figure 6.7. Partial plan of the underground city of Kaymakli (adapted and reproduced from Anahtar, "An Earth Integrated Hotel in Cappadocia," Master's thesis, Middle East Technical U., Ankara, Turkey, 1984, p. 21).

Cappadocia, built toward the end of the Byzantine period, is the only site where underground cities were developed with underground network connections.[5] Within the cities, a network of alleys and passages lead from one neighborhood to another or from one quarter to another. Derinkuyu, an underground city, was discovered in 1963 and is located 29 km from Nevsehir. It is one of the 30 known underground cities. Derinkuyu occupies 1,500 square meters, but only one-fourth of the city has been uncovered so far. It is theorized that this and other underground cities were used primarily when there was a threat of danger, at which time the people abandoned the cities until the surroundings were secure. Before the Hittite period, there were many corridors wide enough for two people to pass. Every corridor had an entrance with a stone door, which was easy to open from the inside and difficult to open from the outside. There were eight levels in this city, which included a missionary school, baptismal pool, a storehouse for wine, and kitchens in the shelters.

Kaymakli, another underground city, is located 20 km south of Nevsehir and 9 km from Derinkuyu. Four levels were opened to visitors in 1974; other floors have not yet been uncovered. Most of the city was used from the fifth to the tenth centuries A.D., and use of the top floor extended back to the Hittite period. The city has a lengthy communication network.[6] The Ozkonak underground city is considered the largest in the region of Cappadocia, and is located 19 km north of Avanos. It has 10 rooms, four large lounges, four graves, three ventilation chimneys, two large round gate stones, and 30 wall recesses. Only a small part of the city has been uncovered.

There are a large number of underground settlements in the region of Kayseri, which was the center of the Cappadocian kingdom. After Nevsehir, this is the second most important center of underground cities. The underground city at Asarkaya is located east of Kayseri. It is speculated that Asarkaya was inhabited between 4000 B.C. and 800 A.D. It has eight levels; three of the levels are above ground and four are below. Other underground cities are Kayseri (3 km from Kayseri City), Agirnas (25 km east of Kayseri), and Doganli (in the Kayseri region).

Underground settlements in Cappadocia were used primarily in the Hellenistic and Roman periods, and many of them were built during this time. The Hittites also used this belowground space extensively and painted many drawings on the rocks. They also built new caves, tunnels, and rooms.

Tunisia

Tunisia has the largest concentration of belowground space used for dwellings in North Africa. Other countries that have similar usage are Egypt, Libya, Morocco, and Algeria. Tunisian subterranean housing originated with the Berber tribes of North Africa, and is thought to have existed for many centuries in the Matmata plateau, in the southern part of Tunisia bordering the northern Sahara region. The plateau stretches from north to south about 500 meters above sea level. It is 200 km long and has an average width of 25 km. With the swift Islamic invasion of North Africa during the seventh century, the Berber tribes were forced to flee by way of the narrow plain thoroughfare located between the Mediterranean Sea and the Matmata plateau (Fig. 6.8). They established their system of belowground usage as a matter of defense and as a response to stressful climatic conditions. The plateau itself is made of light brown desert soil, which holds itself firmly over the cavities. Toward the southern part of the plateau, limestone is most prevalent and determines the form of the belowground dwellings. However, water is very limited within the plateau and, with rainfall averaging less than 100 mm per year, there is not a sufficient supply to support farming. In addition to the Matmata plateau case, subsurface housing was constructed by the Romans in the city of Bulla Regia in the northern part of Tunisia (Fig. 6.9).

At the Matmata plateau, there are around 20 communities where housing is exclusively below ground.[7] The largest

Figure 6.8. Bird's-eye view of Matmata's subterranean village community, northern Sahara Desert, southern Tunisia.

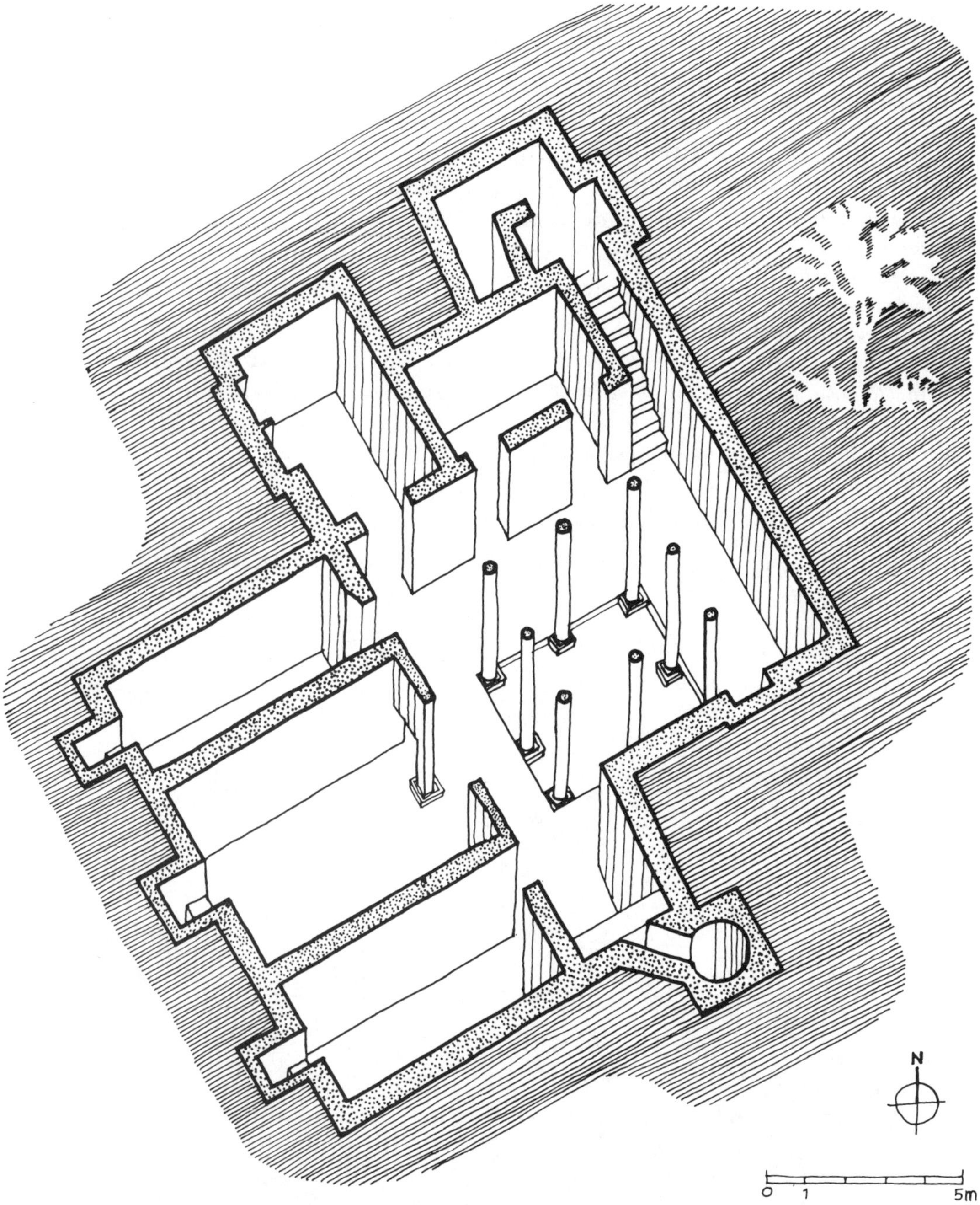

Figure 6.9. Roman subsurface villa, named the Hunt House, constructed within Bulla Regia city, northern Tunisia (Golany, Earth Sheltered Dwellings in Tunisia: Ancient Lessons for Modern Design, *U. of Delaware Press, 1988, p. 118).*

Figure 6.10. Schematic field survey drawing of the community pattern in Matmata village: 1. Well 2. Mosque 3. Souk (marketplace) 4. Post Office 5. School 6. Revenue Office 7. Government Offices 8. Sidi Driss Hotel 9. Marhala Hotel 10. Ksar 11. Sidi Mosa Tomb 12. National Guard (Reproduced from hand-recorded notes of the Institut Technologique d'Art, d'Architecture et d'Urbanisme de Tunis, (n.d.), and also from Ben-Hamza, "The Cave-Dwellers of Matmata: Ritual and

among these communities is the village of Matmata, located at the northern part of the Matmata plateau, some 30 km south of the shore city of Gabes (Fig. 6.10). There are two types of subterranean dwellings in the Matmata plateau. One is the pit type, primarily found in the Matmata village. The other is the cliff type, such as the mountainous cliff villages of Chenini de Tataouine and Douiret. The latter are located approximately 80 km south of Matmata village. Douiret and Chenini are the extreme southern villages of Tunisia, penetrating into the Sahara Desert.

Pit-type subterranean dwellings take the form of a large square or rectangular hole of around 8 or 10 meters in width or length, and 7 to 10 meters in depth. The soil is relatively soft and free of stones. Indigenous people adopted the construction method of cutting the soil and immediately using the space created. They carved vaulted rooms out of each of the pit walls perpendicular to the patio wall, averaging 4

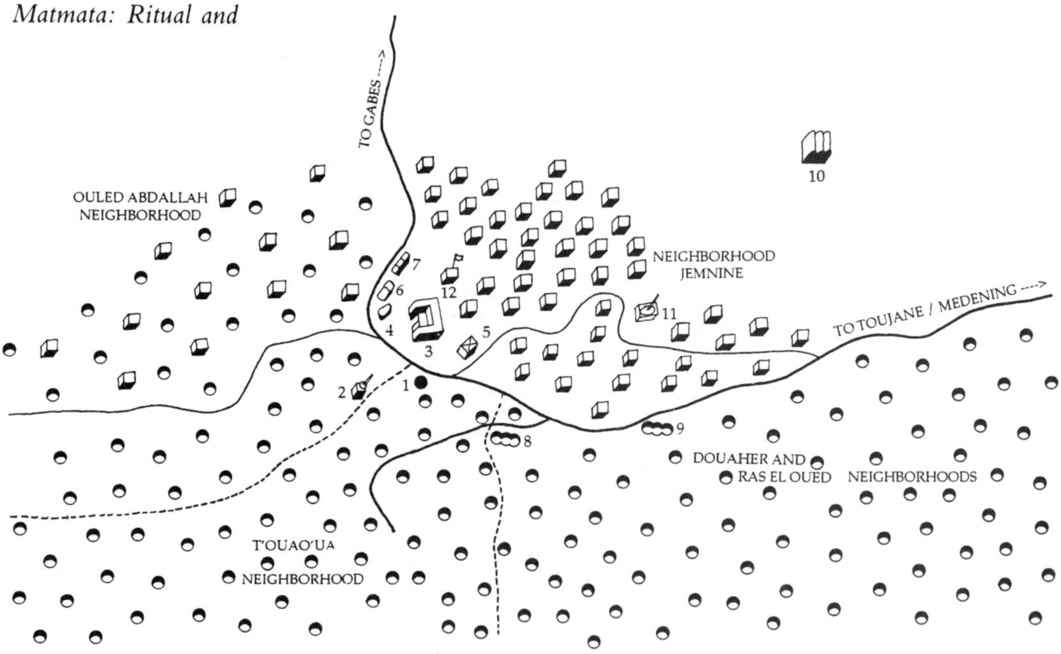

meters in width and 6 meters in depth. The height of each room is between 2.5 and 3 meters. Light shines into the room through the doorway and one small window. No chimneys were constructed for air ventilation or circulation. Within the main room, however, some niches were made for the placement of candles or for closet space. In some cases, the main room was connected to another room constructed deeper into the ground and used mostly for storage. Sometimes, the level of the main room was the same as that of the patio. The floor was around 10 meters from the surface of the soil. In other cases, there was another small room dug horizontally at a depth of 5 or more meters from the surface level. The deeper room has stable diurnal temperatures, while the other experiences some temperature fluctuations. Within the pit-type dwelling, there can be four to eight vaulted rooms. The entrance to the unit from the soil surface is by underground stairways, beginning at the soil level and gradually curving to reach the patio level. In the middle of the patio floor was a large cistern, mortared to store the collected rainwater, with its small opening covered with stones. The water was drawn by a bucket.

The cliff-type belowground dwellings are typically found in Douiret, Chenini de Tataouine, Ghoumrassen, and elsewhere in the southern Matmata plateau. The Chenini de Tataouine village is located on a site composed of alternating soft and hard limestone (Fig. 6.11). Geologically, all layers are horizontally stratified and are 2 meters thick. The dwellers created living spaces by extracting a soft layer, leaving the upper hard layer as a ceiling and the lower hard layer as a floor. Each space was around 3 meters wide and 4 meters or more in depth. The opening was partially closed, leaving space for a door, and the extracted soft layer was used for the construction of a terrace in the front of the dwelling unit. A distant view of the village reveals a horizontal setting of defensible cliff dwellings arranged in a regular, compact form overlooking the lowland.

The subsurface Roman houses built in Bulla Regia are different from the belowground dwellings developed in the Matmata plateau. We tend to believe that the Romans were

Figure 6.10. (Continued) *Economic Decision Making in a Changing Community," Ph.D. diss., Indiana U., 1977, p. 34, and this author's field survey).*

(a)

(b)

aware of the significant advantages of the belowground housing built by the Berbers in North Africa before the Roman occupation of the region. Also, our research findings support the idea that the Romans used the subsurface space in the summer to escape the intense heat of the hot-dry climate of the city of Bulla Regia. The city itself was located in a large, closed valley used for wheat production. The Berbers dug their homes deep below ground, leaving a mass of soil about 5 meters thick above the vaulted room.

In comparison, the Romans left only about one-half meter of soil above their rooms, creating subsurface, not subterranean, dwellings. The Romans also introduced advanced and modern technology to the construction of belowground dwellings. The walls were covered by stones, all the floors were paved, and a light shaft was constructed at the back of the room to allow sunshine and air circulation. The Romans also introduced peristyles into the dwelling, which consisted of Roman stone columns surrounding an open-air patio and encircled by a corridor. A wide guest room or dining room opened to the patio, and a well was constructed on the site of the dwelling. These homes served as villas for the Roman officers.

In northwestern Libya, there are some subterranean villages and cities in the mountains, not far from the Tunisian border, named Gharyan, Nalut, and Ghadamis. These belowground settlements are still populated by Berbers and Arabs.

India

India is another country that experimented with belowground space for many centuries. The focus there has been not cities or villages, but primarily temples and monasteries. The belowground space of India is unique in that it is carved in rock and cliffs, enabling it to last for centuries. The sites selected were usually far from populated areas, isolated in the mountains, close to water, and in tranquil and contemplative environments needed for meditation and spiritual inspiration. Most are Buddhist sites.

Among the sites are:

Figure 6.11. (facing page) Terraced indigenous cliff villages of Chenini de Tataouine and Douiret south of Matmata plateau, Tunisia. Note that the subterranean dwellings are constructed in horizontal layers, following the geological stratification of the mountain. The soft layers of the limestone were excavated for the dwellings.

1. **Aurangabad.** The caves are located north of Aurangabad (400 km east of Bombay) along 3 km from the north to the east. These caves are full of sculptures of goddesses and other deities. The site includes nine caves.
2. **Ellora.** Located about 28 km northwest of Aurangabad, Ellora is famous for its temple dedicated to the god Shiva. All of its spaces are carved from solid red stone. There are 34 caves. The first 12 are Buddhist, the next 17 are Brahmanical, and the last 5 are Jain, all situated along a major ancient highway. Their architecture is a combination of Indo-Aryan and Dravidian. Of Ellora's 34 caves, the Buddhist ones are the oldest.
3. **Ajanta.** About 108 km north of Aurangabad is Ajanta. There are 30 caves in Ajanta; all of them are Buddhist.[8] The caves form a semicircular scarp of rock overlooking a narrow valley and the Waghora stream. The entire façade of Ajanta is rock-cut. The caves beautifully combine painting, sculpture, and architecture. The caves include viharas (monasteries) and five meeting halls. The monasteries reflect the plan of other monasteries throughout India, consisting of a square hole surrounded by cells on three sides, while the front part has a broad veranda. At the back is a shrine, which dominates the structure. The meeting halls are oblong, with apsidal ends and pillared aisles. There is a dome-shaped memorial shrine in the center of the apse. There is only one cave with a two-storied monastery at Ajanta. Cave number 6 has 16 pillars arranged in four rows with 16 cells in the walls. The image of Buddha is separated from the rear wall.[9] The oldest cave probably dates from the first century B.C. to the third century A.D. There is also another group of caves, which dates from the sixth to the seventh centuries. The caves are unique for their quality and quantity of fresco paintings,[10] which form a poem of religious art.[11] Some of the paintings illustrate episodes from the life

of Buddha, especially his incarnation. The Ajanta caves are located in an area which has played an important role in the history of Deccan.[12] Ajanta, like Ellora, is one of the most common types of rock-cut temples, consisting of the shrine proper and great courtyards surrounded by cells for monks. The Dravidian and Jain temples exude power and massiveness and have amazing carved details.[13]

4. **Bedsa.** The small group of caves located at Bedsa are not well known, and are located along the road from Bombay to Poona, near Kamshet.[14]
5. **Karla.** The Karla caves are located 5 km north of the Malavli railway station, 91 km from Bombay. These are Buddhist caves. They consist of one large meeting hall and several monasteries. Designed in great scale, the meeting hall is considered to be one of the most important monuments in the world.[15]

In conclusion, indigenous use of belowground space was limited to dwellings, storage, catacombs, religious centers, sewage, water supply, and tunnels for extraction of natural resources. The dwellings offered a low living standard because technology and safe construction techniques were lacking. These habitats had little appeal, probably due to the serious gap between the design-construction implementation and the required norms for healthy living conditions.

The historical experience of belowground space usage has been widespread throughout the world since early human civilization. With the limited technology available in the past, the construction of such belowground space did not support an acceptable quality of living by our contemporary standards. Yet, the concept of using the belowground space was positive. It provided space for expansion, energy efficiency, food storage, and protection. The study of past experiences like these would certainly introduce some lessons to be applied to modern usage of these spaces. Just as in the past, geo-space use has arisen out of necessity in contemporary times and will continue to do so in the future.

CONTEMPORARY PRACTICE

Belowground space usage, in recent years, has become a movement of almost international proportions. In most contemporary cases, belowground space has been introduced using modern technology and in accordance with modern norms. Technologically advanced countries have become leaders in exploring geo-space for modern land uses, as has been discussed previously in the case of Japan.

Japan

The most substantial strides in geo-space utilization have been in the area of underground shopping centers in Japan. Japan has already constructed 76 belowground shopping centers throughout its large- and medium-sized cities.[16] A good portion of those shopping centers are technologically advanced in design, are operationally sophisticated, and generate a good deal of admiration.

The United States

The U.S. experience is unique.[17] Land pressure and land price increases did not constitute a pressing factor, as in other countries, to accelerate the use of belowground space.[18] Rather, it is for aesthetic reasons that most U.S. public utilities are extended below ground. In the United States there are hundreds of earth-sheltered habitats and more than 90 schools, some university libraries, offices, and agricultural storage spaces located underground.[19] The phenomenon started primarily in the 1970s, at the time of the energy crisis motivated by environmental concerns, and focused on the development of earth-sheltered structures.[20] The buildings were covered by a thick layer of earth around two to three feet wide in order to create effective insulation. One of the most common uses of underground space was for residences of the earth-sheltered type and a few other land uses.[21] Nonresidential belowground land use was more steadily developed, focusing on commercial, storage, offices, parking, and religious, recreational, and agricultural facilities. New York City, and later, Washington, DC, are exceptional in that they also developed underground transportation networks. Recently, a 2.1 km bus tunnel in Seattle was developed.

Since the energy crisis, there has been a substantial lull in geo-space activity in the United States. However, we anticipate the continuation of belowground construction for usage other than housing—especially in urban areas, where land prices are high and space is at a premium.

France

Although the first project for an underground railway in Paris was outlined in 1854, it was not until 1900 that the first underground line was inaugurated, when the World's Fair took place. The underground system was then expanded to become a large city network.[22] The number of users grew from 55 million the first year to 450 million by 1913. The underground network became part of the city plan and the urban fabric of Paris. Between 1930 and 1952, the network expanded to all of the city suburbs. Later, it was expanded and connected with other regions of France.[23] Today, the system includes:

1. Meteor, which is a regional east-west express railway, an intra-urban underground line that is totally automated and links the various networks and services to the major Seine-Rive-Gauche project.
2. Eole (Est-Ouest Liaison Express), a wide-gauge line providing a direct link between the suburban development centers and La Defense.
3. Orbitale, a bypass cutting through the inner city ring.
4. Icare, a major project, still in its early stages of planning, for 150 km of underground toll motorways. The present size of the city underground network is 199 km. The average number of annual users of the underground in Paris is 1.2 billion.[24]

Finally, the French-British 33 kilometer railway and motor vehicle tunnel under the English Channel was recently opened (mid-November 1994).

In Paris, the Eole Project (east-west express connection) is planned as an underground artery to link railroads. Its first stage includes 4 km of tunnels between Villette's Gate to the North and East Railway Stations, where a new railroad station will be built.[25] Along with these rail systems, there are other plans for underground express highways and pedestrian networks within the city.

The decision to move the Paris Food Market, or the Central Market of Les Halles (Forum des Halles), was reached in 1960, but only in 1968 was an official decision made on the site, and the nature of the design was then formulated. A special company, SEMAH, was created to take responsibility for the project. The design, which covers a 32-hectare renovation zone of the area, is composed of Les Halles (8 hectares) and fully utilized down to 28 meters below the surface. The subspace includes an RER station and an underground Forum to link the RER and tube stations to the surface. It is intended to include commercial, cultural, leisure and sports facilities and allow maximum space for pedestrians. The buildings will house a world trade center, with shops, hotels, and restaurants, but will exclude offices. Architects Claude Vasconi and Georges Pencreach created the design from the idea of a blind underground mall pattern. The architects then introduced different levels lit by large oblique glass windows to form a sunken patio 17.5 meters deep, open to the sky and surrounded on three sides by a garden. The Forum is designed to offer extensive commercial uses and to become the "Paris Gateway" for foreigners.

The Forum was opened in 1979, almost two years after the RER station was opened. It has 10 cinemas, a theater, a wax museum, gourmet restaurants, furniture shops, boutiques, and a large number of other shops. The connection with the RER station is through the third and fourth underground floors. The Forum is also surrounded by underground local street networks and two parking lots, with 1,700 parking places. This street system is at the fourth floor and does not detour cars from the major north-south boulevard above ground.

The new Forum, opened in October 1986, was an extension to the previous mall and replaced the originally planned social, cultural, and sports center. It took 30 years to finish the entire project. Today, it is visited by more than 30 million people a year, which averages to 100,000 a day. Seventy-eight percent of the visitors travel by RER or metro, 17 percent are pedestrians, and less than 10 percent drive. The shopping center has sales of 2.7 billion francs a year.[26]

Canada

The Canadian experience is limited to a very few underground shopping centers developed in Montreal and Toronto.[27] The one in Montreal was built in the 1960s and is considered among the most advanced and largest underground shopping centers in the Western world. It consists of shops, offices, restaurants, theaters, and other services. Geo-space evolved in Toronto for pedestrian circulation using a comprehensive and well-planned network.[28] Quebec City conducted a survey of its existing underground structures and outlined a plan assessing the potential of geo-space development in the city. Much of the old Quebec City was erected along slopes.[29]

Sweden

Most of the belowground Swedish experience is focused on a shopping center in Stockholm, on the car repair industry, and military ports. The use of belowground space in Sweden started during the middle of the 19th century with tunnels, pipes, culverts, tramways, and passages.[30] Modern technology enabled the Swedish to penetrate to greater depths and scale the granite rock. The new development left the deep site as a free zone, although issues of legal ownership were raised. The Royal Swedish Library (a large bookstore under Humbgarden Park) and a deep tunnel under central Stockholm carrying wastewater to a discharge point are two of the underground space uses in Sweden.[31] The winding "snake" overflow reservoir under Stockholm was designed to prevent untreated storm water from reaching Lake Malur. Also developed was a large oil tank to be converted to a coal store, a large underwater sewage treatment plant, a traffic link, a tramway, and a waste energy storage area.[32]

China

China today is following its long historic tradition of employing diversified belowground land use.[33] China has probably the largest and most comprehensive underground defense shelters in the world.[34] This author was informed during his research in China that Beijing's civil defense system alone was capable of accommodating 5 million people at a given time, which constitutes half of the city's population, in case of nuclear attack. The whole network provides subsystems for every neighborhood, which are connected to each other and lead through underground tunnels to a few kilometers outside the city of Beijing.[35]

Other countries using geo-space modern design are the United Kingdom, India, Russia, Norway, Finland, the former Yugoslavia, and Holland.[36]

CONCLUSION

The new international movement toward the use of underground space is steadily growing. Existing and future high-level technology, along with urban growth, suggest that there will be further development of this movement and more diversified underground land uses. Contemporary usage of belowground space is becoming increasingly diversified, yet currently, it is still limited mainly to storage, garages, and utilities. While the numerous practical benefits of geo-space have recently led to the construction of some underground schools, libraries, shopping centers, transportation, art halls, and shelters for defense, there is still a general reluctance toward underground housing due to its negative connotation. Thoughtful design can contribute significantly to altering the psychological constraints and biased negative attitudes that have been inherited from the indigenous communities that used the belowground space in the past.

Although geo-space construction can be more expensive initially than conventional construction above ground, further practical experience, large-scale development, and technological advancements can ensure that the price of construction—both below ground and above ground—will equalize over time. With accelerated urbanization throughout the world, skyrocketing urban land prices, increased physical urban expansion of infrastructure, and other associated factors, the increased usage of belowground space has become inevitable. Technologically, the Japanese, Swedish, Americans, and others have proved that such usage is certainly feasible. However, further research and innovative design concepts, especially in the areas of working conditions, health benefits (physical and mental), sleeping conditions, creative accomplishments, and psychological impact, are needed in order to improve the usage of such space and make geo-space a completely viable option for the future.

NOTES

1. Gideon S. Golany, *Chinese Earth-Sheltered Dwellings: Indigenous Lessons for Modern Urban Design* (Honolulu, HI: University of Hawaii Press, 1992): pp. 41–126.
2. John Carmody and Douglas Derr, "The Use of Underground Space in the People's Republic of China," *Underground Space,* Vol. 7, No. 1 (July/Aug. 1982): pp. 7–15.
3. Gideon S. Gollany, *Urban Underground Space Design in China: Vernacular and Modern Practice* (Newark, DE: University of Delaware Press, 1989): p. 93.
4. Omer Yorukoglu, et al., *Underground Cities in Cappadocia,* unpublished manuscript, 1989, p. 16.
5. Id., pp. 25–27.
6. Id., p. 30.
7. Gideon S. Golany, *Earth Sheltered Dwellings in Tunisia: Ancient Lessons for Modern Design* (Newark, DE: University of Delaware Press, 1988): pp. 26–29.
8. Percy Brown, *Indian Architecture* (*Buddhist and Hindu*) (Bombay, India: D. B. Taraporevala Sons & Co., Private Ltd., 1976): pp. 19–26, 26–30, 56–61, and 71–76.
9. Owen C. Kail, *Buddhist Cave Temples of India* (Bombay, India: D. B. Taraporevala Sons & Co., Private Ltd., 1976): pp. 80–92.
10. "Ajanta," *Encyclopaedia Britannica,* Vol. 1 (Chicago: William Benton, Publisher, 1970): p. 476.
11. S. G. Dawne, *Ajanta Ellora and Environs, Mythological, Historical, Sculptural, Pictorial Review,* 4th Ed. (Aurangabad, India: Mukund Prakashan, 1978): p. 125.

12. T. V. Pathy, *Ajanta, Ellora & Aurangabad Caves: An Appreciation* (Aurangabad, India: Shrimati T. V. Pathy, 1978): pp. 7–10.
13. "Temple Architecture," *Encyclopaedia Britannica,* Vol. 21 (Chicago: William Benton, Publisher, 1970): p. 831b.
14. Kail, p. 93.
15. Kail, p. 108.
16. Tetsuya Hanamura, "Japan's New Frontier Strategy: Underground Space Development," *Tunneling and Underground Space Technology,* Vol. 5, No. 1/2 (1990): pp. 13–21.
17. Patrick Horsbrugh, "Geospace Panorama," *Urban Underground Utilization '91, Proceedings* (Tokyo: Urban Underground Space Center of Japan, 1991): pp. 625–32.
18. Raymond L. Sterling, "Recent Developments in Underground Utilization in the USA," *Urban Underground Utilization '91, Proceedings* (Tokyo: Urban Underground Space Center of Japan, 1991): pp. 77–79.
19. Gideon S. Golany, *Earth-Sheltered Habitat: History, Architecture and Urban Design* (New York: Van Nostrand Reinhold Company, 1983): pp. 23–38.
20. D. A. Gilbert, "Earth-Sheltered Housing in Appalachia," *Underground Space,* Vol. 6, No. 2 (Sept./Oct. 1991): pp. 89–92.
21. John Carmody and Raymond Sterling, "Design Guideline for People in Underground Spaces," *Urban Underground Utilization '91, Proceedings* (Tokyo: Urban Underground Space Center of Japan, 1991): pp. 641–46.
22. Truman Stauffer, Sr., "Subsurface Uses in Sweden and France: A Report," in *Underground Utilization: A Reference Manual of Selected Works, Vol. IV: Human Responses and Social Acceptance of Underground Space,* ed. Truman Stauffer (Kansas City, MO: A Geographic Publication, Department of Geosciences, University of Missouri, 1978): pp. 664–77.
23. D. Farray, "The Architecture of French Subways and Underground Railways," *Underground Space,* Vol. 8, No. 2 (1984): pp. 117–24.
24. François Ascher, "Impact of Land Use Transition by Underground Utilization: The Case of Paris," *Urban Underground Utilization '91, Proceedings* (Tokyo: Urban Underground Space Center of Japan, 1991): pp. 64–70.
25. André Guillerme, "Urban Underground Network and Terminal Facilities," *Urban Underground Utilization '91, Proceedings* (Tokyo: Urban Underground Space Center of Japan, 1991): p. 41.
26. Alain Sonntag, "An Image of an Underground Complex, the Forum des Halles in Paris," *Urban Underground Utilization '91, Proceedings* (Tokyo: Urban Underground Space Center of Japan, 1991): pp. 92–96.
27. A. D. Margison, "Canadian Underground NORAD Economically Achieved," *Underground Space,* Vol. 2, No. 1 (Sept. 1977): pp. 9–18.
28. Daniel J. Boivin, "Underground Space Use and Planning in the Quebec City Area," *Tunneling and Underground Space Technology,* Vol. 5, No. 1/2 (1990): pp. 69–84.

29. Michael B. Barker, "Toronto's Underground Pedestrian System," *Tunneling and Underground Space Technology,* Vol. 1, No. 2 (1986): pp. 145–52.
30. Stauffer (1978), pp. 664–77.
31. Hans Wohlin, "Underground Planning," *Urban Underground Utilization '91, Proceedings* (Tokyo: Urban Underground Space Center of Japan, 1991): pp. 50–52.
32. Bjorn Sellberg, "Swedish R & D Programme and Support on Underground Planning in the 1990s," *Urban Underground Utilization '91, Proceedings* (Tokyo: Urban Underground Space Center of Japan, 1991): pp. 619–24.
33. G. Golany, *Urban Underground Space in China* (Newark, DE: University of Delaware Press, 1989).
34. Eugene Wukasch, "Underground Population Defense Structures in China," *Underground Space,* Vol. 7, No. 1 (July/Aug. 1982): pp. 16–20.
35. Golany (1989), pp. 64–71.
36. Further information on some European countries on the subject are:

    Einar Broch and Jan. A. Rygh, "Permanent Underground Openings in Norway—Design Approach and Some Examples," *Underground Space,* Vol. 1, No. 2 (1976): pp. 87–100.

    V. I. Hokkanen, M. Forssen, K. Niva, Sorjonen, and Lahti Institute of Design. "Four Examples of Subsurface Uses in Finland," *Tunneling and Underground Space Technology,* Vol. 9, No. 3 (July 1994): pp. 385–93.

    P. M. Maurenbrecher and H. J. B. Z. Trimbos, "Preliminary Infrastructural Development Plan for the Resort Town of Valkenburg, SE Netherlands, by Adapting Existing Open and Underground Quarry Spaces—Basis for an International Contest," *Urban Underground Utilization '91, Proceedings* (Tokyo: Urban Underground Space Center of Japan, 1991): pp. 661–70.

    M. P. O'Reilly, "Some Examples of Underground Development in Europe," *Underground Space,* Vol. 2, No. 3 (April 1978): pp. 163–78.

    Arthur Quarmby, "Recent Development in Earth-Sheltered Architecture in the United Kingdom," *Urban Underground Utilization '91, Proceedings* (Tokyo: Urban Underground Space Center of Japan, 1991): pp. 633–40.

    Zlatko Svetlicic, "Uses of Underground Space in Yugoslavia," *Tunneling and Underground Space Technology,* Vol. 1, No. 1 (July 1986): pp. 67–70.

SELECTED BIBLIOGRAPHY

GENERAL

COUNTRIES: INDIGENOUS AND MODERN CASES

China
Tunisia
Turkey
United States
Other Countries

SELECTED TOPICS: DESIGN AND CONSTRUCTION

Construction
Design
Drainage, Waterproofing, and Insulation
Natural Hazards
Roofs
Structure
Soil Properties and Performance
Technology
Thermal Performance and Energy
Ventilation
Windows and Light

URBAN DESIGN: CITY PLANNING AND DEVELOPMENT

Climate
Environmental Issues
Industrial and Commercial Use
Military Use
Parking
Psychology
Public Policy, Legislation, and Zoning
Schools
Social and Economic Issues
Transportation

GENERAL

Aughenbaugh, N. B. "Development of Underground Space." In *Earth-Covered Buildings and Settlements,* edited by Frank L. Moreland, 152–68. Springfield, VA: NTIS, 1979.

Bacon, Vinton W. "Overview of Underground Construction in Congested Areas." In *Underground Utilization,* edited by Truman Stauffer, Sr., 2: 176–79. Kansas City, MO: University of Missouri, 1978.

Birkets, Gunnar. *Subterranean Urban Systems.* Ann Arbor: Industrial Development Division, Institute of Science and Technology, University of Michigan, 1974.

Bligh, Thomas. "Building Underground." *Building Systems Design* (October–November 1976): 1–22.

Brown, G. Z. and B. J. Novitski. "Climate Responsive Earth-sheltered Buildings." *Underground Space* 5 (1981): 299, 305.

Campbell, Stu. *The Underground House Book.* Charlotte, VT: Garden Way Publishing, 1980.

Coogan, Alan H. "Classification and Valuation of Subsurface Space." *Underground Space* 3 (1979): 175–86.

Earth-sheltered Housing Design: Guidelines, Examples, and References. Minneapolis: Underground Space Center, University of Minnesota, 1978; revised edition, 1985. Also published, New York: Van Nostrand Reinhold, 1979.

Fairhurst, Charles. "Going Under to Stay on Top." *Underground Space* 1 (1976): 71–86. Also in *Underground Utilization,* edited by Truman Stauffer, Sr., 7: 942–54. Kansas City, MO: University of Missouri, 1978.

Golany, Gideon S. "An Underground Frontier." In *Research/Penn State: Arts and Humanities.*Vol. II, no. 3, September 1981, p. 17.

———. "Contribution of Geo-front Space to Modern Urban Design." In *Design Journal* (of the Department of Architecture and Building Engineering, Tokyo Institute of Technology), no. 3, 1991, p. 19.

———. "Earth-Sheltered Homes: It's 'Cool' Living Below the Ground." Interview in *Herald* (Shenandoah, PA).November 28, 1985.

———. "Moving Down." Interview in *The Penn Stater.* March/April 1983, p. 15.

———. "Subterranean Settlements for Arid Zones." In *Earth-covered Buildings and Settlements,* edited by Frank L. Moreland, 174–202. Springfield, VA: NTIS, 1979.

Hanamura, Tetsuya. "Alice City Network Concept of Urban Underground Space Use." Tokyo: Taisei Corporation, April 1989.

Horsbrugh, Patrick. *Geotecture, Concept, Design, Construction, and Economy of Geospace—The Creation of Subterranean Accommodation.* Minneapolis: Department of Civil and Mineral Engineering, University of Minnesota, 1973.

"House 'Revealed' Itself on Slope" (Case Study). In *Earth Shelter Digest and Energy Report,* no. 11 (September–October 1980): 11–14.

Jannsson, Birger. "City Planning and the Urban Underground." *Underground Space* 3 (1978): 99–115.

Jansson, Birger and Torbjörn Wingqvist. *Planning of Subsurface Use.* Oxford, England: Pergamon Press, 1977.

Jansson, Birger. "Terraspace—A World to Explore." *Underground Space* 1 (1976): 9–18.

Kommendant, August E. "Earth-Covered Structures." In *Earth-Covered Buildings: Technical Notes,* edited by Frank L. Moreland, F. Higgs, and J. Shih, 1–12. Springfield, VA: NTIS, 1979.

Labs, Kenneth B. "Performance of Earth-Covered Development: Planning Issues." In *Earth-Covered Buildings and Settlements,* edited by Frank L. Moreland, 125–40. Springfield, VA: NTIS, 1979.

Legget, Robert F. "The Underground of Cities." In *Underground Utilization,* edited by Truman Stauffer, Sr., 2: 184–88. Kansas City, MO: University of Missouri, 1978.

Moreland, Frank L. "Earth-covered Habitat—An Alternative Future." *Underground Space* I (1977): 295–307.

———, ed. *Alternatives in Energy Conservation: The Use of Earth-Covered Buildings*. Washington, DC: National Science Foundation/USGPO, 1978.

———, ed. *Earth-Covered Buildings and Settlements.* Springfield, VA: NTIS, 1979.

Moreland, Frank L., F. Higgs, and J. Shih, eds. *Earth-covered Buildings: Technical Notes.* Springfield, VA: NTIS, 1979.

Oliver, Paul, ed. *Shelter, Sign & Symbol.* (London: Barrie & Jenkins), 1975.

Potential Use of Underground Space. Minneapolis: Department of Civil and Mineral Engineering, University of Minnesota, 1975.

Randall, William J. "Space Age Cave Dweller." *Congressional Record—Appendix* 109, pt. 10 (18 July 1973): A4569–A4571.

Rapoport, Amos. *House Form and Culture.* Englewood Cliffs, NJ: Prentice Hall, 1969.

Reyner, J. F. *Analysis of Several Surveys Relative to Problems of Shelter Habitability.* Washington, DC: National Academy of Sciences, 1960.

Roy, Robert L. "How to Build an Underground Home." *Farmstead,* March 1979.

Scalise, James W. *Terratecture: The Environmental Benefits of Earth-Integrating Architectural Design Techniques.* Tempe, AZ: College of Architecture, Arizona State University, 1977.

———, ed. *Earth-Integrated Architecture: An Alternative Method for Creating Livable Environments with an Emphasis on Arid Regions.* Tempe, AZ: Architecture Foundation, Arizona State University, 1975.

Stauffer, Truman, Sr., ed. *Underground Utilization: A Reference Manual of Selected Works.* 8 vols. Vol. 1: *Historical Perspective;* Vol. 2: *Uses for Underground Space;* Vol. 3: *Space Construction Underground;* Vol. 4: *Human Response and Social Acceptance of Underground Space;* Vol. 5: *Advantages in Underground Space Use;* Vol. 6: *Regulations and Policy in the Use of Underground Space;* Vol. 7: *The Future of Underground Development;* Vol. 8: *Index.* Kansas City, MO: University of Missouri, 1978.

Sterling, Ray, et al. "Site Constructions for Earth-Sheltered Structures." *Concrete Construction* 25 (1980): 653.

"Study Report by the Subterranean City Concept Subcommittee, 1988–

1990." In *"Proceedings of the Fourth International Conference on Underground Space and Earth Sheltered Buildings, December 3–5, 1991."* Tokyo, Japan: Sponsor—Urban Underground Space Center of Japan, 1991.

van der Meer, Sybe J. "Underground and Earth-covered Housing Deserves Consideration." In *Proceedings IAHS Symposium,* edited by Parviz Rad et al., 2: 1137–50. Coral Gables, FL: International Association for Housing Science, 1976.

Wohlin, Hans. "Underground Planning." In *"Proceedings of the Fourth International Conference on Underground Space and Earth Sheltered Buildings, December 3–5, 1991."* Tokyo, Japan: Sponsor—Urban Underground Space Center of Japan, 1991.

COUNTRIES: INDIGENOUS AND MODERN CASES

China

Andreadaki-Chronaki, Eleni. "Earth Sheltered Architecture as an Alternative for Environmental Planning: The Experience of Santorini. In *Earth Sheltered Buildings Conference* (1986): 6–11.

Aoki, Shiro, and Toshimasa Konishi. "Types and Locations of Earth Construction. In *Symposium on Earth Architecture* (1985): 17–19.

Architectural Journal, ed. "First symposium on cave dwellings and buildings in earth (Zhongguo jianzhu xuehui yaodong ji shengtu jianzhu: Di yi ci xueshu taolun huiyi jiyao)." *Architectural Journal* (Jianzhu xuebao) 10 (1981): 29–30.

Architectural Society of China and Tianjin University, Department of Architecture, eds. *Earth Sheltered Architecture in China* (Zhongguo shengtu jianzhu). Tianjin, China: Tianjin Science and Technology Press, 1985.

Architectural Society of China, ed. *Proceedings of the International Symposium on Earth Architecture* (1–4 November 1985), Beijing.

Arledge, Pat H., and Cora F. McKown. "Innovativeness and Risk-taking Characteristics of Owners of Earth Sheltered Houses." In *Symposium on Earth Architecture* (1985): 41–46.

Asanga, Cletus T., and Robert B. Mills. "Changes in Environment, Grain Quality, and Insect Populations in Pearl Millet Stored in Underground Pits." *Underground Space* 9.5–6 (1985): 316–321.

Attewell, P. B., and A. R. Selby. "The Effects of Large Ground Movements Caused by Tunnelling in Compressible Soils." In *Conference on Underground Space* (1988): 316–326.

Baggs, S. A., and C. F. Wong. "Survey of Radon in Australian Geotecture and Architecture." In *Earth Sheltered Buildings Conference* (1986): 111–116.

Baggs, Sydney A. "Environmental Factors Possibly Influencing Attitude in Australians Living in Above- and Below-ground Dwellings in an Arid Region Mining Town." In *Symposium on Earth Architecture* (1985): 47–54.

———. "Survey and Analysis of Interior/Exterior Environments of Nine

Case Studies of 'Dugout' Dwellings, Coober Pedy, South Australia." In *Conference on Underground Space* (1988): 441–446.
———. "The First International Earth Shelter Building Conference." *Underground Space* 8.3 (1984): 149.
Bai Mingxue. "Earthquake Hazards to the Earth Buildings in Ningxia." In *Symposium on Earth Architecture* (1985): 55–63.
Banks, H. J. "Effects of Controlled Atmosphere Storage on Grain Quality: A Review." *Food Technology in Australia* 83 (1981): 335–340.
Bansal, N. K., and M. S. Sodha. "An Earth-Air Tunnel System for Cooling Buildings." *Tunnelling and Underground Space Technology* 1.2 (1986): 177–182.
Bartz, Janet G. "Attitudes and Residential Satisfaction of Earth Sheltered Housing Consumers." In *Earth Sheltered Buildings Conference* (1986): 137–142.
Beijing City Office of Air Defense. "The Use of Air Raid Shelters in Beijing (Beijing renfang gongcheng de liyong)." *Architectural Journal* (Jianzhu xuebao) 11 (1982): 40–41.
Blick, Edward F., and J. G. Eyerman. "R Factor for Earth Bermed Buildings." In *Earth Sheltered Buildings Conference* (1986): 175–182.
Bowers, Douglas A. "Geocomposites: A Systems Approach." In *Earth Sheltered Buildings Conference* (1986): 103–106.
Boyd, Andrew C. H. *Chinese Architecture and Town Planning, 1500 B.C.–A.D. 1911*. Chicago: University of Chicago Press, 1962.
Boyer, Lester L. "Daylighting Performance of an Earth Covered Dwelling at High Altitude. In *Earth Sheltered Buildings Conference* (1986a): 252–256.
———. "Design and Construction of a Remote, Fully Passive, Two-Story Earth Covered Residence at High Altitude." In *Earth Sheltered Buildings Conference* (1986b): 230–233.
Boyer, Lester L., and Morad R. Atif. "Comparison of Daylighting Distribution in an Underground Dwelling High in the Rockies Using Computer Algorithms, Scale Models, and On-Site Measurements." In *Conference on Underground Space* (1988): 470–475.
Boyer, Lester L., and W. T. Grondzik. "Sunlighting, Daylighting, and Passive Solar Heating for Earth Covered Dwellings." In *Earth Sheltered Buildings Conference* (1986): 220–223.
Britz, Richard D. "Modular Method of Making a Building 'earthship' Structure." In *Earth Sheltered Buildings Conference* (1986): 263–266.
Broch, Einar, and Jan A. Rygn. "Recreational Facilities in Rock in Norway." In *Conference on Underground Space* (1988): 502–508.
Brown, R. L. "Japan: Under Land and Under Water." *Building* 226 (1974): 87–88.
Cai Xiaohong and Lu Younian. "Calculation of Elastic-Plastic Stress for Hydrotechnical Pressure Tunnel (Shuigong yali suidong tansuo xingying li jisuan)." *Underground Space* (Dixia kongjian) 1 (1987): 29–38.
Cai Xiuyue. "The Utilization and Development of the Space in Subway Stations—Some Ideas on the Plan of Xi Dan Station." In *Conference on Underground Space* (1988): 110–115.
Cao Guiquan. "Application of Installation with Heating Pipes and Heat Air

Supply in Underground Construction (Refeng reguan zhuangzhi zai dixia gongcheng de yingyong)." *Underground Space* (Dixia kongjian) 8.3 (1988): 68–72.

Carmody, J. C., and R. L. Sterling. "Design Strategies to Alleviate Negative Psychological and Physiological Effects in Underground Space." In *Earth Sheltered Buildings Conference* (1986): 127–136.

Carmody, J. C., S. L. Shen, Y. J. Huang, L. G. Goldberg, and George D. Meixel, Jr. "A Comprehensive Method for Predicting Residential Energy Savings Due to Basement Insulation." In *Conference on Underground Space* (1988): 404–409.

Carmody, John, and Douglas Derr. "The Use of Underground Space in the People's Republic of China." In *Underground Space* 7.1 (1982): 7–15.

Cave Dwelling Survey and Research Group. *Sunken Cave Dwellings in Gong County* (Gong xian de xiachen shi yaodong jianzhu). Construction Committee of Kaifeng Region, Henan, 1982.

Cena, K., L. Sliwowski, and J. R. Spotila. "Thermal Comfort of Humans in Normal and Special Environments." In *Conference on Underground Space* (1988): 492–495.

Chan Bingshan, Jian Wenbin, and Wu Qiong. "The Resource Significance of the Karst Underground Space." In *Conference on Underground Space* (1988): 280–282.

Chang, Kwang-Chih. *The Archaeology of Ancient China.* 3d ed. New Haven: Yale University Press, 1977.

Chatani, Masahiro, and Koji Yagi. "Analysis of Hillside and Pit Type Dwellings." In *Symposium on Earth Architecture* (1985): 20–23.

Chatani, Masahiro, K. Yagi, T. Nakazawa, and K. Yashiro. "Underground Dwellings in the Loess Land of China. In *Earth Sheltered Buildings Conference* (1986): 36–40.

Chatani, Masahiro, Koji Yagi, Kiyoshi Ishikawa, Naomi Ando, Katsuhiko Yashiro, and S. Nonaka. "Analysis of Recognition of Route in Underground Shopping Areas of Tokyo and Yokohama." In *Conference on Underground Space* (1988a): 433–436.

Chatani, Masahiro, Koji Yagi, Kiyoshi Ishikawa, T. Nakazawa, Katsuhiko Yashiro, and T. Taniguchi. "Research on Spatial Perception in Underground Shopping Areas of Tokyo and Yokohama." In *Conference on Underground Space* (1988b): 437–440.

Chen Benrong and Zhang Yushu. "Glimpse into the Underground City of Beijing (Beijing dixiacheng yi pie)." *Beijing Daily* 21 (December 1983).

Chen Lidao, Zhang Dongshan, and Shu Yu. "The Evaluation of Demand for Underground Space in Urban Areas and Optimal Distribution." In *Conference on Underground Space* (1988): 165–171.

Chen Qigao. "A Program on Wastes Control for a Heat Power Plant by an Underground System." In *Conference on Underground Space* (1988): 416–421.

———. "Research on the Thermal Performance of a Sweet Orange Cellar." In *Earth Sheltered Buildings Conference* (1986a): 190–194.

———. "Theory of Heat Transfer in a Tunnel with Natural Ventilation." In *Earth Sheltered Buildings Conference* (1986b): 205–209.

Chen Yanxun, Nie Minggang, and Xiao Mingyan. "Investigation on Acoustic Environment of Underground Space (Dixia kongjian de shenghuanjing diaocha)." *Underground Space* (Dixia kongjian) 8.3 (1988): 38–50.

Chen Zhanxiang. "Cave Dwellings and Architecture (Yaodong yu jianzhu)." *Architectural Journal* (Jianzhu xuebao) 10 (1981): 31–33.

Chen Ziqiang. "Construction and Management of an Underground Market in the Department Store Nanfang Dasha (Nanfang Dasha dixia shangchang de shigong yu guanli)." *Underground Space* (Dixia kongjian) 4 (1986): 49–55.

Chen, B. J., Q. Y. Chi, and X. K. Li. "Environment and Health in [an] Underground Hospital." In *Conference on Underground Space* (1988): 399–403.

Chinese Academy of Sciences, Institute of Geology. *Chinese Loess Soil Distribution Map* (Zhongguo huangtu fenbu tu). Beijing: Cartographic Press, 1965.

Chongqing University, Tongji University, Harbin Architectural Engineering Institute, and Tianjin University, eds. *Underground Architectural Structure in Rocks* (Yanshi dixia jianzhu jiegou). Beijing: China Building Industry Press, 1928.

Cui Zhensheng, and Gu Zuming. "The Ep-pur Elastic Sealer—A Sealing Material for Underground Engineering." In *Conference on Underground Space* (1988): 208–210. Central Meteorological Department, ed. *Climate of China, Diagrams and Illustrations*. Beijing: Maps Publishing Co., 1978.

Deng Qisheng. "Traditional Measures of Moistureproof in Raw Soil Architecture in China." In *Symposium on Earth Architecture* (1985): 64–68.

Di Cristofalo, S., S. Orioli, G. Silvestrini, and S. Alessandro. "Thermal Behavior of 'Scirocco Room' in Ancient Sicilian Villas. " In *Conference on Underground Space* (1988): 45–47.

Duddeck, Heinz, Deiter Rabe, and Dieter Winselmann. "Design Criteria for Immersed Tube Tunnels: Experiences Gained by the Autobahn-Tunnel under the River Ems." In *Conference on Underground Space* (1988): 172–179.

Duddeck, Heinz. "Guidelines for the Design of Tunnels." In *Conference on Underground Space* (1988): 180–188.

Dunkel, F. V., R. L. Sterling, G. D. Meixel, and L. D. Bullerman. "Underground Structures for Grain Storage: Inter-relatedness of Biologic, Thermal, and Economic Aspects." In *Conference on Underground Space* (1988): 1–6.

Dunkel, F., R. Sterling, and G. Meixel. "Underground Bulk Storage of Shelled Corn in Minnesota." In *Earth Sheltered Buildings Conference* (1986): 78–85.

Dunkel, Florence V. "Grain Storage in South China." *Cereal Foods World* 27 (1982): 409–414.

———. "Underground and Earth-Sheltered Food Storage: Historical,

Geographic, and Economic Considerations." In *Underground Space* 9.5–6 (1985): 310–315.

Dunkel, Florence V., Zhe Lung Pu, Chuan Liang, and Fan-yi Huang. "Insect and Fungal Response to Sorbic Acid-treated Wheat During Storage in South China." *Journal of Economic Entomology* 75 (1982): 1083–1088.

Eberspacher, Warren. "The Earth Systems Structural Forming System." In *Earth Sheltered Buildings Conference* (1986): 257–262.

Editing and Writing Group, ed. *Handbook for the Design of Heating and Air Conditioning in Underground Buildings* (Dixia jianzhu nuan tong kongtiao sheji shouce). Beijing: China Building Industry Press, 1983.

El Bartali, Houssine, and Said Afif. "Temperature Variations in Underground Grain Storage Structures." In *Conference on Underground Space* (1988): 488–491.

El Bartali, Houssine. "Underground Storage Pits in Morocco." In *Earth Sheltered Buildings Conference* (1986): 68–78.

Esaki, Yoichiro, and Ryuzo Ohno. "A Morphological Analysis of Underground Rooms." In *Symposium on Earth Architecture* (1985): 24–26.

Fairhurst, Charles, and Raymond Sterling. "Session Report: Subsurface Storage of Food." *Underground Space* (Dixia kongjian) 7.4–5 (1983): 249–250.

Fan Jiancheng and Kang Ning. "Analysis of Probability of Existence for Underground Construction (Dixia gongcheng de shengcun gailu qianxi)." *Underground Space* (Dixia kongjian) 4 (1986): 19–40.

Feng Zhongwen. "Scientific Management of Utilization of Underground Tunnel Air (Didaofeng liyong de kexue guanli)." *Underground Space* (Dixia kongjian) 1 (1987): 55–63.

Freeman, S. Thomas, et al. "Tunnelling in the People's Republic of China." *Underground Space* 7.1 (1982): 24–30.

Fujita, K., K. Ueda, and Y. Shimomura. "An Empirical Proposal on Stability of Rock Cavern Wall During Construction in Japan." In *Storage in Excavated Rock Caverns, Rockstore* 77.2, Magnus Bergman, ed., 309–314. Oxford, England: Pergamon Press, 1978.

Fukuchi, G. "Guest Editorial." *Tunnelling and Underground Space Technology* 1.3–4 (1986): 223–224.

Fuller, Myron L., and Fredrick G. Clapp. "Loess and Rock Dwellings at Shensi." *Geographical Review* 14 (1924): 215–226.

Gansu Cave Dwelling and Soil Construction Survey and Research Group, ed. *A Collection of Theses on Surveys of Cave Dwellings and Buildings in Earth in China* (Zhongguo yaodong ji shengtu jianzhu diaoyan lunwen xuanji). Lanzhou: Architectural Society of China, 1982.

Gao Boyang and Geng Yonghang. "Research on the Psychological Condition of People Working and Living in Underground Circumstances." In *Conference on Underground Space* (1988): 428–432.

Gao Hewei and Zhang Changquan. "Shanghai People's Square Underground Parking." In *Conference on Underground Space* (1988): 133–137.

Gilman, G. A., and R. A. Boxall. "The Storage of Food Grain in Traditional Underground Pits." *Tropical Stored Products Information* 28 (1974): 19–38.

Goding, Lloyd A. "Ferro Cement as a Building Material for Earth Shelter." In *Earth Sheltered Buildings Conference* (1986): 100–102.
Golany, Gideon S. "Chinese Belowground Dwellings: Cliffside, Pit (Sunken Courtyard), and Earth-Sheltered Dwellings; Home to 40 Million Inhabitants of Northern China". In *Adobe Journal,* Issue 10, 1994, pp. 8–12.
———. *Chinese Earth-Sheltered Dwellings: Indigenous Lessons for Modern Urban Design.* Honolulu: University of Hawaii Press, 1992.
———. *Design and Thermal Performance: Below-Ground Dwellings in China.* Newark, Delaware: University of Delaware Press, 1990.
———. "The Pit and Cliff Houses of China." In *Research/Penn State: Arts and Humanities.*Vol. VI, no. 3, September 1985, pp. 18–19.
———. *Urban Underground Space Design in China: Vernacular and Modern Practice.* Newark, Delaware: University of Delaware Press, 1989.
Granit, Michael. "Underground Design in Urban Areas." In *Conference on Underground Space* (1988): 116–121.
Gu Fubao. *Cave Dwellings in West Henan* (Yuxi yaodong minju). Zhengzhou: Civil Engineering Department, Zhengzhou Engineering College, 1982.
Guan Lijun. "Research on Thermal Environments Inside Loess Cave Dwellings in China." In *Symposium on Earth Architecture* (1985): 82–92.
Gudehus, G. "Geotechnical Innovations for an Underground Garage in Soft Soil." In *Conference on Underground Space* (1988): 86–89.
Guo Yunfu, Ysang Nanfang, and He Tienan. "Design and Economical Benefit of Cavern Oil Storage with Double Approach (Danshuang yindao shantong shiyouku sheji yu jingjixiao yi fenxi)." *Underground Space* (Dixia Kongjian) 1 (1987): 5–16.
Hait, John N. "Passive Annual Heat Storage." In *Conference on Underground Space* (1988): 476–481.
Hanamura, Tetsuya. "Japan's New Frontier Strategy: Underground Space Development." *Tunnelling and Underground Space Technology* 5.1–2 (1990): 13–22.
Hane, Tadashi. "Application of Solar Daylighting System to Underground Space." In *Conference on Underground Space* (1988): 464–469.
———. "Application of Solar Daylighting Systems to Underground Space." *Tunnelling and Underground Space Technology* 4.4 (1989): 465–470.
Hanegreefs, P., S. Clarke, and F. Dunkel. "Underground Bag Storage of Dry Edible Beans in Rwanda (East Africa)." In *Earth Sheltered Buildings Conference* (1986): 74–77.
Hasegawa, F., H. Yoshino, and S. Matsumoto. "Optimum Use of Solar Energy Techniques in a Semi-underground House: First Year Measurement and Computer Analysis." *Tunnelling and Underground Space Technology* 2.4 (1987): 429–436.
History of Chinese Architecture. Group compilation. Beijing: China Building Industry Press, 1979. In Chinese.
Hou Ji-Yao. "Cave Dwellings in North Shaanxi Province." *Knowledge of Architecture* 2 (n.d.). In Chinese.
Inaba, Kazuya, et al. "Investigation and Improvement of Living Environ-

ment in Cave Dwellings in China." *Proceedings of the International Symposium on Earth Architecture, 1–4 November, 1985, Beijing.* Beijing: Architectural Society of China, 1985.

Liu Duanzheng. *Chinese Housing Concepts.* Beijing: Architectural Construction Press, 1957. In Chinese.

Qian Fu-yuan and Su Yu. "Research on the Indoor Environment of Loess Cave Dwellings." *Proceedings of the International Symposium on Earth Architecture, 1–4 November, 1985, Beijing.* Beijing: Architectural Society of China, 1985.

Tunisia

Balland, Viviane. "Morphogenese quaternaire dans les monts de Matmata." In *Maghreb et Sahara, études géographiques offertes à Jean Despois,* 45–57. Special issue of *Acta Geographica.* Paris: Société de Géographie, 1973.

Bardin, P. "Les terres collectives et le paysannat dans le sud de la Tunisie." *La France Méditerranéenne et Africaine* 2 (1938): 65–72.

Ben-Hamza, Kacem. "The Cave-Dwellers of Matmata: Ritual and Economic Decision-Making in a Changing Community." Ph.D. diss., Indiana University, 1977.

Bertholon, L. "Les Ghoumracen, Douiret et Chenine." *Etude . . . Ardh* (1894): 189–91.

———. "Le plateau de Demer et les Ksours des Aouiya." *Etude . . . Ardh*(1894): 185–88.

Beschaouch, Azedine, Roger Hanoune, and Yvon Thébert. *Les ruines de Bulla Regia.* Rome: Ecole Francaise de Rome, Palais Farnese, 1977.

Blanc, E. "Les oasis du sud de la Tunisie." *Bulletin de la Société de Géographie Commerciale de Paris* (1889): 117–24.

Bruun, Daniel. *The Cave Dwellers of Southern Tunisia. Recollections of a Sojourn with the Khalifa of Matmata.* London: W. Thacker and Co., 1898.

Carton, L. "Ksours et troglodytes de la Tunisie." *Bulletin de la Société de Géographie de Lille* (1913): 128–34.

Déambroggio [Kaddour Larby]. "Législation et coutumes des Berbères du Sud Tunisien." *Revue Tunisienne* (1913): 97–103.

Despois, J. "Les greniers fortifiés de l'Afrique du Nord." *Les Cahiers de Tunisie* 1 (1953): 38–60.

Dupas, P. "Note sur les magasins collectifs du Haut-Atlas occidental." *Hespéris* 9 (1929): 303–21.

Duveyrier, H. "Excursion dans le sud de la Tunisie." *Revue Algérienne et Coloniale* (1860): 413–96.

el-Fechfach, Muhamed, et al. *Atlas of Tunisia.* Paris: Jean Afrique, 1980. Printed in Arabic.

Forest, G. "Coutumes des populations de la circonscription de Tataouine." *Mémoire du CHEAM* (Centre de Hautes Etudes d'Administration Musulmane), no. 573 (1942).

Golany, Gideon S. *Earth-Sheltered Dwellings in Tunisia: Ancient Lessons for*

Modern Design. Newark, Delaware: University of Delaware Press, 1988.

Haan, Herman. "Matmata." In *Architects Yearbook,* edited by David Lewis, 2: 126–228. London: Elek Books, 1965.

Hallet, Stanley. "Mountain Villages of Southern Tunisia." *Journal of Architectural Education* 29 (1975): 22–25.

Ibn Khaldun. *Histoire des Berbères et des dynasties musulmanes de l'Afrique septrionale*. Translated by Baron de Slane. 4 vols. Paris: P. Geuthner, 1968.

Idoux, M. "Un été dans le Sud Tunisien." *Mémoires de la Société Bourguignonne de Géographie et d'Histoire* 16 (1990): 20–22.

Jacques-Meuiné, Mme. D. *Greniers-citadelles au Maroc*. In *Publications de l'Institut des Hautes Etudes du Maroc,* pts. 1 and 2, 52 (1951).

Johnston, Harry T. "A Journey through the Tunisian Sahara." *The Geographical Journal,* June 1898.

Joly, A. "Notes géographiques sue le Sud Tunisien." *Bulletin de la Société de Géographie d'Alger* (1907): 281–301; (1909): 223–50, 471–508.

Le Boeuf, J. *La colonisation de l' Extrême Sud Tunisien*. Tunis: Imprimerie Rapide, 1903.

Le Houerou, H. N. "Contribution à l'étude des sols du Sud Tunisien." *Annales Agronomiques* (Paris) (1960): 241–308.

Lelainville, J. "Les troglodytes du Matmata." *Bulletin de la Société Normande de Géographie* (1909): 119–42.

Louis, André, and M. M. Sironval. "Le mariage traditionnel en milieu berbère dans le sud de la Tunisie." *Revue de l'Occident Musulman et de la Méditerranée* (Aix-en-Provence), no. 12 (1972): 93–121.

Louis, André, and Stanley Hallet. "Evolution d'un habitat: Le monde 'berbère' du Sud-Tunisien." *IBLA* (Institut des Belles Lettres Arabes) 42 (1972): 249–67.

Louis, André. "Aus Matmatas et dans les ksars du sud, l'olivier et les hommes." *Cahiers des Arts et Traditions Populaires; Revue due Centre des Arts et Traditions Populaires* (Tunis: Ministère des Affaires Culturelles), no. 3 (1969): 41–66.

———, comp. *Bibliographie ethno-sociologique de la Tunisie*. Tunis: Institut des Belles Lettres Arabes, 1977.

———. "Aux Nefzaouas: Le palmier et les hommes. Du semi-nomadisme à la sédentarisation." Lecture given at the First National Festival of the Sahara, Douz, Tunisia, 17 November 1967. Published in *IBLA* (Institut des Belles Lettres Arabes) 31 (1968): 315–46.

———. "Contacts entre culture 'berbère' et culture arabe dans le Sud-Tunisien." In *Actes du Premier Congrès d'Etudes des Cultures Méditerranéennes d'Influence Arabo-Berbère,* Malta, 3–6 April 1972, 394–405. Algiers: Société Nationale d'Edition et de Diffusion, 1972.

———. "Curieusement adossé à une falaise abrupte, un très vieux village: La Kessera." *IBLA* (Institut des Belles Lettres Arabes) 42 (1979): 235–47.

———. "Evolution des modes de vie au Sahara Tunisien." Lecture given at the Second National Festival of the Sahara, Douz, Tunisia, 9 November 1968. Published in *IBLA* (Institut des Belles Lettres Arabes) 32 (1969): 71–102.

———. "Greniers fortifiés et maisons troglodytes: Ksar-Djouama." *IBLA* (Institut des Belles Lettres Arabes) 28 (1965): 373–400.

———. "Habitat et habitations autour des ksars de montagne dans le Sud Tunisien." *IBLA* (Institut des Belles Lettres Arabes) 34 (1971): 123–48.

———. "Kalaa, ksour de montagne et ksour de plaine dans le Sud-Est Tunisien." In *Maghreb et Sahara, études géographiques offertes à Jean Despois, 257–70.* Special issue of *Acta Geographica.* Paris: Société de Géographie, 1973.

———. "L'habitation troglodyte dans un village des Matmata." *Cahiers des Arts et Traditions Populaires; Revue du Centre des Arts et Traditions Populaires* (Tunis: Secretariat d'Etat aux Affaires Culturelles et à l'Information), no. 2 (1968): 33–60.

———. "Le monde 'berbère' de l'Extrême-Sud Tunisien." *Revue de l'Occident Musulman et de la Méditerranée* (Aix-en-Provence), no. 11 (1972): 107–25.

———. "Sur un piton de l'Extrême-Sud, une étrange cité berbère: Douiret." *IBLA* (Institut des Belles Lettres Arabes) 27 (1964): 381–91.

———. *Carte pedologiue de la Tunisie.* Tunis, 1973.

———. *Etrange cité berbère du Sud Tunisien: Douiret.* Tunis: Société Tunisienne de Diffusion, 1975.

———. *Tunisie du Sud: Ksars et villages de Crêtes.* Paris: Editions du Centre National de la Recherche Scientifique, 1975.

Mathieu, G. "Contribution a l'étude des monts troglodytes, dans l'Extrême Sud Tunisien." *Annales des Mines et de Geology de Tunis,* no. 4 (1949).

Metal, Denyse. "Gravures pariétales d'une habitation du Segdel." *Revue Tunisienne* (1936): 273–341.

Montagne, R. *Un magasin collectif de l'Anti-Atlas. L'Agadir des Ikounka.* Paris: Larose, 1930.

Moreau, Lt. "Le Castellum de Ras Oued el Gordab, près de Ghoumrassen." Paris: Imprimerie Nationale, 1905. Also published in *Bull. Arch. Sousse* (1904): 369–76.

Moreau, P. "Le probléme du nomadisme dans le Sud Tunisien." *L'Afrique et l'Asie* (1950): 42–50.

Peltier, F., and F. Arin. "Les modes d'habitat chez les Djebaliya du Sud-Tunisien." *Revue du Monde Musulman* 8 (1909): 1–28.

Pervinquière, L. "Rapport sur une mission scientifique dans l'Extrême Sud Tunisien." Tunis: Guinle, 1912. Also published in *Mémoires et Documents de la Direction de l'Agriculture du Commerce et de la Colonisation (Tunis),* no. 2 (1912): 51–58 and plates.

Prost, G. "Les migrations des populations du Sud Tunisien." *Bulletin Economique et Social de la Tunisie,* no. 43 (1950): 71–83.

Sladen, Douglas. *Carthage and Tunis, the Old and New Gates of the Orient.* 2 vols. London: Hutchinson and Co., 1906.

Suter, Karl. "Die Wohnhöhlen and Speicherburgen des Tripolitanisch-Tunesischen Berglandes." *Zeitschrift für Ethnologie* 89 (1964): 216–75.

Thomas, J. *A travers le Sud-Tunisien*. Paris: Société d'Editions Géographiques Maritimes et Colonailes, 1930.

Tunisia. Bureau of Native Affairs. *Historique du Bureau des Affaires Indigènes de Tataouine*. Bourg, France: V. Berthod, 1931.

Tunisia. Ministry of Agriculture. Department of Water and Soil Resources. Soil Division. *Carte pedologique de la Tunisie*. Tunis, 1971.

Ward, Philip. *Touring Libya: The Western Provinces*. London: Faber and Faber, 1967.

Warthilâni. "A travers la berbèrie orientale du XVIIIe siècle avec le voyageur al-Warthilâni." *Revue Africaine* 95 (1951): 315–99.

Williams, Maynard O. "Time's Footprint in Tunisian Sands." *National Geographic* 71 (1937): 344–86.

Turkey

Anahtar, Ali Yavuz. "An Earth-Integrated Hotel in Cappadocia." Ankara, Turkey, Master's thesis in Architecture, Middle East Technical University, July 1984, p. 21.

Erguvalani, A. K., and A. E. Yüzer. "Past and Present Use of Underground Openings Excavated in Volcanic Tuffs at Cappadocia Area." In *Storage in Excavated Rock Caverns, Rockstore 77,* edited by Magnus Bergman, 1: 31–36. Oxford, England: Pergamon Press, 1978.

Fahrner, Rudolf. *The Face of Anatolia: Caves and Khans in Cappadocia*. Vienna: Austrian National Printing Press, 1995.

Giovannini, Luciano. "The Rock Settlements." In *Arts of Cappadocia,* edited by Luciano Giovannini, 67–83. Geneva: Nagel, 1971.

Hazer, Fahriye. "Cultural-Ecological Interpretation of the Historic Underground Cities of Göreme, Turkey." In *Alternatives in Energy Conservation, the Use of Earth-Covered Buildings,* edited by Frank L. Moreland, 21–36. Washington, DC: National Science Foundation/USGPO, 1978.

Kostof, Spiro K. *Caves of God: The Monastic Environment of Byzantine Cappadocia*. Cambridge, MA: MIT Press, 1972.

Ozkan, Suha, and Selahattin Onur. "Another Think Wall Pattern: Cappadocia." In *Shelter, Sign & Symbol*. (London: Barrie & Jenkins), 1975, pp. 95–106.

Yorukoglu, Omer, et al. *Underground Cities in Cappadocia* (Ankara: not pub., 1989).

United States

Bennett, David. "University of Minnesota Bookstore." In *Alternatives in Energy Conservation, the Use of Earth-Covered Buildings,* edited by Frank L. Moreland, 117–30. Washington, DC: National Science Foundation/USGPO, 1978.

Boyer, L. L., ed. *An Earth-Sheltered Guest House*. Stillwater, OK: Oklahoma State University, 1979.

Boyer, L. L., W. T. Grondzik, and T. N. Bice, "Energy Usage in Earth-Covered Dwellings in Oklahoma." In *Proceedings: Earth-Sheltered*

Building Design Innovations, edited by Lester L. Boyer, 4.17–4.31. Stillwater, OK: Oklahoma State University, 1980.

Boyer, L. L., W. T. Grondzik, and M. J. Weber. "Passive Energy Design and Habitability Aspects of Earth-Sheltered Housing in Oklahoma." *Underground Space* 4 (1980): 333–39. Also published in *Preconference Proceedings of the Solar Heating & Cooling Systems Operational Results,* 183–89. Colorado Springs, CO: Department of Energy, 1979.

Boyer, L. L., M. J. Weber, and W. T. Grondzik. "Energy and Habitability Aspects of Earth-Sheltered Housing in Oklahoma." Project Report, Presidential Challenge Grant. Stillwater, OK: Oklahoma State University, 1980.

"Largest Earth Shelter in San Francisco." *Earth Shelter Digest and Energy Report,* no. 9 (May–June 1980): 37.

MacCurdy, George G. "American Caves and Cave Dwellers." *American Journal of Archaeology* 4 (1937): 383–87.

Machowski, Barb. "Earth Shelters Enrich American History." *Earth Shelter Digest and Energy Report,* no. 11 (September–October 1980): 48–51.

"Saving by Going Underground." *AIA Journal* 61 (1974): 48–50.

Sterling, Raymond L. "Recent Developments in Underground Utilization in the USA." In *Proceedings of the Fourth International Conference on Underground Space and Earth Sheltered Buildings, December 3–5, 1991.* Tokyo, Japan: Sponsor—Urban Underground Space Center of Japan, 1991.

Other Countries

Ascher, Francois. "Impact of Land Use Transition by Underground Utilization: The Case of Paris." In *Proceedings of the Fourth International Conference on Underground Space and Earth Sheltered Buildings, December 3–5, 1991.* Tokyo, Japan: Sponsor—Urban Underground Space Center of Japan, 1991.

Autio, Jorma. "Economic and Functional Operating Characteristics of Rock Caverns Used for Various Purposes in Finland." In *Conference on Underground Space* (1988): 31–36.

Baggs, Sydney A. "The Dugout Dwellings of an Outback Opal Mining Town in Australia." In *Underground Utilization,* edited by Truman Stauffer, Sr., 4: 573–99. Kansas City, MO: University of Missouri, 1978. Also in mimeo, University of New South Wales, Sydney, Department of Landscape Architecture, 1977.

———. "The Lithotecture of Australia: With Specific Reference to User Health Factors." In *Proceedings: Earth-Sheltered Building Design Innovations,* edited by Lester L. Boyer, 2:19–2:26. Stillwater, OK: Oklahoma State University, 1980.

———. "Underground Architecture." *Architecture Australia* (December 1977–January 1978): 62–69.

Bergman, Sten G. A. *Underground Construction in Sweden.* Stockholm: Swedish Underground Construction Mission, 1976.

Brown, Percy. *Indian Architecture (Buddhist and Hindu),* Bombay: D.B. Taraporevala Sons & Co. Private Ltd., 1976.

Clausen, O. "Sweden Goes Underground." *New York Times Magazine,* 22 May 1966, 23–25.

Crawford, Harriet, ed. *Subterranean Britain—Aspects of Underground Archeology.* New York: St. Martin's Press, 1980.

Dawne, S. G. *Ajanta Ellora and Environs, Mythological, Historical, Sculptural, Pictorial Review.* India: Mukund Prakashan, 4th Ed., 1978.

Duffaut, Pierre. "Ground Pressure and Tunnelling from the Nineteenth Century to the Present." *Underground Space* 1 (1977): 185–200.

———. "Past and Future of the Use of Underground Space in France and Europe." *Underground Space* 5 (1980): 86–91.

Fujita, K., K. Ueda, and Y. Shimomura. "An Empirical Proposal on Stability of Rock Cavern Wall during Construction in Japan." In *Storage in Excavated Rock Caverns, Rockstore* 77, edited by Magnus Bergman, 2: 309–14. Oxford, England: Pergamon Press, 1978.

"Invisible Architecture in the Paris Underground; UNESCO Three-Story Underground Buildings." *Fortune* 73 (1966): 176.

"Japan, Underground Movement." *The Economist* 221 (1966): 1323.

Kovac, Milan. "Subterranean Museum for Pharaoh's Funeral Vessel at Giza, Egypt." *Underground Space* 3 (1979): 303–8.

Labs, Kenneth. "Terratecture: The Underground Design Movement of the 1970's." *Landscape Architecture* 67 (1977): 244–49.

———. "The Use of Earth-Covered Buildings through History." In *Alternatives in Energy Conservation, the Use of Earth-Covered Buildings,* edited by Frank L. Moreland, 7–19. Washington, DC: National Science Foundation/USGPO, 1978.

LaNier, Royce. *Geotecture, Subterranean Accommodation and the Architectural Potential of Earthworks.* South Bend, IN: Privately printed, 1970.

Lang, Thomas A. "Underground Experience in the Snowy Mountains—Australia." In *Protective Construction in a Nuclear Age,* edited by J. J. O'Sullivan, 2: 688–765. New York: Macmillan Co., 1961.

Legget, Robert F. "75-Year-Old Underground Compressed Air Plant Still in Use in Canada." *Underground Space* 4 (1979): 29–31.

O'Reilly, M. P. "Some Examples of Underground Development in Europe." *Underground Space* 2 (1978): 163–78.

Owen, Christopher R. L. "Lalibala: The Rugged Highlands of Ethiopia Hold a Cache of Extraordinary Rockhewn Churches." *Architecture Plus* 2 (1974): 44–49.

Pathy, T. V. *Ajanta, Ellora & Aurangabad Caves: An Appreciation.* Aurangabad, India: Shrimati T. V. Pathy, 1978.

Ponte, Vincent, "Montreal High on Its Underground." *Kansas City Star,* 19 December 1976.

Rudofsky, Bernard. "Troglodytes." *Horizon* 9 (1967): 39.

Wolfe, Nancy H. *The Valley of Bamiyan, Kabul.* Kabul, Afghanistan: Afghanistan Tourist Organization, 1963.

Woodcock, George. "Cave Temples of Western India." *Arts Magazine* 36 (1962): 74–81.

SELECTED TOPICS: DESIGN AND CONSTRUCTION

Construction

Anagnostou, G., and K. Kovári. "The Face Stability of Slurry-Shield-Driven Tunnels." *Tunnelling and Underground Space Technology* 9, no. 2 (1994): 165–174.

Biggart, A. R., J. P. Rivier, and R. Sternath. "Storebælt Railway Tunnel—Construction." In *Proceedings of the International Symposium on Technology of Bored Tunnels Under Deep Waterways*. Copenhagen: Teknisk Forlag A/S, 1993.

Dahlø, T. S., and B. Nilsen. "Stability and Rock Cover of Hard Rock Subsea Tunnels." *Tunnelling and Underground Space Technology* 9, no. 2 (1994): 151–158.

de Ronde, S. F., G. J. Koo, and G. T. Visser. "The Design of the Piet Hein Tunnel, Amsterdam." In *Options for Tunnelling 1993*. Edited by H. Burger. Amsterdam: Elsevier, 1993.

Eisenstein, Z. D. "Large Undersea Tunnels and the Progress of Tunnelling Technology." *Tunnelling and Underground Space Technology* 9, no. 3 (1994): 283–292.

Fairhurst, Charles, et al. "The International Stripa Project: An Overview." *Tunnelling and Underground Space Technology* 8, no. 3 (1993): 315–343.

Grantz, Walter C. "Catalog of Immersed Tunnels." *Tunnelling and Underground Space Technology* 8, no. 2 (1993): 175–263.

———. "Waterproofing and Maintenance." *Tunnelling and Underground Space Technology* 8, no. 2 (1993): 161–174.

Gravesen, Lars, and Nestor S. Rasmussen. "A Milestone in Tunnelling: Rotterdam's Maas Tunnel CelBaxter, D.A. Construction of the Brunswick Street Rail Tunnels, Brisbane." In *Proceedings of the VIIIth Australian Tunnelling Conference, Sydney*. Parkville, Victoria: The Australasian Institute of Mining and Metallurgy, 1993.

Kashima, S., et al. "Study of an Underground Physical Distribution System in a High-Density, Built-Up Area." *Tunnelling and Underground Space Technology* 8, no. 1 (1993): 53–59.

Malmberg, Bo. "Shotcrete for Rock Support: A Summary Report on the State of the Art in Fifteen Countries." *Tunnelling and Underground Space Technology* 8, no. 4 (1993): 441–470.

Nilsen, B. "Empirical Analysis of Minimum Rock Cover for Subsea Rock Tunnels." *Options for Tunnelling 1993*. Edited by H. Burger. Amsterdam: Elsevier, 1993.

Nishida, Y., and N. Uchiyama. "Japan's Use of Underground Space in Urban Development and Redevelopment." *Tunnelling and Underground Space Technology* 8, no. 1 (1993): 41–45.

Nuclear Waste Technical Review Board. "Underground Exploration and Testing at Yucca Mountain." *Tunnelling and Underground Space Technology* 9, no. 3 (1994): 353–363.

Odgård, Anders, David G. Bridges, and Steen Rostam. "Design of the Storebælt Railway Tunnel." *Tunnelling and Underground Space Technology* 9, no. 3 (1994): 293–307.

Pelizza, Sebastiano, and Daniele Peila. "Soil and Rock Reinforcements in Tunnelling." *Tunnelling and Underground Space Technology* 8, no. 3 (1993): 357–372.

Pusch, Roland, and Christer Svemar. "Influence of Rock Properties on Selection of Design for a Spent Nuclear Fuel Repository." *Tunnelling and Underground Space Technology* 8, no. 3 (1993): 345–356.

Rosengren, Lars, and Ove Stephansson. "Modeling of Rock Mass Response to Glaciation at Finnsjon, Central Sweden." *Tunnelling and Underground Space Technology* 8, no. 1 (1993): 75–82.

Rostam, A. "Service Life Design—The European Approach." *Concrete International* 15 (1993): 24–33.

Saveur, Jan, and Walter C. Grantz. "Structural Design of Immersed Tunnels." *Tunnelling and Underground Space Technology* 8, no. 2 (1993): 123–139.

van der Schrier, J. S., et al. "Tunnelling in Urban Areas, Interaction With Loaded Foundation Piles." In *Options for Tunnelling 1993*. Edited by H. Burger. Amsterdam: Elsevier, 1993.

Zhang, Qing, et al. "An Expert System for Prediction of Karst Disaster in Excavation of Tunnels or Underground Structures through a Carbonate Rock Area." *Tunnelling and Underground Space Technology* 8, no. 3 (1993): 373–378.

Design

Bligh, Thomas P. "Building Underground." *Building Systems Design* (October–November 1976): 1–22.

Duke, Buford W., Jr. "Incorporating Earth Shelter Considerations into the Design Process for Nonresidential Building." In *Proceedings: Earth-sheltered Building Design Innovations,* edited by Lester L. Boyer, 5:23–5:30. Stillwater, OK: Oklahoma State University, 1980.

Golany, Gideon S. *Earth-Sheltered Habitat: History, Architecture, and Urban Design*. New York: Van Nostrand Reinhold Co., 1983.

———. "Design of the Earth-Sheltered Habitat: A Challenging Issue in Arid Lands." *ITTC Review* 12 (1983): 75–86.

———. "Subterranean Settlements for Arid Zone." In *Housing in Arid Lands: Design and Planning,* edited by Gideon S. Golany, 109–22. London: Architectural Press, 1980.

Gorman, James. "Underground Architecture." In *Underground Utilization,* edited by Truman Stauffer, Sr., 3: 329–31. Kansas City, MO: University of Missouri, 1978.

Jia Kunnan and Dong Yikang. "Sintered Cave for Living Room." *Proceedings of the International Symposium on Earth Architecture, 1–4 November, 1985, Beijing*. Beijing: Architectural Society of China, 1985.

Labs, Kenneth. "The Architectural Underground." Parts 1, 2. *Underground Space* 1 (1976): 1–8, 135–56.

———. "The Architectural Use of Underground Space: Issues and Limitations." New Haven, CT: privately printed, 1975.

Langley, John B. "The Barrel Shell—Structural Rethinking in Earth-Sheltered Design." *Underground Space* 5 (1980): 92–101.

Newman, Jerry, and Luther C. Godbey. *Design Considerations for Below-Grade-Level Houses.* Saint Joseph, MI: American Society of Agricultural Engineers, 1978.

Pennsylvania State University. Shelter Research and Study Program. *Planning, Analysis, and Design of Shelters.* 5 vols to date. University Park, PA: College of Engineering and Architecture, Pennsylvania State University, 1961.

Powell, Evelyn J. "Design Considerations for the Elderly and Handicapped." In *Earth-Covered Buildings and Settlements,* edited by Frank L. Moreland, 149–51. Springfield, VA: NTIS, 1979.

Preliminary Design Information for Underground Space. Minneapolis: Department of Civil and Mineral Engineering, University of Minnesota, 1975.

Shih, Jason C. "The Optimization of Earth-Covered Shell Structures." In *Earth-Covered Buildings: Technical Notes,* edited by Frank L. Moreland, F. Higgs, and J. Shih, 92–114. Springfield, VA: NTIS, 1979.

Sonntag, Alain. "An Image of an Underground Complex, the Forum des Halles in Paris." In *Proceedings of the Fourth International Conference on Underground Space and Earth Sheltered Buildings, December 3–5, 1991.* Tokyo, Japan: Sponsor—Urban Underground Space Center of Japan, 1991.

Wells, Malcolm B. *Underground Designs.* Brewster, MA: privately printed, 1977.

Drainage, Waterproofing, and Insulation

Bligh, Thomas P., Paul Shipp, and George Meixel. "Where to Insulate Earth-Protected Buildings and Existing Basements." In *Earth-Covered Buildings: Technical Notes,* edited by Frank L. Moreland, F. Higgs, and J. Shih, 251–72. Springfield, VA: NTIS, 1979.

Edvardsen, Knut I. *New Method of Drainage of Basement Walls.* Ottawa: National Research Council of Canada, 1972.

Foute, Steven J., and Douglas B. Cargo. "Earth-Covered Housing: Hydrologic and Pollution Considerations." In *Earth-Covered Buildings and Settlements,* edited by Frank Moreland, 108–24. Springfield, VA: NTIS, 1979.

Gill, G. W. "Waterproofing Buildings Below Grade." *Civil Engineering* 29 (1959): 3–5.

Gratwick, Reginald T. *Dampness in Buildings.* London: Lockwood Press, 1966.

Hagerman, Tor H. "Groundwater Problems in Underground Construction." In *Underground Utilization,* edited by Truman Stauffer, Sr., 3: 451–52. Kansas City: University of Missouri, 1978. Also in *Large Permanent Underground Openings,* edited by Tor Brekke and Finn Jorstad, 319–21. Oslo: Scandinavian University Books, 1970.

Klepper, M. R. "Underground Space: An Unappraised Natural Resource. Some Geologic and Hydrologic Considerations." In *Underground Utilization,* edited by Truman Stauffer, Sr., 1: 31–33. Kansas City: University of Missouri, 1978.

Lane, Charles A. "Waterproofing Earth-Sheltered Houses." *Fine Homebuilding*, no. 2 (April–May 1981): 35–37.

McGroarty, Bryan. "Waterproofing: Design to Work." *Earth Shelter Digest and Energy Report*, no. 12 (November–December 1980): 23–26.

Moyers, John C. *Value of Thermal Insulation in Residential Construction*. Oak Ridge, TN: Oak Ridge National Laboratory, 1971.

Randall, Frank A., Jr. "Construction Recommendations for Moisture Control." *Concrete Construction* 25 (September 1980): 675.

Remson, I., G. M. Hornberger, and F. J. Molz. *Numerical Methods in Subsurface Hydrology*. New York: John Wiley & Sons, 1971.

Natural Hazards

"Disasters Point Way to Earth Shelters." *Earth Shelter Digest and Energy Report*, no. 8 (March–April 1980): 11–14.

Dowding, Charles H. "Seismic Stability of Underground Openings." In *Storage in Excavated Rock Caverns, Rockstore 77*, edited by Magnus Bergman, 2: 231–38. Oxford, England: Pergamon Press, 1978.

Fukushima, Kunio, M. Hamada, and T. Hanamura. "Earthquake Resistant Design of Underground Tanks." In *Storage in Excavated Rock Caverns, Rockstore 77*, edited by Magnus Bergman, 2: 315–20. Oxford, England: Pergamon Press, 1978.

Pratt, H. R., W. A. Hustrulid, and D. E. Stephenson. *Earthquake Damage to Underground Facilities*. Washington, DC: Department of Energy Report Prepared by E. I. Du Pont De Nemours & Co., 1978.

Tamura, C. "Design of Underground Structures by Considering Ground Displacement during Earthquakes." *Proceedings of U.S.–Japan Seminar on Earthquake Engineering Research with Emphasis on Lifeline Systems*. Tokyo, 1976.

Yamahara, Hiroshi, Y. Hisatomi, and T. Morie. "A Study on the Earthquake Safety of Rock Cavern." In *Storage in Excavated Rock Caverns, Rockstore 77*, edited by Magnus Bergman, 2: 377–82. Oxford, England, Pergamon Press, 1978.

Roofs

Bargabus, Dave, and D. Laubach. "Covered vs. Conventional Roofs: What's Best?" *Earth Shelter Digest and Energy Report*, no. 10 (July–August 1980): 26–28.

Behr, Richard, Ernst W. Kiesling, and Gary Boubel. "Thin Shell Roof Systems and Construction Techniques for Earth Sheltered Housing." In *Proceedings: Earth-sheltered Building Design Innovations*, edited by Lester Boyer, 7:17–7:27. Stillwater, OK: Oklahoma State University, 1980.

Gordon, Ann. "Rooftop Plantings for Earth-Covered Buildings in Temperate Climates." In *Proceedings: Earth-sheltered Building Design Innovations*, edited by Lester Boyer, 3:11–3:16. Stillwater, OK: Oklahoma State University, 1980.

Roy, Robert L. "Earth Roof Thickness: The Tradeoffs." *Earth Shelter Digest and Energy Report*, no. 13 (January–February 1981).

Structure

Bello, Arturo A. "Simplified Method for Stability Analysis of Underground Openings." In *Storage in Excavated Rock Caverns, Rockstore 77,* edited by Magnus Bergman, 2: 289–94. Oxford, England: Pergamon Press, 1978.

Bodonyi, J., and S. Fekete. "Calculation and Design of Large Underground Openings." In *Storage in Excavated Rock Caverns, Rockstore 77,* edited by Magnus Bergman, 2: 295–300. Oxford, England: Pergamon Press, 1978.

Chicago Urban Transportation District. "Underground Construction Problems, Techniques & Solutions." In *Proceedings of a Seminar,* U.S. Department of Transportation, 20–22, October 1975, Chicago. Springfield, VA: NTIS, n.d.

Darvas, Robert M. "Structural Problems in Earth Covered Buildings." In *Earth-Covered Buildings: Technical Notes,* edited by Frank L. Moreland, F. Higgs, and J. Shih, 70–73. Springfield, VA: NTIS, 1979.

Earth-Sheltered Construction: A Special Issue of Concrete Construction 25 (September 1980).

Getzler, Z., Menachem Gellert, and Ruben Eitan. "Analysis of Arching Pressures in Ideal Elastic Soil." *Journal* (Soil Mechanics and Foundations Division, American Society of Civil Engineers) 96 (1970): 1357–72.

Getzler, A., and L. Lupu. "Experimental Study of Buckling of Buried Domes." *Journal* (Soil Mechanics and Foundations Division, American Society of Civil Engineers) 95 (1969): 605–24.

Gill, H. L. "Model Study of Arching Above Buried Structures." *Journal* (Soil Mechanics and Foundations Division, American Society of Civil Engineers) 95 (1969): 1120–22.

———. "Static Loading of Small Buried Arches." Port Hueneme, CA: U.S. Naval Civil Engineering Laboratory, 1965.

Lenzini, Peter A. "Ground Stabilization: Review of Grouting and Freezing Techniques for Underground Openings." *Underground Space* 1 (1977): 227–40.

Shih, Jason C. "Literature Review on the Structural Optimization for the Possible Application of Earth-Covered Buildings." In *Earth-Covered Buildings: Technical Notes,* edited by Frank L. Moreland, F. Higgs, and J. Shih, 115–22. Springfield, VA: NTIS, 1979.

Sterling, Ray. "Structural Systems for Earth Sheltered Housing." In *Earth-Covered Buildings: Technical Notes,* edited by Frank L. Moreland, F. Higgs, and J. Shih, 60–69. Springfield, VA: NTIS, 1979. Also in *Underground Space* 3 (1978): 75–81.

Vadnais, Kathleen. "Double Envelope Concept Applied to Earth-Sheltering." *Earth Shelter Digest and Energy Report,* no. 13 (January–February 1981).

Soil Properties and Performance

Ashbel, D., et al. *Soil Temperature in Different Latitudes and Different Climates.* Jerusalem: Hebrew University, 1965.

Bergman, S. Magnus. "Geo-Planning—the Key to Successful Underground Construction." *Underground Space* 2 (1977): 1–7.

Blick, Edward F. "A Simple Method for Determining Heat Flow through Earth Covered Roofs." In *Proceedings: Earth-sheltered Building Design Innovations,* edited by Lester L. Boyer. 3: 3–23. Stillwater, OK: Oklahoma State University, 1980.

Bligh, Thomas P. "Thermal Energy Storage in Large Underground Systems and Buildings." In *Storage in Excavated Rock Caverns, Rockstore 77,* edited by Magnus Bergman, 1: 63–71. Oxford: Pergamon Press, 1978.

Bliss, Donald E., et al. "Air and Soil Temperatures in a California Date Garden." *Soil Science* 53 (1942): 55–64.

Bodonyi, J. "Engineering Geology Related to Underground Openings." *UNESCO International Post-Graduate Course on the Principles and Methods of Engineering Geology.* Budapest, 1975.

Boileau, G. G., and J. K. Latta. "Calculation of Basement Heat Losses." In *Underground Utilization,* edited by Truman Stauffer, Sr., 5: 754–64. Kansas City, MO: University of Missouri, 1978.

Carson, James E. "Analysis of Soil and Air Temperatures by Fourier Techniques." *Journal of Geophysical Research* 68 (1963): 2217–32.

———. *Soil Temperature and Weather Conditions.* Lemont, IL: Argonne National Laboratory, 1961.

Crabb, G. A., Jr., and J. L. Smith. "Soil-Temperature Comparisons under Varying Covers." In *Soil Temperature and Ground Freezing,* 32–80. Washington, DC: National Academy of Sciences, National Research Council, 1953.

Davies, G. R. "Thermal Analysis of Earth-Covered Buildings." In *Proceedings of the 4th National Passive Solar Conference,* edited by G. Franta, 744. Newark, DE: American Section of the International Solar Energy Society, 1979.

Davis, William B. "Earth Temperature: Its Effect on Underground Residences." In *Earth-Covered Buildings: Technical Notes,* edited by Frank L. Moreland, F. Higgs, and J. Shih, 205–9. Springfield, VA: NTIS, 1979.

de Vries, D. A. "Thermal Properties of Soils." In *Physics of Plant Environment,* edited by Willem R. van Wijk, 210–35. New York: John Wiley & Sons, 1963.

Drapeau, F. J. *Mathematical Analysis of Temperature Rise in the Heat Condition Region of an Underground Shelter.* Washington, DC: U.S. Department of Commerce, National Bureau of Standards, 1965.

Druker, E. F., and J. T. Haines. "A Study of Thermal Environment in Underground Survival Shelters Using an Electronic Analog Computer." *ASHRAE Transactions* 70 (1964): 7–20.

Flathau, W. J., and J. P. Balsara. "Soil-structure Interaction—An Overview." In *Earth-Covered Buildings: Technical Notes,* edited by Frank Moreland, F. Higgs and J. Shih, 13–33. Springfield, VA: NTIS, 1979.

Fluker, B. J. "Soil Temperatures." *Soil Science* 86 (1958): 35–46.

Givoni, Baruch. "Modifying the Ambient Temperature of Underground Buildings." In *Earth-covered Buildings: Technical Notes,* edited by Frank

L. Moreland, F. Higgs, and J. Shih, 123–38. Springfield, VA: NTIS, 1979.

Golany, Gideon S. "Soil Thermal Performance and Geo-Space Design." In *Geotechnical Engineering, Japan Society of Civil Engineers*. Proceedings of JSCE, No. 445/III-18, March 1992–3, pp. 1–8.

Jumikis, A. R. *Thermal Soil Mechanics*. New Brunswick, NJ: Rutgers University Press, 1966.

Kusada, T. and F. J. Powell. *Heat Transfer Analysis of Underground Heat Distribution Systems*. Report 10-194. Washington, DC: National Bureau of Standards, 1970.

Kusada, T., and P. R. Achenbach. "Comparison of Digital Computer Simulations of Thermal Environment." In *Occupied Underground Protective Structure with Observed Conditions*. Report 9473. Washington, DC: National Bureau of Standards, 1966.

———. "Numerical Analyses of the Thermal Environment of Occupied Underground Spaces with Finite Cover Using a Digital Computer." *ASHRAE Transactions* 69 (1963): 439–52.

Kusada, T. *Earth Temperature Beneath Five Different Surfaces*. Report 10373. Washington, DC: U.S. Department of Commerce, National Bureau of Standards, 1971.

———. "Least Squares Technique for the Analysis of Periodic Temperature of the Earth's Surface Region." *Journal of Research of the National Bureau of Standards* 71C, no. 1 (1967): 43–50.

———. "The Effect of Ground Cover on Earth Temperature." In *Alternatives in Energy Conservation, the Use of Earth-Covered Buildings,* edited by Frank L. Moreland, 279–303. Washington, DC: National Science Foundation/USGPO, 1978.

Labs, Kenneth. "The Underground Advantage: Climate of Soils." In *Proceedings of the 4th National Passive Solar Conference,* edited by G. Franta. Newark, DE: American Section of the International Solar Energy Society, 1979.

Langbein, W. B. "Computing Soil Temperatures." *Transactions of the American Geophysical Union* 30 (1949): 543–47.

Lytton, Robert L. "Soil and Ground Water Considerations." In *Alternatives in Energy Conservation, the Use of Earth-Covered Buildings,* edited by Frank L. Moreland, 257–62. Washington, DC: National Science Foundation/USGPO, 1978.

Maxwell, Robert K. "Temperature Measurements and the Calculated Heat Flux in the Soil." Master's thesis, University of Minnesota, 1964.

McBride, M. F., et al. "Measurement of Subgrade Temperatures for Prediction of Heat Loss in Basements." *ASHRAE Transactions* 85, pt. 1 (1979).

Penrod, E. B. "Variation of Soil Temperature at Lexington, Kentucky, from 1952–1956." Bulletin 57. Lexington, KY: Engineering Experiment Station, University of Kentucky, 1960.

Peters, A. *Soil Thermal Properties: An Annotated Bibliography*. Philadelphia: Franklin Institute, 1962.

Pollack, H. A., and D. S. Chapman. "The Flow of Heat from the Earth's Interior." *Scientific American* 237 (1977): 60–76.

Shipp, P. H. "Thermal Characteristics of Large Earth Sheltered Structures." Ph.D. diss., University of Minnesota, 1979.
Singer, Irving A., and R. M. Brown. "The Annual Variations of Sub-soil Temperatures about a 600-Foot Circle." *Transactions of the American Geophysical Union* 37 (1956): 743–48.
Smith, A. "Diurnal, Average, and Seasonal Soil Temperature Changes at Davis, California." *Soil Science* 28 (1929): 457–68.
Smith, W. O. "The Thermal Conductivity of Dry Soils." *Soil Science* 53 (1942): 435–59.
"Survival Shelters." In *ASHRAE Guide and Data Book Applications, 1971*, 177–202. New York: American Society of Heating, Refrigerating, and Air-Conditioning Engineers, 1971.
Van Straaten, J. F. *Thermal Performance of Buildings.* Amsterdam: Elsevier Publishing Co., 1967.
Willi, W. O. "Bibliography on Soil Temperature." Mandan, ND: U.S. Department of Agriculture, Agricultural Research, Northern Great Plains Field Station, 1964.

Technology

Dunnicliff, John. *Geotechnical Instrumentation for Monitoring Field Performance.* New York: John Wiley & Sons, Inc., 1993.
Kämppi, A. "Viikinmäki: Helsinki's New Central Wastewater Treatment Plant." *Tunnelling and Underground Space Technology* 9, no. 3 (1994): 373–377.
Kovari, K., R. Fechtig, and Ch. Amstad. "Experience with Large-Diameter Tunnel Boring Machines in Switzerland." *Options for Tunnelling 1993.* Edited by H. Burger. Amsterdam: Elsevier, 1993.
Kramer, Steven R., William J. McDonald, and James C. Thompson. *An Introduction to Trenchless Technology.* Arlington, VA: National Utility Contractors Association, 1994.
Molenaar, V. L. "Construction Techniques." *Tunnelling and Underground Space Technology* 8, no. 2 (1993): 141–159.
Murray, C., S. Courtley, and P. F. Howlett. "Developments in Rock-Breaking Techniques." *Tunnelling and Underground Space Technology* 9, no. 2 (1994): 225–231.
Robbins, Richard, and Martin Kelley. "Tunnelling Machine Development for Undersea Projects—A Review of the Issues." *Tunnelling and Underground Space Technology* 9, no. 3 (1994): 323–328.
Soliman, E., H. Duddeck, and H. Ahrens. "Two- and Three-Dimensional Analysis of Closely Spaced Double-Tube Tunnels." *Tunnelling and Underground Space Technology* 8, no. 1 (1993): 13–18.

Thermal Performance and Energy

Allen, F. C. "Ventilating and Mixing Processes in Nonuniform Shelter Environments." Springfield, VA: National Technical Information Service, 1970.
Barnes, Paul R., and Hanna Shapira. "Passive Solar Heating and Natural

Cooling of an Earth-Integrated Design." In *Proceedings: Earth-sheltered Building Design Innovations,* edited by Lester L. Boyer, 7:29–7:36, Stillwater, OK: Oklahoma State University, 1980.

Bedell, Berkley. "Earth-sheltered Homes Help to Achieve Energy Independence." *Earth Shelter Digest and Energy Report,* no. 9 (May–June 1980): 48–49.

Bennett, David, and Thomas P. Bligh. "The Energy Factor—A Dimension of Design." *Underground Space* 1 (1977): 325–32.

Bligh, Thomas P. "A Comparison of Energy Consumption in Earth-Covered vs. Non-Earth-Covered Buildings." In *Alternatives in Energy Conservation, the Use of Earth-Covered Buildings,* edited by Frank L. Moreland, 85–105. Washington, DC: National Science Foundation/USGPO, 1978.

Chester, C. V., et al. "An Earth-Covered Residential Concept for the Humid Continental Region." In *Earth-covered Buildings: Technical Notes,* edited by Frank L. Moreland, F. Higgs, and J. Shih, 171–94. Springfield, VA: NTIS, 1979.

Coffee and Crier, Architects. "An Underground Solar-Heated Residence (Plan)." In *Alternatives in Energy Conservation, the Use of Earth-Covered Buildings,* edited by Frank L. Moreland, 149. Washington, DC: National Science Foundation/USGPO, 1978.

Cunningham, Kim. "Underground Architecture: An Alternate Approach to Architecture and Energy Conservation." Master's thesis, Oklahoma State University, 1977.

Drucker, Eugene E., and Herbert S. Y. Cheng. "Analog Study of Heating in Survival Shelters." In *Symposium on Survival Shelters,* 35–80. New York: American Society of Heating, Refrigerating, and Air-conditioning Engineers, 1963.

Ducar, G. J., and G. Engholm. "Natural Ventilation of Underground Fallout Shelters." *ASHRAE Transactions* 71, pt. 1 (1965): 88–100.

"Earth-Sheltered Colorado House is 100% Solar Heated." *Underground Space* 4 (1980): 403–4.

Eckert, E. R. G., T. P. Bligh, and E. Pfender. "Energy Exchange between Earth-Sheltered Structures and the Surrounding Ground." In *Earth-Covered Buildings: Technical Notes,* edited by Frank L. Moreland, F. Higgs, J. Shih, 226–50. Springfield, VA: NTIS, 1979.

Emery, A. F., D. R. Heerwagen, and C. J. Kippenhan. "Earth-Sheltered Passive Solar Structures—Occupant Comfort and Energy Use." In *Earth-Sheltered Building Design Innovations,* edited by Lester L. Boyer, 4:33–4:40. Stillwater, OK: Oklahoma State University, 1980.

Golany, Gideon S. "Free-Energy Cooling Systems for Houses in the Desert." In *Innovations for Future Cities,* edited by Gideon S. Golany, 246–62. New York: Praeger Publications, 1976. Also in *Proceedings of the International Conference of Energy Use Management,* edited by Rocco A. Fazzolare and Craig B. Smith, 2: 339–46. New York: Pergamon Press, 1977.

———. "Urban Design Morphology and Thermal Performance." In *Atmospheric Environment,* Vol. 29, 1995, pp. 1–11.

Hylton, Joe. "Underground-Solar House Wind Tunnel Test." In *Earth-*

Covered Buildings: Technical Notes, edited by Frank L. Moreland, F. Higgs, and J. Shih, 195–204. Springfield, VA: NTIS, 1979.

Jackewicz, Shirley A. "New Cave Dwellers Save on Energy Bills, Keep Comfortable." *Wall Street Journal,* 24 May 1979.

Labs, Kenneth. "Earth Tempering as a Passive Design Strategy." In *Earth-Sheltered Building Design Innovations,* edited by Lester L. Boyer, 3.3–3.10. Stillwater, OK: Oklahoma State University, 1980.

Nordell, Bo, et al. "The Combi Heat Store—A Combined Rock Cavern/Borehole Heat Store." *Tunnelling and Underground Space Technology* 9, no. 2 (1994): 243–249.

Orlowski, Henry. "Thermal Chimneys and Natural Ventilation." In *Earth-Covered Buildings: Technical Notes,* edited by Frank L. Moreland, F. Higgs, and J. Shih, 220–25. Springfield, VA: NTIS, 1979.

Rice, Irvin. "Heating, Ventilating, and Air Conditioning of Underground Installations." In *An Introduction to the Design of Underground Openings for Defense,* edited by Le Roy Goodwin, *Quarterly of The Colorado School of Mines* 46: 259–304. Golden: Colorado School of Mines, 1951.

Schutrum, L. F., and N. Ozisik. "Solar Heat Gains through Domed Skylights." *ASHRAE Journal* 3 (1961): 51–60.

Sohar, Ezra. "Man in the Desert." In *Arid Zone Settlement Planning: The Israeli Experience,* edited by Gideon S. Golany, 447–518. New York: Pergamon Press, 1979.

"The Energy Initiative Program: An Earth-Covered Military Plant." *Underground Space* 3 (1979): 269–70.

Thompson, Robert. "Air Quality Maintenance in Underground Buildings." *Underground Space* 1 (1977): 355–64.

van Hove, J. "Productivity of Aquifer Thermal Energy Storage (ATES) in The Netherlands." *Tunnelling and Underground Space Technology* 8, no. 1 (1993): 47–52.

Ventilation

"Energy Conservation by Building Underground." *Underground Space* 1 (1976): 19–33.

"Underground Building Climate." *Solar Age* 4 (1979): 44–50.

Windows and Light

Boyer, Lester L. "Subterranean Designs Need Daylighting." *Earth Shelter Digest and Energy Report,* no. 4 (July–August 1979): 32–34.

Burts, E., and Eva McDonald. "Opinions Differ on Windowless Classrooms." *NEA Journal* 50 (October 1961): 13–14.

Chambers, J. A. "A Study of Attitudes and Feelings toward Windowless Classrooms." Ed.D. diss., University of Tennessee, 1963.

Collins, Belinda Lowenhaupt, ed. *Windows and People: A Literature Survey. Psychological Reaction to Environments With and Without Windows.* Gaithersburg, MD: U.S. Department of Commerce, National Bureau of Standards, 1975.

Karmel, L. J. "Effects of Windowless Classroom Environment on High School Students." *Perceptual and Motor Skills* 20 (1965): 277–78.

URBAN DESIGN: CITY PLANNING AND DEVELOPMENT

Climate

Barber, E. M. "Adequacy of Evaporative Cooling and Shelter Environmental Prediction." Report 10689. Washington, DC: National Bureau of Standards, 1972.

Behr, Richard, Ernst Kiesling, and Gary Boubel. "Earth-Sheltered Housing Potentials for West Texas." *Earth Shelter Digest and Energy Report,* no. 5 (September–October 1979): 24–25.

Gelder, John. "Underground Desert Shelter." B. Arch. thesis, University of Adelaide, Australia, 1977.

Golany, Gideon S. "Subterranean Settlements for Arid Zones." In *Earth-Covered Buildings and Settlements,* edited by Frank L. Moreland, 174–202. Springfield, VA: NTIS, 1979. Also in *Housing in Arid Lands: Design and Planning,* edited by Gideon S. Golany, 109–22. London: Architectural Press, 1980.

Green, Kevin W. "Passive Cooling: Designing Natural Solutions for Summer Cooling Loads." *Research and Design* 2 (Fall 1979): 4.

Hummell, John D., David E. Bearint, and James A. Eibling. "Survival Shelter Cooling: Conventional and Novel Systems." *ASHRAE Transactions* 71, pt. 1 (1965): 125–33.

Kiesling, E. W., and Gary A. Boubel. "Ashford Earth-Covered Residence, Muleshoe, Texas." In *Earth-Covered Buildings and Settlements,* edited by Frank Moreland, 273–77. Springfield, VA: NTIS, 1979.

Kiesling, E. W., Gary A. Boubel, and R. Behr. "Earth Forms Texas House." In *Earth Shelter Digest and Energy Report,* no. 4 (July–August 1979): 8–11.

Knauer, Virginia, and Lewis M. Branscomb. *Eleven Ways to Reduce Energy Consumption and Increase Comfort in Household Cooling*. Washington, DC: National Bureau of Standards, 1971.

Labs, Kenneth. "Terratypes: Underground Housing for Arid Zones." In *Housing in Arid Lands: Design and Planning,* edited by Gideon S. Golany, 123–40. London: Architectural Press, 1980.

Langley, John B. "Sun Belt Subsoil Studied." *Earth Shelter Digest and Energy Report,* no. 12 (November–December 1980): 20–21.

Langley, John B., and James L. Gray. *Sun Belt Earth-Sheltered Architecture.* Winter Park, FL: Sun Belt Earth-Sheltered Research, 1980.

Shih, J. C., and S. Kumar. "A Case for Subsurface and Underground Housing in Hot-arid Lands." Washington, DC: U.S. Army Research Office, 1976.

van der Meer, Wybe J. "The Possibility of Subterranean Housing for Arid Zones." In *Housing in Arid Lands: Design and Planning,* edited by Gideon S. Golany, 141–150. London: Architectural Press, 1980.

Environmental Issues

Carleton, Joseph G. "An Environmentalist's Views on Underground Construction." In *Underground Utilization,* edited by Truman Stauffer, Sr., 6: 894–99. Kansas City, MO: University of Missouri, 1978.

Dasler, A. R., and D. Minard. "Environmental Psychology of Shelter Habitation." *Transactions. Proceedings of the ASHRAE Semiannual Meeting,* Vol. 71, pt. 1, 115–24. Chicago: American Society of Heating, Refrigerating, and Air-Conditioning Engineers, 1965.

LaNier, Royce. "Assessing Environmental Impact of Earth-Covered Buildings." *Underground Space* 1 (1977): 309–15.

———. "Earth-Covered Buildings and Environmental Impact." In *Alternatives in Energy Conservation, the Use of Earth-Covered Buildings,* edited by Frank L. Moreland, 269–78. Washington, DC: National Science Foundation/USGPO, 1978.

Lasitter, H. A. *Acoustics Noise Reduction in Shielded Enclosures.* Port Huemene, CA: Naval Civil Engineering Laboratory, 1967.

Wells, Malcolm. "To Build Without Destroying the Earth." In *Alternatives in Energy Conservation, the Use of Earth-Covered Buildings,* edited by Frank L. Moreland, 211–32. Washington, DC: National Science Foundation/USGPO, 1978. Also in *Underground Utilization,* edited by Truman Stauffer, Sr., 4: 546–52. Kansas City, MO: University of Missouri, 1978.

Industrial and Commercial Use

"Building Underground: Factories and Offices in a Cave." *Engineering News-Record* 166 (18 May 1961): 58–59.

Callahan, S. J. "The Development of Limestone Mines into Underground Warehouses and Manufacturing Plants." In *Underground Utilization,* edited by Truman Stauffer, Sr., 1: 22–26. Kansas City, MO: University of Missouri, 1978.

Carlier, R. "Design of Windowless Factories in Belgium." *Annales, Institut Technique du Bâtiment et des Travaux Publics* 22 (1969): 1441–83.

Chryssafopoulos, Nicholas. "Employee Attitudes." In *Underground Utilization,* edited by Truman Stauffer, Sr., 4: 530–31. Kansas City, MO: University of Missouri, 1978.

Fairhurst, Charles. "U.S. Develops New, Commercial Uses for Earth-Sheltering." *Underground Space* 5 (1980): 31–35.

Keighley, E. C. "Visual Requirements and Reduced Fenestration in Office Buildings—A Study of Window Shape." *Building Science* 8 (1973): 311–20.

Lonsdale, Richard E. "Underground Space as a Locational Consideration in Industry." *Underground Space* 1 (1976): 59–60. Also in *Underground Utilization,* edited by Truman Stauffer, Sr., 5: 687–88. Kansas City, MO: University of Missouri, 1978.

Military Use

Blaschke, Theodore E. "NORAD's Underground Command Center." *Civil Engineering* 34 (1964): 36–39.

Jones, Lloyd S. "Non-traditional Military Uses of Underground Space." *Underground Space* 2 (1978): 153–58.

Lally, Elaine K. "The Strange Language of Cheyenne Mountain: The

NORAD Underground." In *Underground Utilization,* edited by Truman Stauffer, Sr., 7: 965–74. Kansas City, MO: University of Missouri, 1978.
Margison, Arthur D. "Canadian Underground NORAD Economically Achieved." *Underground Space* 2 (1977): 9–17.
Olson, J. J., et al. *Ground Vibrations from Tunnel Blasting in Granite: Cheyenne Mountain (NORAD), Colorado.* Minneapolis: Bureau of Mines, Twin Cities Mining Research Center, 1972.
Shih, Jason, and S. Kumar. *Optimization of Subsurface and Underground Shells.* Washington, DC: U.S. Army Research Office, 1970.

Parking

Igarashi, K., and S. Okumura. "Conception of an Underground Parking System." *Tunnelling and Underground Space Technology* 8, no. 1 (1993): 61–64.
Matsushita, K., S. Miura, and T. Ojima. "An Environmental Study of Underground Parking Lot Developments in Japan." *Tunnelling and Underground Space Technology* 8, no. 1 (1993): 65–73.
Narvi, S., et al. "Legal, Administrative, and Planning Issues for Subsurface Development in Helsinki." *Tunnelling and Underground Space Technology* 9, no. 3 (1994): 379–384.
National Research Council. *In Our Own Backyard: Principles for Effective Improvement of the Nation's Infrastructure.* Edited by Albert A. Grant and Andrew C. Lemer. Washington, DC: National Academy Press, 1993.
Pells, P. J. N., R. J. Best, and H. G. Poulos. "Design of Roof Support of the Sydney Opera House Underground Parking Station." *Tunnelling and Underground Space Technology* 9, no. 2 (1994): 201–207.
Pells, P. J. N., P. A. Mikula, and C. J. Parker. "Monitoring of the Sydney Opera House Underground Parking Station." In *Proceedings of the VIIth Australian Tunnelling Conference, Sydney.* Parkville, Victoria: The Australasian Institute of Mining and Metallurgy, 1993.
Pickell, Mark B., ed. *Pipeline Infrastructure II.* New York: American Society of Civil Engineers, 1993.

Psychology

Bechtel, Robert B. "Psychological Aspects of Earth Covered Buildings." In *Earth-Covered Buildings and Settlements,* edited by Frank L. Moreland, 71–77. Springfield, VA: NTIS, 1979.
Collins, Belinda L. "Review of the Psychological Reaction to Windows." In *Underground Utilization,* edited by Truman Stauffer, Sr., 4: 532–40. Kansas City, MO: University of Missouri, 1978.
Paulus, Paul B. "On the Psychology of Earth-Covered Buildings." In *Alternatives in Energy Conservation, the Use of Earth-Covered Buildings,* edited by Frank L. Moreland, 65–69. Washington, DC: National Science Foundation/USGPO, 1978. Also printed in *Underground Space* I (1976): 127–30.

Seybert, Jeffrey A. "Psychological Factors and Future Expansion of Underground Space Utilization." In *Underground Utilization,* edited by Truman Stauffer, Sr., 7: 1018–19. Kansas City, MO: University of Missouri, 1978.

Wunderlich, Elizabeth. "Psychology and Underground Development." In *Underground Utilization,* edited by Truman Stauffer, Sr., 4: 526–29. Kansas City, MO: University of Missouri, 1978.

Public Policy, Legislation, and Zoning

Barker, Michael B. "Earth-Sheltered Construction: Thoughts on Public Policy Issues." *Underground Space* 4 (1980): 283–88.

Colvin, Brenda. *Land and Landscape: Evolution, Design, and Control.* 2d ed. London: J. Murray, 1973.

Duffaut, Pierre. "Site Reservation Policies for Large Underground Openings." *Underground Space* 3 (1979): 187–93.

Fischer, Hans C. "National and International Cooperation in Underground Construction." Stockholm: Swedish Underground Construction Mission, 1976.

Green, Melvyn. "Building Codes and Underground Buildings." In *Earth-Covered Buildings and Settlements,* edited by Frank L. Moreland, 25–29. Springfield, VA: NTIS, 1979.

Guinnee, John W., and William G. Gunderman. "Terraspace: In Consideration of a Policy." In *Underground Utilization,* edited by Truman Stauffer, Sr., 6: 834–36. Kansas City, MO: University of Missouri, 1978.

Hamburger, Richard. "Public Policy Considerations and Earth-Covered Settlements." In *Earth-Covered Buildings and Settlements,* edited by Frank L. Moreland, 1–6. Springfield, VA: NTIS, 1979.

———. "Strategies for Legislative Change." In *Alternatives in Energy Conservation, the Use of Earth-Covered Buildings,* edited by Frank L. Moreland, 243–46. Washington, DC: National Science Foundation/USGPO, 1978.

Harza, Richard D. "National Commitment to Better Urban Life Assures a Rapid Growth in Underground Construction." Paper presented at the annual meeting of the Geological Society of America, Washington, DC, 1971.

Higgs, Forrest S. "Integrating Earth-Covered Housing into Existing Energy-Efficiency Code Structures." In *Earth-Covered Buildings and Settlements,* edited by Frank L. Moreland, 44–53. Springfield, VA: NTIS, 1979.

Labs, Kenneth. "Underground Development, Zoning, and You." *Earth Shelter Digest and Energy Report,* no. 4 (July–August 1979): 4–7.

LaNier, Royce, and Frank L. Moreland. "Earth-Sheltered Architecture and Land-Use Policy." *Underground Space* 1 (1977): iii–iv.

Legal, Economic, and Energy Considerations in the Use of Underground Space. Washington, DC: National Academy of Sciences, 1974.

Schools

Allen, Phelan G. "Fremont Elementary School Air Conditioning." In *Underground Utilization,* edited by Truman Stauffer, Sr., 4: 562–63. Kansas City, MO: University of Missouri, 1978.

Bailey, Edwin R. "Alternatives to Educational Space Needs." In *Underground Utilization,* edited by Truman Stauffer, Sr., 2: 180–83. Kansas City, MO: University of Missouri, 1978.

Carter, Douglas N. "Community and Building Official Reaction to Earth-Covered Buildings: A Case Study, Terraset Elementary School, Reston, Virginia." In *Earth-Covered Buildings and Settlements,* edited by Frank L. Moreland, 78–81. Springfield, VA: NTIS, 1979.

———. "Terraset Elementary School, Reston, Virginia." In *Alternatives in Energy Conservation, the Use of Earth-Covered Buildings,* edited by Frank L. Moreland, 135–49. Washington, DC: National Science Foundation/USGPO, 1978.

———. "Terraset School." *Underground Space* 1 (1977): 317–23.

Cooper, James G., and Carl H. Ivey. *A Comparative Study of the Educational Environment and the Educational Outcomes in an Underground School, a Windowless School, and Conventional Schools.* Washington, DC: Defense Civil Preparedness Agency, 1964.

———. *Final Report of the Abo Project.* Santa Fe: New Mexico Department of Education, 1964.

Demos, G. D. "Controlled Physical Classroom Environments and Their Effect upon Elementary School Children." Windowless Classroom Study. Riverside County, CA: Palm Springs School District, 1965.

Larson, C. Theodore. *The Effect of Windowless Classrooms on Elementary School Children.* Ann Arbor, MI: Architectural Research Laboratory, University of Michigan, 1965.

Lutz, Frank W. "Studies of Children in an Underground School." In *Alternatives in Energy Conservation, the Use of Earth-Covered Buildings,* edited by Frank L. Moreland, 71–77. Washington, DC: National Science Foundation/USGPO, 1978. Also in *Underground Space* 1 (1976): 131–34.

Lutz, Frank W., and Susan B. Lutz. *The Preliminary Report on the Abo Project.* Washington, DC: Defense Civil Preparedness Agency, 1964.

Lutz, Frank W., Patrick D. Lynch, and Susan B. Lutz. *Abo Revisited: An Evaluation of the Abo Elementary School and Fallout Shelter (Final Report).* Washington, DC: Defense Civil Preparedness Agency, 1972.

Platzker, J. "The First 100 Windowless Schools in the U.S.A." Paper presented at the twenty-eighth annual convention of National Association of Architectural Metal Manufacturers, 1966.

Tikkanen, K. T. "Window Factors in Schools." In *Overhead Natural Lighting, Part 2.* Helsinki, Finland: University of Technology, 1974.

"Three Underground Schools." *Progressive Architecture* 56 (1975): 30.

Social and Economic Issues

Davidoff, Linda. "Social Issues in Community Planning for Earth-Covered Shelters." In *Earth-Covered Buildings and Settlements,* edited by Frank L. Moreland, 17–24. Springfield, VA: NTIS, 1979.

Dorum, Magne. "Energy Economy in Rock Stores. A Study of Heat Requirement in Air-conditioned Stores, Freestanding and in Rock Caverns." In *Storage in Excavated Rock Caverns, Rockstore 77,* edited by Magnus Bergman, 1: 73–77. Oxford, England: Pergamon Press, 1978.

Foster, Eugene L. "Social Assessment—A Means of Evaluating the Social and Economic Interactions between Society and Underground Technology." *Underground Space* 1 (1976): 61–63.

Fox, Greg. "Low Cost Earth-Sheltered Housing." *Alternative Sources of Energy,* no. 48 (March–April 1981).

Garrison, W. L. "Social, Economic, and Planning Impacts of Rapid Excavation and Tunneling Technology." In *Underground Utilization,* edited by Truman Stauffer, Sr., 6: 854–58. Kansas City, MO: University of Missouri, 1978.

Golany, Gideon S. *Ethics and Urban Design: Culture, Form, and Environment.* New York: John Wiley & Sons, 1995.

Hoch, Irving. "Economic Trends and Demand for the Development of Underground Space." In *Legal, Economic, and Energy Considerations in the Use of Underground Space,* 68–86. Washington, DC: National Academy of Sciences, 1974.

Isakson, Hans R. "Institutional Constraints on the Marketing and Financing of Earth-Covered Settlements." In *Earth-Covered Buildings and Settlements,* edited by Frank L. Moreland, 7–11. Springfield, VA: NTIS, 1979.

Korell, Mark L. "Financing Earth-Sheltered Housing: Issues and Opportunities." *Underground Space* 3 (1979): 297–301.

McKown, Cora, and K. Kay Steward. "Consumer Attitudes Concerning Construction Features of an Earth-Sheltered Dwelling." *Underground Space* 4 (1980): 293–95.

McWilliams, Donald B., and Stephen M. Findley. "A Life Cycle Cost Comparison Between a Conventional and an Earth-Covered Home." In *Earth-Covered Buildings and Settlements,* edited by Frank L. Moreland, 94–107. Springfield, VA: NTIS, 1979.

Muller, C. A., and R. A. Taylor. "No Cause for Apprehension about Costs of Insuring Earth-Sheltered Homes." *Underground Space* 5 (1980): 28–30.

Newcomb, Richard T. "Dynamic Analyses of Demands for Underground Construction." In *Legal, Economic, and Energy Considerations in the Use of Underground Space,* 87–103. Washington, DC: National Academy of Sciences, 1974.

Parker, Albert D. *Planning and Estimating Underground Construction.* New York: McGraw-Hill, 1970.

Rockaway, John D., and N. B. Aughenbaugh. "Go Underground for

Low-Cost Housing." In *Underground Utilization,* edited by Truman Stauffer, Sr., 4: 614–18. Kansas City, MO: University of Missouri, 1978.

Roy, Robert L. *Underground Houses: How to Build a Low-Cost Home.* New York: Sterling Publishing, 1979.

Smay, V. E. "Underground Living in this Ecology House Saves Energy, Cuts Building Costs, Preserves the Environment." *Popular Science,* June 1974, 88–89+.

Williams, John E. "The Application of Life-Cycle Cost Techniques for Earth-Covered Building Analysis: A Preliminary Evaluation." In *Underground Utilization,* edited by Truman Stauffer, Sr., 5: 765–80. Kansas City, MO: University of Missouri, 1978.

———. "Comparative Life-Cycle Costs." In *Alternatives in Energy Conservation, the Use of Earth-Covered Buildings,* edited by Frank L. Moreland, 43–64. Washington, DC: National Science Foundation/USGPO, 1978.

Transportation

Alders, Charles. "A Study of Earth-Covered Buildings by an Electric Utility Company." In *Alternatives in Energy Conservation, the Use of Earth-Covered Buildings,* edited by Frank L. Moreland, 107–15. Washington, DC: National Science Foundation/USGPO, 1978.

Anttikoski, U., et al. "Bedrock Resources and Their Use in Helsinki." *Tunnelling and Underground Space Technology* 9, no. 3 (1994): 365–372.

Asche, H. R., and D. A. Baxter. "Design and Ground Support Monitoring of the Brunswick Street Rail Tunnels, Brisbane." *Tunnelling and Underground Space Technology* 9, no. 2 (1994): 233–242.

Berg, N. "Aspects on Underground Location of Urban Facilities—Power Supply, Oil Storage, and Sewage Treatment." In *Geological and Geographical Problems of Areas of High Population Density.* Sacramento, CA: Association of Engineering Geologists, 1970.

Branyan, S. G. "Operation and Maintenance of Underground Storage." In *Symposium on Salt,* 609–15. Cleveland: Northern Ohio Geological Society, 1963.

Brown, Verne. "Underground Location of Utilities as Part of Underground Development." In *Underground Utilization,* edited by Truman Stauffer, Sr., 2: 254–56. Kansas City, MO: University of Missouri, 1978.

Espedal, Tor Geir, and Gunnar Nærum. "Norway's Rennfast Link: The World's Longest and Deepest Subsea Road Tunnel." *Tunnelling and Underground Space Technology* 9, no. 2 (1994): 159–164.

Hoffman, George A. "Urban Underground Highways and Parking Facilities." Memorandum RM-3680-RC. Santa Monica, CA: RAND Corporation, 1963.

Homes and Narver, Inc. *Engineering Study of Underground Highway & Parking Garage & Blast Shelter for Manhattan Island.* Oak Ridge, TN: Oak Ridge National Laboratory, 1966.

Hunt, F. R. "Underground Power Transmission." *Underground Space* 3 (1978): 19–33.

Meyer, Kirby. "Utilities for Underground Structures." In *Alternatives in Energy Conservation, the Use of Earth-Covered Buildings,* edited by Frank L. Moreland, 165–81. Washington, DC: National Science Foundation/USGPO, 1978.

Mitchell, Ansel. "Operation and Maintenance of Underground Facilities." In *Underground Utilization,* edited by Truman Stauffer, Sr., 1: 74–76. Kansas City, MO: University of Missouri, 1978.

Odgård, Anders, David G. Bridges, and Steen Rostam. "Design of the Storebælt Railway Tunnel." *Tunnelling and Underground Space Technology* 9, no. 3 (1994): 293–307.

Paulson, Boyd C., Jr. "Research and Development Needs for Systems and Management in Underground Transportation Construction." *Underground Space* 2 (1977): 81–89.

Rosander, A. "Underground Sewage Treatment Plant." *Underground Space* 2 (1977): 39–45.

Rosengren, Lars. "Preliminary Analysis of the Dynamic Interaction Between Norra Länken and a Subway Tunnel for Stockholm, Sweden." *Tunnelling and Underground Space Technology* 8, no. 4 (1993): 429–439.

Sherman, Richard G., et al. "Boston Harbor Outfall Tunnel: An Environmental Imperative." *Tunnelling and Underground Space Technology* 9, no. 3 (1994): 309–322.

Sikkel, H. A. "An Underground Bus Terminal in Amsterdam: Legal and Policy Issues." *Tunnelling and Underground Space Technology* 8, no. 1 (1993): 31–36.

Smith (Wilbur) and Associates. *Transportation and Parking for Tomorrow's Cities.* New Haven, CT: Automobile Manufacturers Association, 1966.

van Hengstum, L. A., et al. "Railway Projects Move Dutch Transport into the Twenty-First Century." *Tunnelling and Underground Space Technology* 9, no. 3 (1994): 343–351.

INDEX